MICROBIOLOGY

An environmental
perspective

MICROB

IOLOGY

An environmental perspective

Paul Edmonds

Associate Professor of Microbiology and Public Health
University of Wisconsin—Oshkosh

Macmillan Publishing Co., Inc.
New York

Collier Macmillan Publishers
London

Macmillan Publishing Co., Inc.
866 Third Avenue, New York, New York 10022

Collier Macmillan Canada, Ltd.

Library of Congress Cataloging in Publication Data

Edmonds, Paul, (date)
  Microbiology : an environmental perspective.

  Includes bibliographies and index.
  1. Microbial ecology. I. Title
QR100.E35   576'.15   76-58450
ISBN 0-02-333580-7

Printing: 2   3   4   5   6   7   8   Year: 8   9   0   1   2   3   4

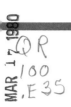
Preface

Environmental biology is one of many popular courses in environmental science programs that are being initiated and/or expanded on college campuses throughout the nation. Yet, there is a paucity of information in these programs that concerns the activities of microorganisms in nature. Traditional books on microbial ecology characterize these activities for microbiologists and students with advanced training in the biological sciences, but I know of no other book that has made this attempt for beginning students.

Microbiology: An Environmental Perspective is *not* intended to be a textbook for use in an introductory course for microbiology majors. The aim of this book is to provide students with an understanding of vital activities that microorganisms perform in nature. Although this book is written for the nonspecialist in microbiology, it is assumed that readers will have some command of basic biological concepts. In order to accommodate students with a diversity of backgrounds, the basic chemical concepts that are required for understanding environmental relationships are discussed in Chapter 2. The diversity of microbial habitats is discussed in Chapter 3, after which emphasis is placed on physiological activities of microorganisms in relation to broad-scale interactions that occur in natural ecosystems. Subsequent discussions deal with the epidemiology of some common human and animal diseases. Then, specific examples of influences that microorganisms have on the quality of the environment—human health, pollution (air, water, and land), and welfare (economic losses, etc.) are presented. The final chapter delineates specific roles that microorganisms play in polluted ecosystems and emphasizes the importance of controlling such activities.

I have taken a biased approach in the development of this book. It is not intended to be all-inclusive in the coverage of microbial activities in nature. Cell biology and basic physiology will be deemphasized except where necessary to clarify a functional concept. Also, culture methods and details of techniques used for studying microorganisms are deliberately omitted, because these aspects are covered thoroughly in many excellent general microbiology books.

In order to facilitate reading, a list of key words is provided at the end of each chapter. Pertinent periodicals and monographs are listed under

v

''Selected Readings'' for those who may desire more details of a particular topic. A comprehensive glossary is provided at the end of the book.

I believe that *Microbiology: An Environmental Perspective* will provide the nonmicrobiologist with a broader dimension of organismic activities in the total ecosystem.

Finally, I wish to acknowledge my colleagues, friends, and relatives for their support during the preparation of this book. The names of those individuals who generously provided photographs for use as illustrations appear throughout the book in figure captions that acknowledge their contributions.

My special appreciations are expressed to those who read parts or all of the manuscript, and who took the time to offer constructive criticisms. Among this group are Dr. Daniel D. Burke, Department of Microbiology, University of Illinois, Urbana; Dr. Karl T. Kleeman, Department of Biology, California State University, Fresno; Dr. Michael G. Petit, Department of Microbiology, Colorado State University, Ft. Collins; Dr. Ramon J. Seidler, Department of Microbiology, Oregon State University, Corvallis; Dr. David L. Wright, Department of Biology, University of Wisconsin–Oshkosh, and my former students: Penny Schiller, Danielle Meyer, Kenneth L. Hobbs, and David M. Janssen.

Lastly, I wish to acknowledge the support of my family, and the cooperation of the staff at Macmillan Publishing Company. Without the courteous cooperation of Charles E. Stewart, Jr., Woodrow W. Chapman, Bruce Bumby, Mollie Horwitz, and Lola Peters this project could not have been completed.

Paul Edmonds

Contents

Introduction

1

Microbial groups 3

Historical perspective   3
Living cells   6
Procaryotic organisms   7
   Bacteria   9
   Blue-green algae   25
   Mycoplasmas   26
Eucaryotic organisms   28
   Fungi   30
     Yeasts   30
     Molds   32
     Slime molds   34
   Algae   35
   Protozoa   36
   Isolated cells of multicellular organisms   39
Viruses (noncellular agents)   40
   Bacterial viruses (bacteriophages)   42
   Animal viruses   44
Selected readings   50

2

Nutrition and metabolism 53

Essential chemical concepts   54
   Atoms and molecules   54
   Chemical bonds   56
   Acids and bases   57

Biological molecules   60
Requirements for growth   70
   Water   71
   Carbon   72
   Energy   72
   Nitrogen   72
   Accessory materials   73
Nutritional types of organisms   74
   Autotrophs   74
      Photoautotrophs   74
      Chemoautotrophs   75
   Heterotrophs   75
      Photoorganotrophs   75
      Chemoorganotrophs   75
Cellular metabolism   76
   Enzymes   76
   Energy-yielding processes   81
      Photosynthesis   82
      Fermentation   83
      Aerobic respiration   85
      Anaerobic respiration   88
   Energy-utilizing processes   88
      Biosynthesis of macromolecules   89
      Biosynthesis of proteins   90
   The molecular basis of mutation   92
Selected readings   95

3

Diversity of microbial habitats   97

Environmental selecting factors   99
   Physical   99
      Solar radiations (temperature and light)   99
      Osmotic pressure   102
      Hydrostatic pressure   103
   Chemical   104
      pH (acidity and alkalinity)   104
      Available gases (O_2 and CO_2)   105
   Biological   105

Types of microbial habitats   106
　Atmospheric environments   106
　　Outdoor air   107
　　Indoor air   107
　Aquatic environments   108
　　Fresh water   108
　　Salt water   110
　Terrestrial environments   112
　Biological environments   116
　Microenvironments   117
Modes of dispersal   117
Selected readings   122

4

Microbial interference   123

Competition for survival in nature   124
　Changes in ecosystems: succession   125
　Competition for nutrients   129
　Competition for oxygen   130
　Competition for space   131
Role of antimicrobials in nature   132
　Antibiosis from antibiotics   133
　Antibiosis from bacteriocins   140
　Antibiosis from other substances   141
　Sensitive and resistant species   145
Selected readings   147

5

Intermicrobial and extramicrobial relationships   149

Types of symbiotic relationships   150
　Mutualism   151
　Commensalism   151
　Parasitism   152
Intermicrobial associations in nature   153

Lichens   154
Lysogeny   155
Miscellaneous associations   157
   Protozoan associations   157
   Bacterial associations: a triad   160
   Mycoviruses   163
   Cyanophages   164
Extramicrobial associations in nature   165
   Microbial associations with animals and insects   165
      Microbial–ruminant associations   165
      Microbial–insect associations   170
   Microbial associations with higher plants   172
      Mycorrhizae   172
      Bacterial–legume associations   174
   Obligate microbial parasites of higher organisms   177
Selected readings   179

6

Transformations in geochemical cycles   181

Dynamic aspects of microbial populations   182
   Determinants of population size   182
      Unicellular growth   183
      Population growth patterns   185
      The significance of measuring microbial activity   187
   Energy conversion and biosynthesis   189
Role of microorganisms in geochemical cycles   193
   Carbon cycle   193
   Nitrogen cycle   195
   Phosphorus cycle   198
   Sulfur cycle   199
Selected readings   201

7

The human body: a natural ecosystem   203

Inhabitable anatomical regions   204
   The skin   204

The gastrointestinal tract   208
Miscellaneous areas   210
The indigenous microflora   211
Normal ecological niches   211
Sterile (forbidden) zones   215
Defense mechanisms   215
Mechanical barriers   216
Immune barriers   216
Phagocytosis   216
Antibodies   217
Types of immunity   220
Natural immunity   220
Acquired immunity   220
Germ-free animals   221
Selected readings   224

8

Epidemiology of human microbial diseases   225

Sources of environmental pathogens   227
Living reservoirs   227
Inanimate reservoirs   228
Modes of transmission for pathogens   229
Direct transmission   229
Indirect transmission   229
Vehicle-borne   229
Vector-borne   230
Airborne   230
Epidemiological investigations   230
Retrospective studies   230
Prospective studies   231
Human infectious diseases   231
Selected bacterial diseases   234
Boils and carbuncles   234
"Strep" sore throat   235
Pneumonia   237
Tuberculosis   239
Typhoid fever   242
Cholera   244

Selected viral diseases   246
   Smallpox   246
   Polio   247
   Measles   249
   Hepatitis   251
   Influenza   253
Selected fungal (mycotic) diseases   256
   Dermatomycoses   256
   Systemic mycoses   257
Selected protozoan diseases   260
   Amebiasis   260
   Giardiasis   260
Selected venereal diseases   262
   Gonorrhea   262
   Syphilis   263
Prevention and control of infectious diseases   265
   Chemotherapy   266
   Prophylactic immunization   269
   Environmental sanitation   269
Selected readings   270

9

Epidemiology of zoonotic diseases   271

Selected types of zoonoses   271
   Bacterial zoonoses   272
     Anthrax   272
     Brucellosis   274
     Bubonic plague   276
     Salmonellosis   277
   Viral zoonoses   279
     Encephalitides   280
     Yellow fever   282
     Rabies   283
   Protozoan zoonoses   284
     Malaria   285
     Toxoplasmosis   287
Prevention and control of zoonoses   288

Prophylactic immunizations   289
Environmental sanitation   289
Selected readings   290

10

Microbial toxins in the environment   293

Types of microbial toxins   293
  Bacterial toxins   294
    Exotoxins   295
    Endotoxins   296
  Algal toxins   297
    Blue-green toxigenic algae   297
    Eucaryotic toxigenic algae   299
  Fungal toxins (aflatoxins)   301
Ecological consequences of microbial toxins   303
  Effect on human beings   303
  Effect on food chains   305
Microbial toxins as insecticidal agents   308
  Bacterial toxins harmful to insects   310
  Fungal toxins harmful to insects   312
Selected readings   313

11

The role of microorganisms in polluted environments   315

Microbiological aspects of air pollution   316
  Nonindustrial microbial aerosols   317
  Industrial microbial aerosols   321
Microbiological aspects of water pollution   323
  Drinking water — problems of purification   324
  Sewage — problems of treatment   327
  Solid wastes — problems of disposal   331

Microorganisms as tools for detecting specific
pollutants   333
    Bacterial and protozoan assays   334
    Tissue culture assays   335
Selected readings   337

Comprehensive glossary   339

Index   359

Introduction

In this era of ecological awareness, maintaining the quality of the environment is a major concern of many college students. Yet, the techniques and approaches of separate disciplines among the physical, social, and biological sciences in our institutions are inadequate to deal justly with the complexity and magnitude of environmental problems. Environmental science programs are numerous, but too often students develop obscure views of the environment as a totality — a functional system of subunits that include physical, chemical, and biological components. Among the biological components, the activities of some microorganisms are vital to the functioning of the system as a whole and to the existence of all other biological species including human beings.

Microbiology: An Environmental Perspective introduces nonmajors in microbiology to the functional aspects of microorganisms in nature. Many individuals harbor vague concepts about microorganisms and recognize them as "germs" that function to produce discomfort and dreaded human diseases; others view them only as complex agents studied by highly trained scientists. Although both views are within the vast scope of microbiology, they fail to convey the broad spectrum of activities in nature that are mediated by microorganisms.

CHAPTER 1

Microbial groups

- **Historical perspective**
- **Living cells**
- **Procaryotic organisms**
    Bacteria
    Blue-green algae
    Mycoplasmas
- **Eucaryotic organisms**
    Fungi
        Yeasts
        Molds
        Slime molds
    Algae
    Protozoa
    Isolated cells of multicellular organisms
- **Viruses (noncellular agents)**
    Bacterial viruses (bacteriophages)
    Animal viruses
- **Key words**
- **Selected readings**

What are microorganisms? Unlike terms used to describe higher forms of life (plants and animals), the term *microorganism* does not describe a unified group or a single biological entity with similar structures and function. Microorganisms are highly diverse, but all are microscopic or submicroscopic, all are undifferentiated unicells, and all share a common feature — smallness. This does not mean that all microorganisms are of an equal size. The size variation will become evident as we examine the types of cells found within the various groups of microorganisms.

Historical perspective

From the earliest recordings of ancient civilizations through the Middle Ages, human infirmities were ascribed to demons, evil spirits, and retributive gods. Such views were inadequate to satisfy the curiosity of a few. Those curious-minded individuals probed into the mysteries of life by collecting and examining all types of living things. At that time the world of microscopic organisms was unknown.

Between the fourteenth and seventeenth centuries, the fields of chemistry, botany, zoology, and medicine evolved from superstitious arts, filled with uncertainty and chaotic techniques, into well-disciplined subjects based on scientific principles. A few workers in the medical profession postulated the existence of invisible agents of disease. For example, Girolamo Fracastoro (1483–1553) published a book in 1546 which described three ways for the transmission of infectious agents: by direct contact, by fomites (inanimate objects), and by the air. Later, the microscope emerged as an important instrument. Athanasius Kircher (1602–1680), Robert Hooke (1635–1703), and Marcello Malpighi (1628–1694) used magnifying lenses in their studies of a variety of biological specimens, but Anton van Leeuwenhoek (1632–1723) is recognized as the first person to see and accurately describe the invisible creatures of the microbial world. Unlike most scholars of that era, Leeuwenhoek was not a trained scientist, but ground lenses and assembled them into microscopes as a hobby. Sketches of creatures that he had observed in saliva, teeth scrapings, and a variety of other aqueous mixtures were presented to the Royal Society of London in 1674. His drawings were highly accurate, because morphological descriptions were made relative to items such as grains of sand and mustard seeds. As a result of such comparisons, some of the organisms that he described are now recognized as bacteria and protozoans.

Although many divergent views relative to the origin of life had been publicized in the scientific community, debate over that controversial issue was intensified by the discovery of microorganisms. Some leading scholars of that time ardently supported the "theory of spontaneous generation" and attempted to prove its validity through experimentation. As a result of much confusion and inaccurate interpretation of experimental data, controversial debate on this subject continued for many years. Finally, through a series of carefully planned experiments, Louis Pasteur (1822–1895) proved that "germs" did not originate spontaneously in sterile solutions, but developed from other "germs" that had entered his flasks on contaminated dust particles; thus ended the debate over the theory of spontaneous generation. This was only one of many great contributions that Pasteur made to science; others were the establishment of concepts for the "germ theory of disease," the characterization of life in the absence of air (anaerobic growth), the development of techniques for pasteurization, and the development of a vaccine for rabies.

While Pasteur's accomplishments in France were unprecedented in the scientific community, a young physician, Robert Koch (1843–1910), was making news in Germany. Koch established himself as a capable scientist at an early age and proved subsequently to be

Pasteur's most ardent critic and competitor. The peak of their rivalry centered on the anthrax problem in Europe. After Pasteur had cultivated the anthrax bacillus in his laboratory, Koch succeeded in isolating the organism *(Bacillus anthracis)* in pure culture and described it as the causative agent of anthrax, a disease common among animals that can be transmitted to human beings. Subsequently, Koch's succession of accomplishments on the development of methods and techniques for the study of different microorganisms in pure culture established him as a renowned scholar.

Koch's successful accomplishments in microbiology can be attributed in part to his clinical background as a physician and to his methodological approach to the study of infectious diseases. To prove that a specific disease was caused by a specific microorganism, he specified a sequence of procedures and conditions that had to be met. Actually, he developed criteria that would support the germ theory of disease. Today, we recognize those criteria as *Koch's postulates:* (1) the infectious microbe must be present in every diseased animal but not present in healthy animals; (2) the infectious microbe must be isolated from the diseased animal and grown in a pure culture; (3) the isolated infectious microbe must provoke the characteristic symptoms of the disease when inoculated into a susceptible (nonimmune) animal; and (4) the infectious microbe must be reisolated from the experimentally diseased animal, grown in pure culture, and demonstrated to be the same as the microbe isolated from the original diseased animal.

By following Koch's postulates, the investigator can discriminate between casual associations (transient contaminants), and the causative agents for many infectious diseases. Even today, with some modifications, those postulates remain useful, but strict adherence to them is inappropriate under certain circumstances. For example, some infectious agents do not grow well when isolated in pure culture, and the causative organisms of human diseases such as leprosy and gonorrhea *(Mycobacterium leprae* and *Neisseria gonorrhoeae,* respectively) do not provoke the characteristic symptoms of the diseases they cause when inoculated into experimental animals. Nevertheless, Koch's postulates contributed significantly to the development of fundamental principles in microbiology.

Together, Pasteur and Koch are credited with laying the foundation for the subsequent development of microbiology into a dynamic field of science. Many nonprofessionals recognize the names of these historical figures, because their major contributions dealt with organisms that cause diseases in human beings and in animals. However, during the same period less familiar figures were making laudable discoveries in areas of applied microbiology. The activities of microorganisms in the soil were exposed to the scientific community by Martinus

W. Beijerinck (1851–1931) and Sergei N. Winogradsky (1856–1953).

Beijerinck isolated the first symbiotic nitrogen-fixing bacterium from structures on the roots of leguminous plants (e.g., soybeans, peas, alfalfa), and described the process of symbiotic nitrogen fixation. During the process legume roots become infected with certain kinds of bacteria, then cells in the infected roots become modified in a manner to convert atmospheric nitrogen into cellular constituents. In subsequent studies he demonstrated that atmospheric nitrogen could be fixed by many organisms, including free-living bacteria (*Azotobacter* species) through a nonsymbiotic process.

Winogradsky was also interested in the action of soil microorganisms. He characterized the cyclical process of microbial conversion of nitrogenous compounds in nature and described specific microbial groups that have the ability to transform ammonia to nitrite, and nitrite to nitrates—processes that are extremely important to the environment as a whole.

Although not a physician, Winogradsky was highly impressed with the work of Pasteur and Koch. Pasteur was equally impressed by Winogradsky's discoveries, and invited him to join his team in Paris. Because of circumstances pertaining to World War I, Winogradsky declined the honor. In 1922, however, in response to an invitation from Roux, one of Pasteur's colleagues, Winogradsky moved to Paris and spent the last 25 years of his life at the Pasteur Institute investigating the ecological relationships of soil microorganisms.

And so, from the pioneering efforts of a quartet of giants—Pasteur, Koch, Beijerinck, and Winogradsky—during their studies of pathogenicity and the cyclical conversion of nitrogenous compounds in nature, the theory and practice of environmental microbiology was born.

Living cells

Prior to our discussion of types of organisms it seems appropriate to specify those functional characteristics which clearly separate living cells from nonliving things, and to describe a minimal structural plan for all cells. A *living cell,* the smallest functional unit in any organism, exhibits the following characteristics or attributes for some period during its life: (1) the ability to acquire and utilize nutrients, (2) the ability to grow, (3) the ability to excrete waste, (4) the ability to respond to stimuli, (5) the ability to adapt, and (6) the ability to reproduce. Such qualities are not exhibited by dead cells, inanimate objects, or virions (extracellular virus particles).

In terms of structure and function, each cell is a separate and different unit of life. There is no such thing as a typical cell, although we continue to use the expression. In unicellular organisms, all activities of life take place in an independent unit. In multicellular organisms, each cell has its own uniqueness, but all cells function cooperatively and contribute to the life-sustaining processes of the organism.

Regardless of cell type or kind of organism, all cells have the same minimal structural plan: a unit or cell membrane, which may or may not be the outer boundary (a rigid layer called the cell wall is the outermost structure in some cells); a cytoplasm, which is the waterlike collection of substances in the cell's interior; and a nucleus or nuclear region (the location of the genetic material).

Each of these components is absolutely essential to a living cell. The *cell membrane* is a selective barrier that functions to regulate the entry of nutrients and the exit of cell products. The cell wall (present in all plants and in some microorganisms) functions to give the cell rigidity and protection. The *cytoplasm,* the compartment in which all vital processes take place, contains a variety of substances in solution. In cells that have a discrete *nucleus,* the genetic material is separated from the cytoplasmic substances by a nuclear membrane, but in other types of cells, the genetic material is not enclosed by a membrane but is distributed throughout the cytoplasm in areas that comprise the *nuclear region.*

Although the minimal structural plan described is adequate to represent some cell types, many other cells have additional internal and external components, some of which are highly complex in both structure and function. A generalized version of the manner in which all living cells are interrelated is shown in Figure 1.1. Although living organisms have been grouped taxonomically into five kingdoms, their bodies are composed of either procaryotic or eucaryotic cells. *Kingdom Monera* is composed of organisms with procaryotic cells (bacteria and blue-green algae), and *Kingdom Prostista* is composed of single-celled organisms with eucaryotic cells. *Kingdoms Plantae, Fungi,* and *Animalia* are composed of multicellular and multinucleate organisms whose bodies are made of eucaryotic cells. These differences will become clearer as we proceed with a discussion of specific types of organisms.

Procaryotic organisms

Procaryons are organisms with simple structures that are characterized predominantly by the absence of certain features. Conspicuous among these are intracellular membranous structures (mitochondria

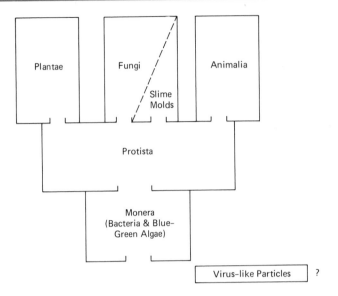

Figure 1.1. Simplified version of the five-kingdom system for classifying organisms. All procaryotic cells are in the Kingdom Monera, all eucaryotic unicellular cells in the Kingdom Protista, and all eucaryotic multicellular cells in the Kingdom Plantae, Fungi, or Animalia; viruslike particles are of uncertain origin.

and chloroplasts), a nucleus (i.e., chromosomes or genes separated from the cytoplasm by a membrane), and structures for sexual reproduction. All procaryons are single-celled organisms, but aggregations of unicells are typical among some groups. Organisms with the characteristics that we have described are known today as bacteria, blue-green algae, and mycoplasmas. The structural simplicity of these groups is probably related to their lack of modification during the course of their evolutionary development.

It is generally agreed that biological diversity results from cellular development at various stages in evolution. Procaryotic cells are considered to be ancestral forms of life. The vast array of eucaryotic cells with varying degrees of complexity in their structures are considered to be descendants of less complex ancestral procaryons, but their development probably occurred along diverging lines, as described previously (see Figure 1.1). Therefore, the least complex and simplest forms of organisms (bacteria and blue-green algae) are at the bottom of the taxonomic ranks, and the most complex forms of life are at the top of the taxonomic ranks. Cells with varying degrees of complexity are distributed between the two extremes and occupy intermediate taxonomic ranks.

Bacteria

Those procaryotic organisms that we designate as *bacteria* are widely distributed in the biosphere. All organisms within this group are unicellular, but a considerable degree of variation can be found among cell types.

Morphologically, bacterial cells are structurally simple (Figures 1.2 A and B). On the basis of cell shapes, three broad groups are recognized: bacilli (rodlike), cocci (spherical or ovoid), and spirilla (spirals and corkscrews). Among such groupings individual cells may appear as single units, pairs, chains, or clusters, but cells in such configurations are usually constant for a particular species. However, we must remember that these groupings are aggregations of undifferentiated unicells, and not cooperative units. Owing to the constancy of cell configurations in such groupings, taxonomic relationships are implied by aggregations of spherical cells: diplococci (pairs), streptococci (chains), tetracocci (four cells arranged in squares), sarcinae (cubical packets), and staphylococci (grapelike clusters). Other designations for cellular arrangements of the bacteria are used less frequently.

The bacteria, like all other kinds of microorganisms, cannot be seen with the unaided human eye; to observe their cell morphology, a microscope must be used. As one might expect, in addition to variations in cell shape, bacteria also exhibit a great deal of variation in cell size.

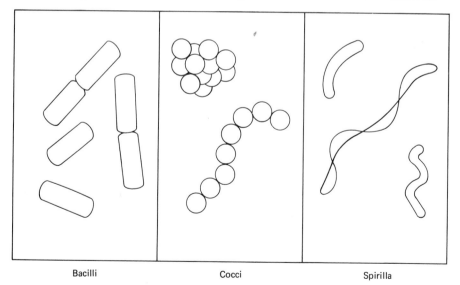

| Bacilli | Cocci | Spirilla |

Figure 1.2A. Schematic view of three basic morphological types of bacterial cells.

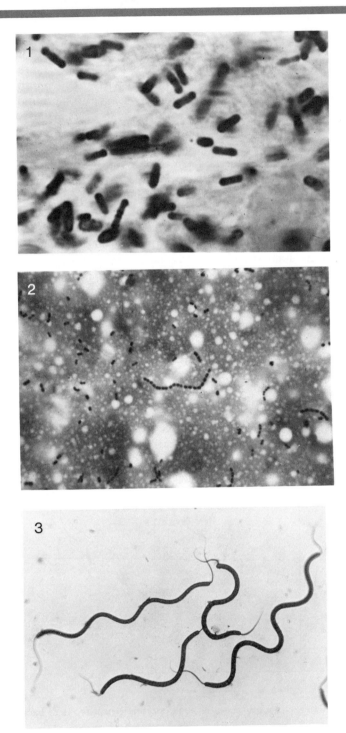

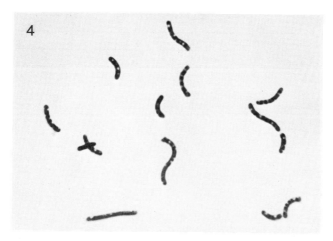

Figure 1.2B. Photo of bacterial cells through a compound microscope: **(1)** bacilli *(Clostridium perfringes in muscle tissue)*; **(2)** coccus *(Streptococcus lactis)*; and **(3, 4)** Spirilla *(Spirillum volutans* and *Spirillum species)*. [Courtesy of the Carolina Biological Supply Company.]

Some examples of cell dimensions for various types of bacteria are shown in Table 1.1. It is important to remember that individual cells of closely related bacteria within the same species are not all equal in size. To allow for individual variations, a range of dimensions is given in the table to show the minimum and maximum size limits within which cells for bacteria of a particular species should fit. Some of the largest bacterial cells are found among the spiral and filamentous forms. For example, *Spirochaeta plicatilis* has helical cells 0.5 to 0.75 micrometer (μm) wide and 100 to 200 μm long, and *Herpetosiphon geysericola* has cells (rods or filaments) 0.5 μm wide and 10 to 150 μm in length. On the other hand, bacteria such as *Coxiella burnetii* have extremely small cells (width 0.2 to 0.4, length 0.4 to 1.0 μm), dimensions near the limits of resolution of light microscopes.

Modern *compound microscopes* (Figure 1.3) are complicated instruments that are the product of optical improvements that took place over a long period. They are called "compound" microscopes because they contain two lenses (the ocular lens and the objective lens). The total magnification of a specimen when viewed with a compound microscope is the product of the magnification of the two lenses (ocular × objective). For example, if the ocular lens has a 10× magnification and the objective lens has a 100× magnification, the total magnification of the specimen is 1,000×. Present instruments bear little resemblance to the simple microscopes that Leeuwenhoek constructed and used to make the first accurate observations of microbial life. His simple

Table 1.1.   Cell Dimensions of Selected Bacteria

Name and Morphological Type	Range of Cell Sizes (μm)		
	Diameter	Width	Length
Cocci (spheres or ovals)			
Staphylococcus aureus	0.8– 1.0		
Streptococcus pyogenes	0.6– 1.0		
Micrococcus roseus	1.0– 2.5		
Neisseria gonorrhoeae	0.6– 1.0		
Bacilli (rod-shaped)			
Bacillus anthracis		1.0– 1.2	3.0– 3.5
Haemophilus influenzae		0.2– 0.3	0.5– 2.0
Clostridium botulinum		0.8– 1.3	4.4– 8.6
Yersinia pestis		0.5– 1.0	1.0– 2.0
Spiralla (spiral or corkscrews)			
Treponema pallidum		0.09– 0.18	6.0– 20
Vibrio cholerae		0.5	1.5– 3.0
Spirillum volutans		1.4– 1.7	14.0– 60

Source: Assembled from R. E. Buchanan and N. E. Gibbons, eds., *Bergey's Manual of Determinative Bacteriology,* 8th ed. (Baltimore, Md.: The Williams & Wilkins Company, 1974).

microscopes contained only a single lens and were capable of magnifying specimens no more than about 300 times. In spite of that low magnification, images projected by Leeuwenhoek's microscopes were extremely clear, superior even to images projected by compound microscopes that other investigators were using during that period, because their lenses were of poor quality.

The ability of light microscopes (either simple or compound) to project clear and distinct images of two separate points in close proximity is called the *resolving power*. Resolving power, an inherent property of every microscope, is determined by the wavelength of light rays being transmitted and the quality of the lens. Consequently, the maximum resolution of modern compound microscopes is approximately 0.2 μm. In other words, cell and/or structures whose diameter is less than 0.2 μm cannot be perceived clearly with a compound microscope. Images become fuzzy or distorted when the total magnifications is increased above 2,000×, because the resolving power is not increased accordingly.

Living bacterial cells in an aqueous suspension are almost colorless. Because of this lack of contrast, they are difficult to see with light microscopes. This fact was realized by Robert Koch while he was working with the anthrax bacillus. Consequently, he experimented with a variety of techniques in an attempt to improve the visibility of bacteria in specimens to be examined microscopically. One technique was to spread on a glass slide a small drop of fluid in which the bacteria were suspended. After allowing the film to dry, they were examined with a light microscope. Such preparations are called *smears*. By subjecting

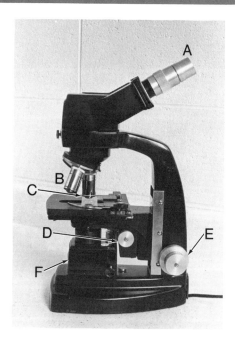

Figure 1.3. One kind of modern light microscope: (A) ocular; (B) objectives; (C) specimen slide on stage; (D) condenser adjusting knob; (E) focusing knob, outer section for coarse adjustment, inner for fine adjustment; (F) built-in lamp.

smears to mild heat for a short time, the cells adhere to the slide and are referred to as *fixed*. When Koch examined such preparations, the cells were not changed, but remained colorless or transparent. Koch and his associates tried staining bacteria with coal-tar or aniline dyes used by histologists to make structures in mammalian tissues easier to observe. This method proved successful and gave rise to the use of stains in bacteriology.

Today, coal-tar dyes are used extensively in bacteriology, and numerous staining methods have been devised for specific purposes Some are used to detect a particular kind of structure (e.g., flagellum, spore, or capsule), and others are used to differentiate one kind of bacteria from another. When a single dye is used merely to increase the contrast in a specimen by making the bacteria darker than the surrounding medium, the process is referred to as a *simple stain*. The major characteristic of simple stains is their ability to react equally with all kinds of bacteria. Methylene blue is a good example of a widely used simple stain. The simple staining procedure is rapid and simple. A dilute solution of methylene blue is applied dropwise to heat-fixed smears for 1 minute. The smear is then rinsed with tap water, blotted dry, and observed with a light microscope (Figure 1.4).

Differential stains, unlike simple stains, do not react equally with all kinds of bacteria. Therefore, they can be used to detect differences in cellular structures, or to distinguish between morphologically simi-

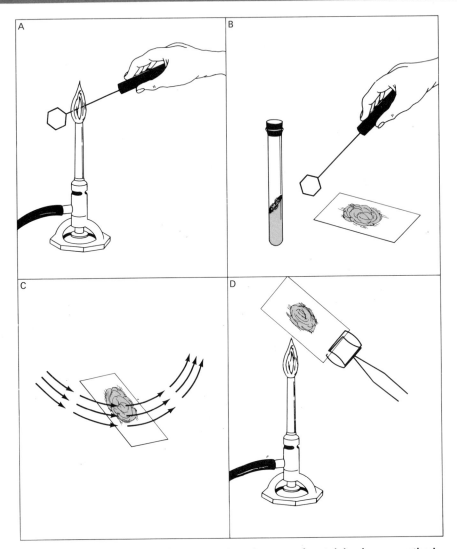

Figure 1.4. Steps in the preparation of a smear for staining by any method. (A) An inoculating loop being sterilized in the flame of a bunsen burner. (B) Bacterial cells being removed from a test tube culture and being spread on a microscope slide. (C) The slide containing the smear of cells is allowed to air dry. (D) The air-dried smear being heat-fixed.

lar bacteria. The *Gram stain* is the most widely used and probably the most important differential stain employed in bacteriology. The procedure, developed by Christian Gram, a Danish physician, in the nineteenth century, is used primarily to divide bacteria into one of two broad groups: *gram positive* and *gram negative*. The differentiation is based upon the color exhibited by bacterial cells after heat-fixed smears have been treated with four reagents, each applied for a specific time and in a sequential manner: a primary stain (crystal violet), a mordant (iodine–potassium iodide solution), a decolorizing agent (95 percent ethyl alcohol), and a red counterstain (Safranin-O). When cells are observed with a light microscope after all reagents have been applied, gram-positive bacteria are stained purple because they retain the crystal violet, and gram-negative bacteria are stained pink or red because they lose the crystal violet and retain the Safranin-O. The sequential manner in which the reagents are applied and the reaction time for each are listed in Table 1.2.

It is important to remember that the Gram staining procedure is not absolute, because certain kinds of bacterial cells exhibit a variable reaction, and others are stained poorly or not at all. Bacteria in the genus *Mycobacterium* are among the types that react poorly with the Gram stain. For their detection, the acid-fast staining method is used (see Chapter 8).

The importance of Gram's staining procedure becomes obvious when one considers the fact that gram-positive and gram-negative bacteria have several fundamental differences. In addition to basic differences in inherent chemical and physical properties, bacteria within each category differ in respect to the diseases they cause and in respect to their susceptibilities to certain kinds of therapeutic antimicrobial agents. Thus, by merely examining a heat-fixed Gram-stained smear with a light microscope, a great deal can be learned about the nature of bacteria. The gram reaction is consistent when applied to young (actively growing) bacterial cells and is recognized as a reliable method for separating the bacteria into two large taxonomic subgroups: gram-positive bacteria and gram-negative bacteria. However,

Table 1.2. Pertinent Aspects of the Gram Staining Procedure*

Step No.	Reagent	Purpose	Time	Result/Function
1	Crystal violet	Primary stain	1 minute	Stains all cells purple
		(Rinse smear with tap water)		
2	Iodine solution†	Mordant	1 minute	Increases affinity between cells and primary stain
		(Rinse with tap water)		
3	Ethyl alcohol (95%)	Decolorizer	30 seconds	Removes primary stain from some cells, while others remain purple
		(Rinse with tap water)		
4	Safranin-O (a red dye)	Counterstain	10–30 seconds	Stains colorless cells pink or red, while others remain purple
		(Rinse with tape water and blot dry)		

*Other versions of the procedure exist, but all employ essentially the same steps, and the end results are the same.
†Iodine–potassium iodide solution.

it is important to remember that the Gram staining procedure is of no taxonomic value when applied to procaryotic organisms that grow normally without cell walls (see the discussion of *Mycoplasmas* later in the chapter) and organisms that are composed of eucaryotic cells.

Although the ability to study bacteria by light microscopy has been enhanced by the use of various stains, the compound microscope itself has undergone some drastic changes in terms of instrument design and optical improvements. A modern compound microscope can be modified in a number of ways to expand its versatility. As a result, several versions of the light microscope have evolved: the dark-field microscope, the phase-contrast microscope, and the fluorescence microscope. Each is modified or equipped to be used for a specific purpose. Collectively, the types of observations that can be made in microbiology with them are innumerable. For details relative to the principles of light microscopy, the reader is referred to one of the basic microbiology books listed at the end of the chapter.

To study microbial anatomy in great detail and to observe extremely small intracellular structures, an *electron microscope* must be used. Using this instrument, scientists have learned a great deal about the structure of various components within bacteria and other cells. The development of the *transmission electron microscope* (Figure 1.5) was one of the great scientific achievements of the twentieth century.

Figure 1.5. Transmission electron microscope. [UW-Oshkosh Electron Microscope Laboratory, courtesy of Dr. Rodney Cyrus.]

Owing to the complex nature of the instrument, a discussion of its principles is beyond the scope of this book. However, we shall present a brief overview of its major features and capabilities. The basis upon which the electron microscope was conceived is the fact that a beam of electrons behaves in a manner similar to light rays and exhibits wavelengths that are much shorter than wavelengths of visible light. In the electron microscope, glass lenses cannot be used because electrons will not pass through glass; instead electromagnets function as the lenses. When an electron microscope is being operated, a beam of electrons is directed from an electron source into the specimen chamber and is focused onto the specimen by electromagnetic lenses. To obtain clear images, the specimen chamber into which electrons are expelled must be maintained under high vacuum, because electrons are scattered by gas molecules present in the air. As electrons impinge on the specimen in the evacuated chamber, they become scattered and produce images which are then projected on a small screen, similar to the screen of an ordinary television set. By use of a built-in camera, the investigator can make photographs of images, as desired, while viewing each specimen.

The transmission electron microscope is used primarily to examine the structure of very small intracellular components. Cells in their living state cannot be observed with this instrument. To be viewed by the transmission electron microscope, specimens must be prepared in a very specific manner. During this treatment, whole cells are chemically fixed and *sectioned*—cut into very thin slices with an ultramicrotome. Both processes are complex and tedious to perform. The thin sections are then inserted into the specimen chamber and observed. By using electrons that exhibit very short wavelengths instead of light, the resolving power of the electron microscope is several orders of magnitude greater than the resolving power of compound microscopes. Furthermore, the magnification capability of the modern transmission electron microscope ranges from approximately 600× to 500,000×. As a result, the transmission electron microscope enables the investigator to see components in bacterial cells with great detail (Figures 1.6A and B).

A distinct characteristic of all bacteria is the absence of a true nucleus. Unlike higher organisms, each bacterial cell contains a single chromosome, and it is not separated from the cytoplasm by a membrane. Within the cytoplasm, the bacterial chromosome is concentrated in areas called nuclear regions. Ribosomes (small dark particles) can also be seen in thin sections of bacterial cells, but discrete organelles (membrane-encircled bodies) are conspicuously absent. However, the cell membrane and the cell wall can be seen as distinct compo-

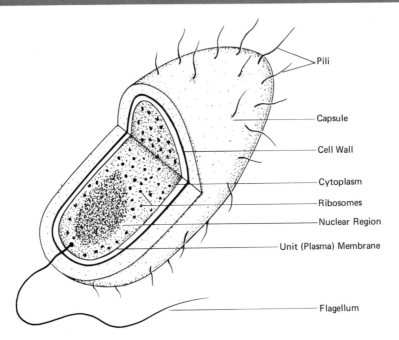

- Pili
- Capsule
- Cell Wall
- Cytoplasm
- Ribosomes
- Nuclear Region
- Unit (Plasma) Membrane
- Flagellum

**Figure 1.6A.   Cutaway view of a generalized bacterial cell show-
ing its structural components.**

nents. It is important to remember that many functions vital to bacteria are carried out by enzymes associated with the cell membrane (see Chapter 2).

The *scanning electron microscope* (Figure 1.7) is a recent modification of the transmission electron microscope. Its capability is unique, because only surface features of structures can be observed. Preparatory procedures for specimens to be examined by the scanning electron microscope are relatively simple. Whole bacterial cells can be coated with a thin film of a heavy metal such as gold, after which they are placed in the specimen chamber. The entire surface of each cell is then scanned with a very narrow beam of electrons that moves back and forth over the specimen. Images of surface structures are then projected on a screen, from which photographs can be taken as desired. A major feature of the scanning electron microscope is its great depth of field. Thus, thick specimens of varying sizes can be examined, but it is important to remember that only surface details can be visualized. Another feature that extends the usefulness of the scanning electron microscope is its magnification capability, ranging from approximately 15× to 100,000×. The surface topography of whole bacterial cells as revealed by this instrument is shown in Figure 1.8.

In addition to the unique capabilities that were stated above for

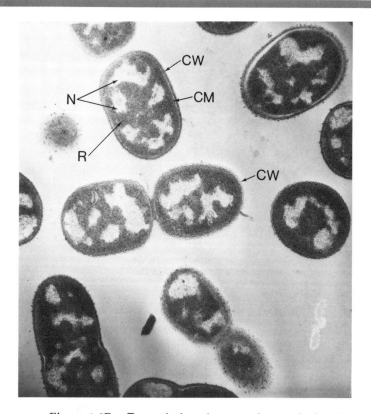

Figure 1.6B. Transmission electron micrograph showing fine structure of bacterial cells (*Achromobacter* sp.). The outer layer, or cell wall (CW), appears as a rough surface, and the inner cell membrane (CM) appears as a smooth surface. The nuclear region (N) appears as white dense areas scattered through the cell, and the ribosomes (R) appear as the dark granulated areas. Note that some cells are in various stages of elongation in preparation for division (about ×17,000). [Courtesy of G. B. Chapman, *J. Bacteriol.* 95:1862, 1968, with permission.]

each type of electron microscope, other facts to remember are: (1) cells in their living state cannot be examined with electron microscopes, (2) electron microscopes are expensive instruments and costly to operate, and (3) the individual who operates them must possess specific skills in the techniques of electron microscopy.

As mentioned previously, some structures present in cells are absolutely essential for life, but other components are variable. A particular kind of variable structure may be found characteristically in some bacterial species but be consistently absent in bacterial cells of a different type; flagella and pili are examples. *Flagella,* appendages that

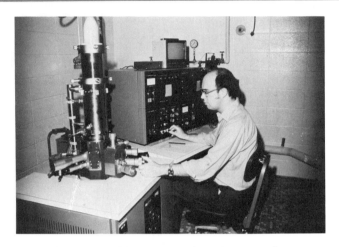

Figure 1.7. Modern scanning electron microscope. [UW-Oshkosh Electron Microscope Laboratory, Courtesy of Dr. Rodney Cyrus.]

enable certain types of bacteria to move independently, are possessed by most kinds of motile bacteria but are not present on nonmotile bacteria or bacteria that move as a result of different mechanisms. *Pili*, which resemble flagella, do not contribute to the motility of cells that possess them. Pili are not present on all bacteria: some cells do not have any, others may have various types. Of special interest is the *sex pilus*, because it plays a role in bacterial conjugation. It is important to remember that certain kinds of bacteria produce variable structures in response to specific changes in the environment that surrounds them. This type of behavior is exhibited by bacteria that are able to form capsules and/or endospores.

A *capsule* is the gelatinous material that is tightly bound to the outermost surface of a microbial cell. When such material is loosely attached to the cell's surface, it is called a *slime layer*. Capsules or

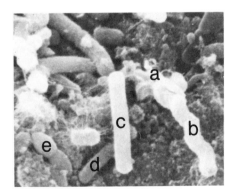

Figure 1.8. Scanning electron micrograph of bacteria on the colonic epithelium of the dog (about ×6,000): (A) and (B) large spiral-shaped bacteria; (C) and (D) bacilli; (E) cocobacilli. [Courtesy of C. P. Davis, unpublished.]

slime layers are produced by many kinds of microbial cells, and they may play important ecological roles. Encapsulated cells of the bacterium *Streptococcus pneumoniae* (formerly called *Diplococcus pneumoniae*) are pathogenic for human beings, because they are able to avoid being destroyed by the body's defense mechanisms. Nonencapsulated cells of pneumococci are harmless. Capsules are often formed in response to changes in the cell's environment. For example, pneumococci will form capsules when injected into the peritoneal cavity of a mouse, but the same cells will not form capsules when grown in some synthetic media under laboratory conditions. Capsules and slime layers may contribute to the survival of other kinds of cells in natural habitats by enabling them to become attached to a wide variety of surfaces.

Endospores (bacterial spores) are unique structures, formed only by certain types of bacteria, and should not be confused with the types of spores formed by other organisms, especially the fungi. The most commonly found, and best studied, types of bacteria that form endospores are found within two genera: *Bacillus* and *Clostridium*. The genus *Bacillus* contains only aerobic species, and the genus *Clostridium* contains only anaerobic species.

Bacterial endospores are extremely resistant to heat, drying, and the action of many chemicals. Thus, they survive in environments that would otherwise be destructive. An endospore is a dormant state of the vegetative cell from which it was formed. Structurally and chemically, endospores are different from vegetative cells. During the formation of an endospore, the vegetative cell undergoes a series of complex changes, among which are the formation of new structures and the synthesis of a chemical substance, *dipicolinic acid,* that was not present in the vegetative cell. The process ends with the development of a mature endospore. Bacterial endospores are structurally complex, composed of several distinct layers. The outermost layer is called the *exosporium,* within which is the *spore coat,* which may consist of several layers. Beneath the spore coat is the *cortex.* The cortex surrounds the *core,* which is not vastly different from the interior of a vegetative cell.

A cell may remain dormant in the spore state for long periods (even years), but when introduced to a favorable environment, it will germinate and reproduce itself vegetatively. The germination and outgrowth of a dormant endospore of *Bacillus subtilis* is shown in Figure 1.9. Note the thick outer spore coat and the dense cortex in the dormant spore as compared to the thinner outer coat and the fibrous cortex just 10 minutes later. Progressively during germination the core swells and pushes outward. As the inner and outer spore coats disintegrate, the new cell begins to emerge, a process known as *outgrowth*. After emergence, the cell elongates and begins to divide.

The preceding discussion is an oversimplification of a complex pro-

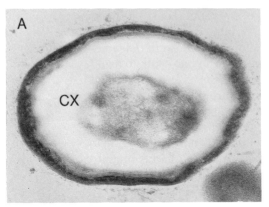

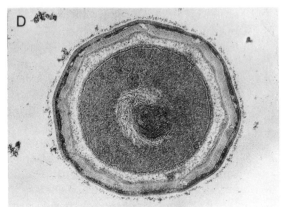

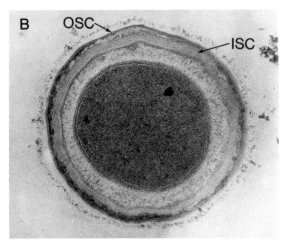

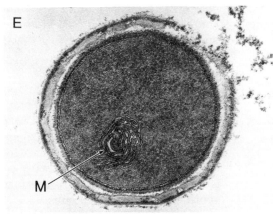

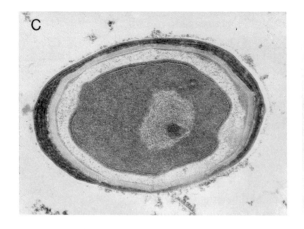

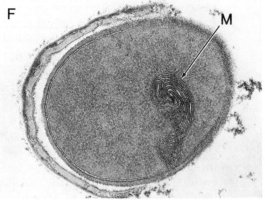

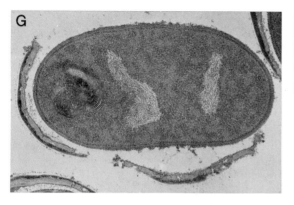

Figure 1.9. Transmission electron micrograph of an endospore of *Bacillus subtilis* showing sequential structural changes that occur during germination and outgrowth: (A) dormant endospore, (B) at 10 minutes, (C) 20 minutes, (D) 30 minutes, (E) 40 minutes, (F) 50 minutes (emerging vegetative cell), (G) 60 minutes. (H) 90 minutes (cell preparing for division by binary fission). CX, cortex; OSC, outer spore coat; ISC, inner spore coat; M, mesosome. [Courtesy of R. H. Doi and L. Y. Santo. *J. Bacteriol.* **120**:475, 1974, with permission.]

cess. However, some knowledge of endospores and of their outgrowth is of practical significance, because sterilization methods in hospitals, in pharmaceutical firms, and in various food-processing industries must be designed to destroy them. When sterilization processes are inadequate in such industries, serious consequences may result. A number of recent human deaths from botulism (an illness that results from the consumption of a potent toxin produced by an anaerobic bacterium, *Clostridium botulinum)* resulted from eating improperly sterilized canned food items.

In terms of ecological significance, organisms that form endospores are more versatile, because such structures are formed when cells are confronted with adverse conditions, such as starvation, when available nutrients become exhausted. Furthermore, spores enhance a cell's survivability while being transported to a suitable environment.

In a suitable environment, bacteria usually reproduce by *binary fission* a process in which a single cell divides transversely into two separate units. Each new cell, often called a *daughter cell,* contains an exact complement of genetic information and cytoplasmic materials from the parent cell. Each daughter cell can then repeat the process of binary fission and produce two additional cells. In other words, bacterial cells increase by geometric progression. Generation time (time required for a given population to double) may be as short as 20 minutes for some species. For this reason, under optimum conditions for growth, bacterial cells may increase to astronomical figures in a very short period (see Chapter 6). Although binary fission is the predominant mode of reproduction among the bacteria, some members of these groups propagate themselves through other processes, such as fragmentation and budding.

An interesting fact to remember is that some species of bacteria are capable of transferring genetic material from cell to cell through processes that do not involve cell division. These mechanisms are transformation, transduction, and conjugation. *Transformation* is a process through which a free (or isolated) fragment of a bacterial chromosome is taken up by a recipient bacterial cell and subsequently becomes integrated into the recipient's chromosome. *Transduction* is the transfer of a gene or several genes from a donor bacterial cell to a recipient through an intermediate agent — a specific type of bacterial virus (an entity that will be described later in the chapter). The third process, *conjugation,* is the transfer of genetic material through a mating process that requires cell-to-cell contact. Bacterial cells that participate in the process of conjugation are of two mating types. One partner acts as the donor of genetic material and the other acts as the recipient. All donor cells harbor a transmissible nonchromosomal genetic element called the *F-factor* within their cytoplasm, and are designated as F^+ cells, or males. Recipient cells, which do not harbor the F-factor, are designated as F^- cells, or females. During conjugation, every F^-, or female, cell is always converted to a F^+, or male, cell because the F-factor carries genes that governs the heritability of maleness. Both mating types may be of the same species, or each partner may belong to a different genus and species.

Previously, we stated that sex pili were among the kinds of variable structures that bacterial cells may possess. It is of interest to note that F^+ cells have sex pili but F^- cells do not. Furthermore, the sex pilus is absolutely essential to the conjugation process, because it is the structure that forms the conjugation bridge through which the genetic material passes from donor cell to recipient cell. The F-factor is one of several kinds of genetic elements called plasmids. A *plasmid* is an autonomous hereditary genetic element that exists within the cytoplasm of a bacterial cell, separate from its chromosome. Several classes of plasmids have been described, and they are known to exist in many types of bacteria. Plasmids are not essential to a cell, but each of them contains genes that confer a new property or selective advantage to the bacterial cell that harbors it. *R-factors* are plasmids that exhibit such characteristics, because they carry genes for resistance to antibacterial drugs. When R-factors are present in disease-producing bacteria, the property of drug resistance can be conferred to other bacteria through conjugation. Instances of this phenomenon often occur in bacteria that are commonly present in the environment of hospitals. The management of hospital-acquired infections caused by R-factor-mediated drug-resistant bacteria has emerged as a serious problem (see Chapter 8).

Blue-green algae

Those procaryons designated as *blue-green algae* are more conspicuous in nature than bacteria are. They are abundant in soil and water habitats, and the range of environmental conditions in which they grow are extremely diverse. Members of this group are found in desert soils, others grow in the acid waters of hot springs, and still others grow abundantly in fresh and marine bodies of water.

Organisms within the blue-green algal group are characterized by their primitive cell structure. Cytoplasmic membrane-enclosed organelles are conspicuously absent, and their genetic material is dispersed throughout the cytoplasm in nuclear regions, as in the bacteria. As a result of their procaryotic nature, many scientists have suggested that the blue-green algae be taxonomically designated as cyanobacteria. Those procaryotic features also clearly differentiate blue-green algae from eucaryotic algae. In spite of their procaryotic nature, blue-green algae differ from bacteria by having more complex cellular structures and by having a different complement of photosynthetic pigments. All the blue-green algae are photosynthetic organisms and have pigment centers located in membrane-bound structures. In some species such membranes are complex and multilayered. Types of photosynthetic pigments found in blue-green algae are chlorophyll a and various kinds of accessory pigments (phycocyanins, phycoerythrins, carotenes, and xanthophylls). The characteristic color exhibited by a particular species results from a specific complement of those pigments. For example, the blue-green color of some species results from the presence of a combination of pigments: chlorophyll a (green) and phycocyanin (blue). However, all members of the group are not blue-green in color; a wide range of colors are exhibited by different species. As a result of the distinct blue, green, or red colorations that certain species produce, their abundant growth in fresh and marine bodies of water, referred to as "algal blooms," is easily recognized. All photosynthetic organisms contain a number of pigments, but the type of chlorophyll pigment varies among organisms within different taxonomic groups. Only one kind of chlorophyll (chlorophyll a) is present in all species of blue-green algae. Eucaryotic algae and photosynthetic higher plants have chlorophyll a, but many species within these groups also have other kinds of chlorophyllous pigments.

Most members of the blue-green algae exist as unicells or as filamentous forms and reproduce primarily by the process of binary fission. However, reproduction in some species occurs by fragmentation of the filaments. Certain structures—notably heterocysts, gas vacuoles, and akinetes—are produced by some filamentous types. These structures appear to have unique functions. For example, species of blue-green algae that possess *heterocysts* also have the ability to fix atmospheric nitrogen (see Chapter 5 for a discussion of the process). Heterocysts are structurally different from adjacent vegetative cells and develop at various intervals along the filament. Although present in blue-green algae that fix nitrogen, the role that heterocysts play in the process is not known.

Gas vacuoles are hollow, rigid structures that develop in certain filamentous and afilamentous species. Gas vacuoles, most prevalent in algae that live in fresh and marine waters, function to increase the buoyancy of the organism. By floating, photosynthetic organisms are maintained in the upper waters, where the availability of light is greatest.

Certain species of blue-green algae form *akinetes,* a resting stage or spore that functions to increase the survivability and dispersal of the species. In a favorable habitat, akinetes germinate and develop into the filamentous vegetative form. In general, considerable diversity of form and cellular arrangement can be found among the various species of blue-green algae. Some common types are shown in Figure 1.10.

Mycoplasmas

A unique group of microorganisms, the *mycoplasmas,* were first recognized in the eighteenth century, although they were then called "pleuropneumonia-like organisms" (PPLO). The group includes both saprophytes and pathogens.

The mycoplasmas are free-living procaryotic cells that do not possess cell walls. As entities without cell walls, they are less complex in chemical structure than the bacteria, and all of them can be cultivated in the laboratory on enriched synthetic media. Some mycoplasmas display an absolute requirement for sterols, a criterion that can be used to differentiate them. Four genera are recognized: (1) those in the genus *Mycoplasma* require sterols for growth; (2) those in the genus *Acholeplasma* do not require sterols for growth; (3) *Thermoplasma acidophilum* (the only species described) does not require sterols for growth, but grows optimally at 59°C; and (4) *Spiroplasma citri* is the most recently proposed species, but its exact position among the mycoplasmas is somewhat questionable.

The mycoplasmas are unique in respect to their growth patterns on solid and in liquid media. On agar plates, the colonies exhibit the distinct appearance of a fried egg. The dark center is more dense than

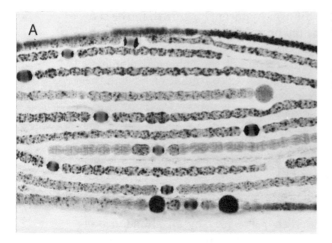

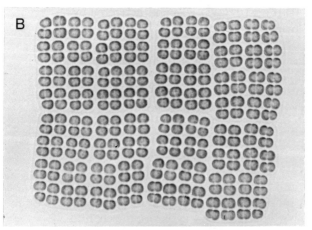

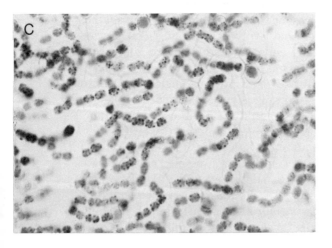

Figure 1.10. Representative types of blue-green algae: (A) *Anabaena;* (B) *Meresmopedia;* (C) *Nostoc.* [Courtesy of the Carolina Biological Supply Company.]

the outer edge, which is rather diffuse. Apparently, growth in the center penetrates down and into the medium, while growth on the periphery extends horizontally. In liquid media, mycoplasmas exhibit both spherical and filamentous elements. Some of the spherical elements have diameters within the range 0.125 to 0.220 μm, at or beyond the range of resolution of light microscopes. The unique growth pattern for one species, *Mycoplasma pneumoniae,* has recently been described (Figure 1.11). Observations with a scanning electron microscope revealed that morphological changes in *M. pneumoniae* occur in an orderly and sequential manner, progressing from small spherical forms to filamentous forms and back to spherical forms larger than those initially observed. Three distinct phases of growth were observed: Phase I (8 hours to 2 days) was characterized by the predominance of spherical forms. Phase II (2 to 6 days) was characterized by an abundance of straight and branching filaments with bulbous elements situated along their lengths, and phase III (6 to 10 days) was characterized by the reappearance of spherical forms somewhat larger than those observed initially.

The mycoplasmas are widespread in nature. Saprophytic species are commonly found in polluted waters and soils, and the pathogenic species are often associated with atypical pneumonia and various other infections in human beings and with pleuropneumonia in cattle.

Eucaryotic organisms

Eucaryotic organisms represent within themselves all the complexities of life—ranging from independent unicellular forms to the human organism, which contains differentiated cells organized into tissues, organs, and systems, all of which function cooperatively. Eucaryons include cells that are present in many biological groups, all of which probably represent types that evolved from ancestral procaryons. Apparently, the less-stable cells responded to environmental pressures and evolved into organisms with a greater degree of complexity. These cells are characterized by a nucleus enclosed within a nuclear membrane, sexual reproductive structures, and cytoplasmic *organelles,* membrane-enclosed bodies that carry out specialized functions. Mitochondria and chloroplasts are examples of the latter.

Mitochondria are found in all eucaryotic cells, but they vary both in number and in kind with type of cell and organism. They have two very distinct components, an outer membrane and an inner membrane. The *outer membrane,* resembles the plasma or unit membrane common to all cells, but the *inner membrane* is greatly folded. The folds of the inner membrane are called *cristae,* and associated with them is

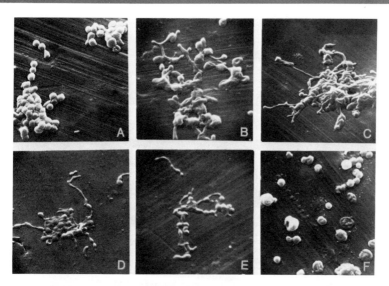

Figure 1.11. Growth cycle of *Mycoplasma pneumoniae* as observed with a scanning electron microscope: (A) phase I (8 hours): spherical forms; (B) early phase II (2 days): spherical cells, short and straight branching filaments; (C) phase II (3 days): microcolony composed of interwined filaments and occasional round forms; (D) mid-phase II (4 days): microcolony composed of both round and filamentous forms with bulbous elements; (E) late phase II (6 days): shorter filaments with both terminal and intervening bulbs; (F) phase III (10 days): larger, asymmetrical round forms (×4,100) [Courtesy of J. D. Pollack, *J. Bacteriol.* **104**:499, 1970, with permission.]

the cell's energy-generating system. Mitochondria are responsible for the production and packaging of energy into high-energy components for cellular functions.

Chloroplasts, unlike mitochondria, are found only in eucaryotic algae and eucaryotic green plants. They are conspicuously absent from animal cells. The function of chloroplasts is to carry out the reactions of photosynthesis. They contain chlorophylls and other accessory pigments required for capturing radiant energy and for transforming it into chemical energy that can be used for cellular processes.

Both mitochondria and chloroplasts are autonomous structures within eucaryotic cells. They contain their own genetic material, distinctly different from that within the cell's nucleus, and they contain apparatus for synthesizing their own proteins. Furthermore, they replicate themselves, independent of their host cell, by binary fission. As a result of such unique features, several theories have been proposed to account for their origin. One hypothesis suggests that mitochondria and chloroplasts may have evolved from free-living procaryotic ancestors and become obligate endosymbionts of eucaryotic cells (see Chapter 5).

The microscopic forms of eucaryotic organisms are distributed among the following groups: fungi (yeasts and molds), algae, protozoa, and isolated single cells of multicellular organisms which can be cultivated as unicellular independent entities under laboratory conditions.

Fungi

The *fungi* represent a diverse group of saprophytes. Organisms within this group obtain and utilize only preformed nutrients, most often from dead or decaying organisms. However, some types are *parasites;* that is, they live in or on tissues of another organism and inflict harm to it while obtaining their nutrients. Other types of fungi are *parasite-saprophytes.* In this type of association, the parasite kills the other organism and lives saprophytically on the decomposing remains.

The fungi include both macroscopic forms of organisms. Chlorophyll and other photosynthetic pigments are conspicuously absent from these cells, and their structures are not differentiated into roots, stems, and leaves. Fungi also differ from higher plants by having cell walls composed of chitin, whereas the cell walls of higher plants are composed of cellulose.

Yeasts Afilamentous unicellular forms of fungi are called *yeasts.* They are generally recognized by their distinct morphology, which

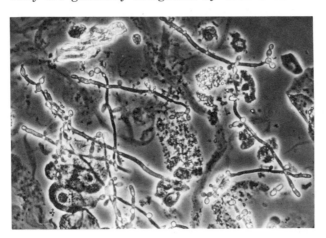

Figure 1.12A. Budding yeast cells of *Candida albicans* isolated from urine, as viewed with a phase-contrast microscope (×1,000). Note the attached hyphae. [Courtesy of G. D. Roberts, *J. Clin. Microbiol.* 2:261, 1975, with permission.]

consists of ovoid, ellipsoidal, or budding cells (Figures 1.12A and B). They can be differentiated from bacteria by their large size, morphological features, and the presence of chitin in their cell walls, from algae by not having photosynthetic pigments, and from protozoans by possessing rigid cell walls.

Yeasts reproduce asexually by budding or fission, and sexually through complex processes that involve both conjugation and sporulation. Patterns of reproduction among these cells are diverse, but the concept can be generalized from the life cycle of baker's yeast, *Saccharomyces cerevisiae*. This organism has heterothallic structures, which, although morphologically similar, produce two different types of sexual gametes (opposite mating types). Mating types are differentiated

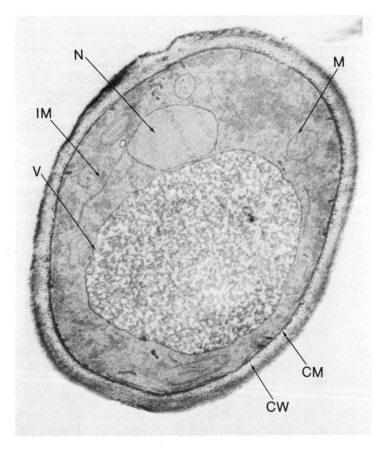

Figure 1.12B. Transmission electron micrograph of a yeast cell, *Candida albicans* (about ×24,000). CW, cell wall; CM, cell membrane; N, nucleus; IM, internal membrane; M, mitochondria; V, vacuole. [Courtesy of E. Balish, unpublished.]

by + and − notations in Figure 1.13. If each mating type is kept separate, homogenous gametes will not mate among themselves but each type will reproduce itself asexually by budding. Mating occurs only between opposite types.

During the sexual phase, opposite mating types conjugate, after which they undergo the process of *mitosis*. This process results in the formation of a cell with a double set of chromosomes, one set from each mating type *(diploid)*. Then the diploid cell undergoes a process of *meiosis*, which results in the formation of four ascospores (two of each mating type). These are *haploid,* because each contains only half of the chromosomes present in the diploid phase. Note that meiosis is the opposite of mitosis and results in reducing the chromosome number of the diploid cell to its haploid number. Some cells do not undergo meiosis after mitosis but instead enter into an asexual cycle (diploid) and reproduce themselves by budding. We have described the reproduction cycle for *Saccharomyces cerevisiae,* which consists of two separate asexual phases or a combination of an asexual and a sexual phase. Remember that variations from this cycle can be found in other yeasts.

Interest in this group of fungi is not new. Yeasts cells were first described by Leeuwenhoek in the seventeenth century and were characterized by Pasteur in the nineteenth century as the agents responsible for alcoholic fermentation. Subsequently, yeasts emerged as industrially important organisms, because of their ability to transform a variety of substances into products of economic interest—wines, alcohols, organic acids, and a variety of other substances. A distinct characteristic of yeasts is the large quantity of gas that is evolved as a product during fermentation. For this reason, baker's yeast is a common household item. When added to freshly prepared dough, yeast cells grow and produce carbon dioxide, a gas, which causes the dough to rise. When such dough is baked in an oven, a light-textured bread or other product results.

In addition to the variety of strains that are of economic importance, other species, such as *Candida albicans,* often cause serious infections in human beings. Many types of yeast infections are superficial, but serious systemic diseases may occur in debilitated patients and in those suffering from such traumas as postoperative wounds and severe burn injuries.

Molds   Fungi in the *mold* group are very conspicuous in nature and are recognized by their filamentous structures and macroscopic size, features that separate them from the afilamentous cells of yeasts. This is not a definitive basis for separating the two groups, because some species of yeast develop pseudomycelia and some species of molds grow as diphasic entities, exhibiting yeastlike cells under one set of

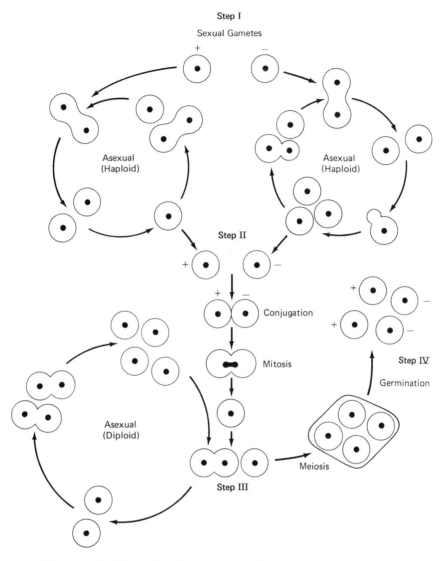

Figure 1.13. Life cycle of baker's yeast, *Saccharomyces cerevisiae.* In step I, each haploid sexual gamete (opposite mating types) reproduces itself asexually. Step II is the sexual phase; opposite mating types conjugate and undergo mitosis. In step III the resulting diploid cell may reproduce itself asexually or undergo meiosis. In step IV the ascospore germinates, releasing four sexual gametes (two of each mating type).

environmental conditions and filamentous growth under other conditions. Mold colonies have a fibrous or cobweblike appearance that results from the aggregation of filaments (hyphae) to form a mass, called a *mycelium*.

Morphological features (type of hyphae, spores, and sexual gametes) are stable characteristics among different groups of molds. Hyphae may be *septate,* with cross-walls, or *nonseptate,* without cross-walls, and may appear as submerged vegetative structures or as aerial spore-bearing filaments (Figures 1.14A and B). Although vegetative structures may exhibit some degree of specialization, molds do not have differentiated systems of roots, stems, and leaves.

Slime Molds The *slime molds* comprise an interesting group of eucaryotic cells, because they have some characteristics that resemble the fungi and others that are distinctly animal-like and similar to the protozoans. Their name derives from a slimy vegetative cell mass, called a *plasmodium,* which glides or crawls over the surface of decaying material. During this process, the organism absorbs nutrients and engulfs bacteria. At some stage in their life cycle, all slime molds produce funguslike fruiting bodies.

Slime molds are divided into two types: acellular and cellular. *Acellular* slime molds have multinucleated vegetative structures that always grow in the jellylike plasmodium form. In highly favorable environments, this structure will continue to enlarge and may weigh many pounds. Under less favorable conditions, the plasmodium will develop into fruiting bodies which are usually brightly colored.

Cellular slime molds have uninucleated vegetative cells which are flagellated. The protozoanlike cells multiply by binary fission and function for a period in this stage but eventually aggregate into the jellylike plasmodium and produce asexual fruiting bodies. In this stage the cells resemble fungi.

Currently, scientists regard the slime molds as possible models for studying the mechanisms of cell differentiation.

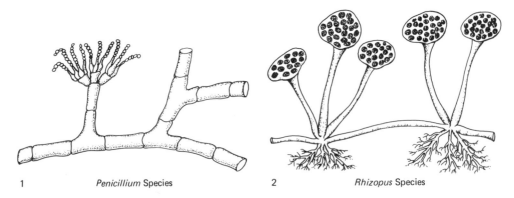

1   *Penicillium* Species   2   *Rhizopus* Species

Figure 1.14A. Types of mold hyphae: (1) septate (*Penicillium* species) and (2) nonseptate (*Rhizopus* species).

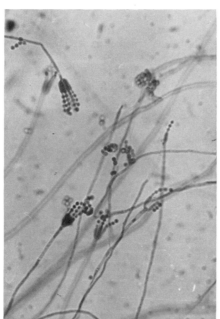

Figure 1.14B. Compound Light microscope photo of molds: (l.) *Penicillium* species and (r.) *Rhizopus* species. [Courtesy of the Carolina Biological Supply Company.]

Algae

The eucaryotic *algae* represent the extremes of diversity among microorganisms. They exhibit a wide variety of structural appearances and considerable heterogeneity can be found among their reproductive structures (sexual gametes and asexual spores). However, the various algal groups can be differentiated by means of the varying biochemical characteristics of their pigments, storage products, and cell-wall components. Sizes range from the unicellular *Chlorella* to the large multicellular kelps, *Macrocystis*. Morphologically, this group has rootlike, stemlike, and leaflike structures, but these lack the capacity to function as true roots, stems, and leaves. All algae are composed of undifferentiated vegetative cells, and they do not have a vascular system comparable to higher plants. Representative species of microscopic eucaryotic algae are shown in Figure 1.15.

Unlike the molds, which are saprophytes or parasites, all eucaryotic algae have photosynthetic pigments contained within chloroplasts (Figure 1.16). These internal structures are self-replicating cytoplasmic units. All photosynthetic algae contain chlorophyll a, but other types of chlorophylls and accessory pigments vary considerably

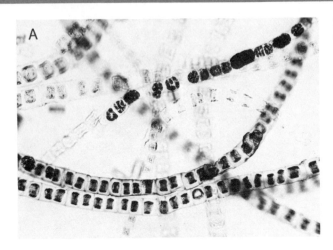

Figure 1.15. Compound light microscope photo of two kinds of microscopic eucaryotic green algae **(A)** *Ulothrix* species; **(B)** *Volvox* species. [Courtesy of the Carolina Biological Supply Company.]

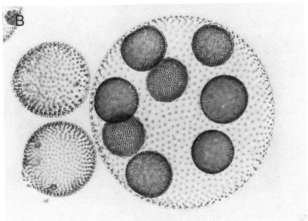

among the various groups. Owing to the difference in ratios of chlorophylls to accessory pigments, the colors displayed by the various algal groups are stable characteristics. In the Crysophyta carotenoid pigments mask chlorophylls to produce a golden-brown color. Similarly, the phycobilin pigments mask the chlorophylls in the Rhodophyta to produce their characteristic red color.

Considering this synoptic view of the algae as a background, our discussion in succeeding chapters will be restricted to selective types that are microscopic in size.

Protozoa

Organisms characterized as *protozoa* are predominately unicellular animals, but some forms possess algalike features. However, they

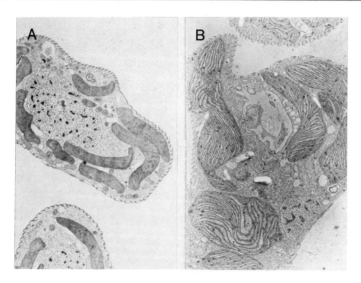

Figure 1.16. Transmission electron micrograph showing *Euglena gracilis* chloroplast membrane structures (×3,400): (A) compact chloroplast; (B) swollen chloroplast. [Reprinted with permission of the Society of Protozoologists from *The Journal of Protozoology*, 1973, Volume 20, pages 652–653.]

do not possess photosynthetic pigments. Some authorities consider the protozoa as regressive forms of algae. Among the known groups of microorganisms, protozoa contain the most complex type of unicells, and all members of this group have some form of motility.

Like other groups of microorganisms, protozoa are a heterogeneous group of organismic forms (Figure 1.17). With few exceptions they are overwhelmingly unicellular entities in which the outer covering is a nonrigid pellicle. Rigid plantlike cell walls are conspicuously absent.

Protozoa exist in nature both as free-living entities and as obligate parasites. Their form of nutrition may be *saprozoic* (absorptive) or *holozoic* (ingestive). The latter process is usually considered to be a distinctive trait of animals.

Reproduction among the protozoa may involve either sexual or asexual processes. Many of the parasitic protozoa have complex life cycles in which stages must pass through a specific host, usually mammals or arthropods.

We usually associate the protozoa with a variety of parasitic diseases, but many free-living forms play important ecological roles. They contribute significantly to soil fertility by decomposing organic matter, and they aid in maintaining an ecological balance in the biosphere by consuming bacteria and other microscopic organisms as food.

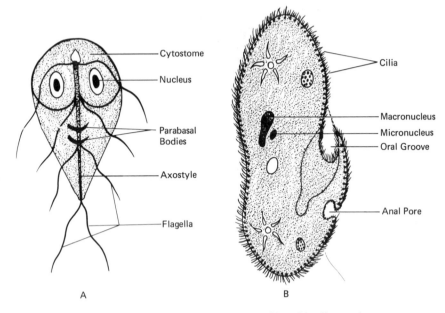

Figure 1.17. Representative protozoans: (A) a *Giardia* species, a common inhabitant of human gastrointestinal tract; (B) a *Paramecium* species, a free-living inhabitant of organically rich waters.

Although the protozoa comprise a heterogenous group of unicellular organisms, some of the most interesting members are found among the ciliates. Organisms within this group are unicells that exhibit a great deal of differentiation in cellular structures. All members of the ciliates are characterized by having hairlike structures, *cilia* on their cells during some period of their life. Cilia are organs of locomotion. They beat in a rhythmic and coordinated manner and propel the organism from one place to another. Organisms within this group also possess two nuclei: a macronucleus and a micronucleus. The *macronucleus* plays an important role in regulating growth and cellular activities but does not function in sexual reproduction. The *micronucleus,* on the other hand, plays a functional role in sexual reproduction. The ciliates obtain food by ingesting bacteria and other materials through an oral groove (mouth) connected to a food vacuole where digestion occurs. These food vacuoles then move within the cytoplasm of the cell as a result of cytoplasmic streaming. Soluble nutrients are absorbed, and undigested waste is excreted through the anal pore. *Paramecium* is a genus of ciliates that exhibit the characteristics mentioned above. See Chapter 5 for some interesting ecological roles that various *Paramecium* species play in nature.

Isolated cells of multicellular organisms

Many types of cells from tissues of higher organisms will function as independent units of living matter when cultivated in a medium apart from the organism *(in vitro)*. In such environments, cells may lose their ability to carry out specialized functions but will grow as undifferentiated unicellular entities. The laboratory system for cultivating cells of higher organisms in artificial environments is called *tissue* or *cell culture*.

During the process of adapting to growth in tissue culture, most cells undergo transformation. Subsequently, transformed cells are unable to function in a specialized manner, but, instead, grow indefinitely in an undifferentiated state with characteristics of unicellular microorganisms.

Perusal of the literature in the field of tissue culture will quickly reveal a confusing and somewhat unique terminology. Therefore, special attention is given here to the use of a few basic terms. During the development of tissue culture as a specialized area of microbiology, many terms were acquired from the related field of bacteriology, but different meanings were given to other terms. One such term is "transformation"; in bacteriology, the term refers to a process in which a portion of a bacterial chromosome (a single gene or several genes) existing in a free state is taken up by a recipient bacterial cell and incorporated into the chromosome of the recipient bacterial cell, whereas in tissue culture "transformation" refers to an adaptive change in cellular function as a result of continued cultivation *in vitro* or in response to an infection by a virus such as SV-40 (see the next section for a discussion).

Cell cultures prepared directly from tissues of higher organisms are called *primary cultures,* and the cells are referred to as a *primary cell line.* Cell cultures prepared from a primary cell line are called *secondary cultures.* Subsequently, cells from secondary and succeeding cultures can be used to prepare new tissue culture systems. This process is called *serial passage.* However, primary cell lines are usually characterized by a limited life span when cultivated *in vitro*. On the contrary, continuous cell cultures can be propagated by serial passage *in vitro* for an indefinite period, and cells with this capability are referred to as *established cell lines.*

Unquestionably, the rapid expansion of virology as a basic science is directly related to, if not dependent upon, the development of tissue culture techniques. Since viruses are incapable of self-duplication in a free-living state (see the next section), tissue culture techniques provide us with a system for the *in vitro* cultivation, outside the host, of many virus types. In addition to studies on the basic biology of viruses, valuable information has been obtained from the use of tissue culture in the study of host–parasite relationships at the cellular level. In essence, tissue culture has emerged as a powerful biochemical tool with unlimited potential. Its techniques are useful to scientists concerned with the development of chemotherapeutic agents and to those involved with research into virus diseases and cancer. It also appears promising as a tool for the assessment of cellular injury from environmental pollutants.

Viruses (noncellular agents)

Although much younger than the other areas of microbiology, the field of virology has emerged as a dynamic and fascinating area of study. Detailed study of the bacteria preceded that of the viruses, not because viruses were less interesting, but because techniques for studying biological systems at the subcellular level is of recent origin when compared with techniques for studying bacteria.

Viruses differ markedly from organisms previously described in other microbial groups. Most viruses are submicroscopic in size and can only be visualized with an electron microscope. Viruses are structurally simple. The virus particle, or *virion,* consists of a core that is composed of only one type of nucleic acid, either *deoxyribonucleic acid* (DNA) or *ribonucleic acid* (RNA). The nucleic acid core is surrounded by a protein coat (called a *capsid).* Protein subunits from which the capsid is constructed are called *capsomers.* In many viruses, the capsid is the outermost structure, but the nucleocapsid (core + capsid) of certain other viruses is surrounded by a membranous layer called an *envelope.* Extracellular viruses are lacking in the ability to acquire and utilize nutrients, the ability to reproduce, and the ability to grow. In other words, the extracellular virus particle is unable to function as an independent unit of life. All living cells are composed of both DNA and RNA and have the ability to carry out life processes as independent self-sufficient units. In contrast, the virion is inert and is no more alive than the organic chemicals that we may find in any biology or chemistry stockroom.

The characteristics described preclude the viruses from being considered as either procaryotic or eucaryotic cells. However, extracellular virus particles have the ability to invade and replicate themselves within susceptible living cells. A susceptible cell is referred to as a host, and the virus is called a parasite. Actually, viruses are obligate intracellular parasites, because they cannot function as independent units of life when existing as extracellular entities. In the intracellular state viruses cannot reproduce themselves from preexisting viruses, but genes released from the virus particle assume control of the host cell's biosynthetic capacity. Then, while under the control of viral genes, the host cell synthesizes viral structural components, which are then assembled intracellularly into new viruses and released.

The relationship of viruses to the various groups of microorganisms is questionable (see Figure 1.1). Some authorities consider viruses to be regressive forms of preexisting parasitic cells. Supporting this concept is the fact that viruses are unable to replicate themselves when in the extracellular state. Consequently, they could not have existed as primitive living systems prior to the evolution of procaryotic cells. In relation to other microbial groups, viruses may be considered as unique infectious agents that bridge the gap between inanimate materials and the simplest forms of living cells.

Viruses are a heterogenous group of submicroscopic agents that selectively parasitize living cells of all biological groups. Viruses are highly specific in the selection of cells that they infect and are commonly categorized, on the basis of host-range specificity, as animal viruses, plant viruses, algal viruses, bacterial viruses (bacteriophages), and so on. Classification systems for viruses within a category are extremely complex and are based on both structure and chemical composition.

When a susceptible cell is infected by a virus, a series of complex interactions take place between the host cell and the infecting virus. The host–parasite interactions are initiated by the virus genetic information that enters the host cell during the invasion process. The replication cycle for most viruses usually involves at least five steps: (1) attachment of the extracellular virus particle to the host cell, (2) penetration or entry of the virus or its nucleic acid into the host cell, (3) synthesis or production of essential viral components, (4) assembly of components into new virus particles, and (5) release of virus progeny (with or without lysis of the host cell). Variations in reproductive cycles are determined, to a degree, by the type of virus and the type of host cell.

Since structural features and chemical properties of a particular kind of virus are constant and stable, the following criteria are generally used to differentiate them into classes:

1. The type of nucleic acid present in the core of the virus particle
2. Whether the nucleic acid is single- or double-stranded
3. The number and arrangement of capsomers in the capsid
4. The size and shape of the capsid
5. The presence or absence of an envelope

In addition to those criteria, various specialized tests have been developed for the definitive identification of viruses.

In the sections to follow we shall present some general aspects of bacterial viruses (bacteriophages) and of animal viruses.

Bacterial viruses (bacteriophages)

The heterogenous group of viruses that have the ability to invade and replicate themselves within bacterial cells are called *bacterio-phages*. Viruses within this group are extremely diverse. Of equal importance is the degree of diversity that exists among the various kinds of cells that function as hosts. In fact, many microbiologists believe that most species of bacteria are capable of serving as host for one or more kinds of bacteriophages.

Structurally, some bacteriophages are extremely complex; they also vary in terms of chemical composition. Bacteriophages, like all other kinds of viruses, contain only one type of nucleic acid, either DNA or RNA (although bacteriophages with a core of DNA, and bacteriophages with a core of RNA, have both been described). Furthermore, the nucleic acid molecule in both groups may be either single- or double-stranded. Most DNA bacteriophages are double-stranded, and most RNA bacteriophages are single-stranded. Although less common, single-stranded DNA bacteriophages and double-stranded RNA bacteriophages have also been described.

A schematic diagram of one type of bacteriophage (phage) and the events that may occur when a bacterial cell is invaded are shown in Figure 1.18. During the infection process, this type of phage particle becomes attached to the host bacterial cell by tail fibers. Then, through a syringelike action, the phage injects its nucleic acid into the bacterial cell, but the protein shell of the phage remains outside the bacterium.

Subsequent interactions within the phage-infected bacterial cell may follow either one of two paths. When the interactions result in the vegetative production of phage progeny and the ultimate destruction *(lysis)* of the host bacterial cell, the process is called the *lytic cycle* and the infecting agent is referred to as a *virulent phage*. In some phage-infected bacterial cells, the phage nucleic acid does not lead to the production of phage progeny but becomes incorporated into the bacterial chromosome. This process is called *lysogeny,* and the infecting agent is called a *temperate phage*. In this state the host bacterial cell is not destroyed but carries the newly acquired piece of phage nucleic acid (a *prophage)* as an integral part of the bacterial chromosome. Under some environmental conditions, temperate phages undergo spontaneous vegetative development through the lytic cycle. However, temperate phages usually transfer genetic information to bacterial cells without destroying them and so are also called *transducing phages*. Often, lysogenic bacterial cells exhibit characteristics that are different from

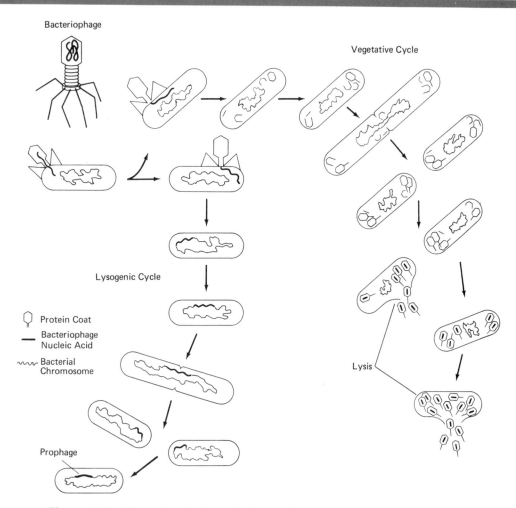

Figure 1.18. Life cycle of a bacteriophage (phage). The phage infects a susceptible nonlysogenic bacterial cell and injects its genes. Subsequently, the process may follow either of two paths. The vegetative cycle leads to the production of new phages and ultimately to lysis of the bacterial cells. The lysogenic cycle does not produce new phage progeny, but the virus genes are incorporated into the bacterial chromosome as a prophage.

their nonlysogenic counterparts. For example, the bacterium *Corynebacterium diphtheriae* causes the disease diphtheria when in the lysogenic state with a specific prophage, but the nonlysogenic cells of this bacterium are harmless.

Currently, a great deal of research is being conducted on various kinds of bacteriophages and on the nature of their interaction with various hosts (bacterial cells). Extensive use of bacteriophages as research tools can be attributed in part to the simplicity of the techniques used to grow bacterial host cells in the laboratory. On the other hand, it is more difficult to propagate animal viruses in the laboratory, because of the meticulous techniques required to grow host mammalian cells in tissue culture. Thus, bacteriophages have emerged as invaluable research tools for molecular biologists. Knowledge gained from research studies that employ bacteriophages has contributed significantly to the understanding of how all cells (procaryotic and eucaryotic) function at the molecular level. Bacteriophages also play important ecological roles in nature (see Chapter 5).

Animal viruses

Viruses that are able to invade and replicate themselves in mammalian cells are called *animal viruses*. This group contains a large number of viruses, with great diversity in terms of morphological features and chemical composition. Animal viruses, like bacteriophages, contain either DNA or RNA in their core, and the nucleic acid molecule may be either single- or double-stranded. Replication of animal viruses in mammalian cells follows the general pattern of replication described previously for the replication of bacteriophages in bacterial cells. Yet, the replication process for animal viruses differs considerably from that of bacteriophages with respect to the specific mechanisms involved. Some animal viruses replicate themselves within the cytoplasm of host mammalian cells, whereas other animal viruses replicate themselves within the host-cell nucleus. Furthermore, animal viruses tend to be released from host cells by either of two major processes. In some cases, the host cell ruptures in a manner that allows mature virions to be released. This process is lethal to the host cell, and the newly released virions are referred to as *naked* or *nonenveloped*. In many ways, this process resembles the lytic destruction of bacterial cells by virulent bacteriophages. Other kinds of animal viruses are released from the host cell gradually through a *budding-off* process, and each newly formed virion is released surrounded by a host-derived membrane called an *envelope* (Figure 1.19). This process does not result in immediate death to the host cell.

Taxonomically, the animal viruses have been divided into several broad categories on the basis of chemical and physical properties or on the basis of epidemiological peculiarities (Table 1.3). On the basis of the latter, the *arbovirus group* contains all animal viruses that are transmitted naturally from a reservoir in the environment to a susceptible host by an arthropod vector (e.g., mosquito, tick, mite). Furthermore, within these heterogenous groups are specific agents that produce important communicable diseases in humans and animals. As a result of their disease-producing potential, many types of animal vi-

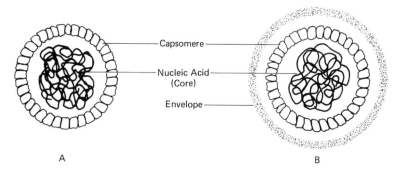

Figure 1.19. Generalized representation of structural components of an animal virus: (A) naked (nonenveloped) virion; (B) enveloped virion.

ruses are familiar to nonscientists because their names correlate with the diseases they produce. The polio, rabies, and influenza viruses are widely known. The epidemiology of diseases caused by these viruses is discussed in Chapters 8 and 9.

In our previous discussion of the cultivation of mammalian cells *in vitro,* we characterized those cells that exhibit the property of indefinite growth when propagated by serial passage in tissue culture systems as established cell lines. Such cells are referred to as "transformed" because they exhibit an adaptive change in cell function as a

Table 1.3. Some General Characteristics of Selected Animal Viruses*

Type of Nucleic Acid in Core	Taxonomic Group	Presence of an Envelope	Site of Replication in Host Cell	Representative Disease Produced
DNA	Adenoviruses	Naked	Nucleus	Respiratory illness
	Herpes viruses	Enveloped	Nucleus	Fever blisters
	Papoviruses	Naked	Nucleus	Tumors in animals
	Pox viruses	Enveloped	Cytoplasm	Smallpox
RNA	Arboviruses	Enveloped	Cytoplasm	Equine Encephalitides
	Leukoviruses	Enveloped	Cytoplasm	Tumors in animals
	Myxoviruses	Enveloped	Nucleus	Influenza
	Paramyxoviruses	Enveloped	Cytoplasm	Measles
	Picornaviruses	Naked	Cytoplasm	Polio
	Reoviruses	Naked	Cytoplasm	Nonspecific
	Rhabdoviruses	Enveloped	Cytoplasm	Rabies virus

*Only a partial representation of those known is given; the characteristics listed are not intended to be inclusive.

result of being cultivated *in vitro.* Normal mammalian cells, when grown *in vitro,* exhibit the phenomenon of *contact inhibition,* a process characterized by repeated cell division and random movement of cells until they come in contact with one another. As a result of surface contact between cells, both cell division (growth) and motility are inhibited. Transformed cells, on the contrary, do not exhibit the phenomenon of contact inhibition but continue to divide in an unregulated manner. Such continuous or established mammalian cell lines have emerged as invaluable biological tools, serving as *in vitro* hosts in which many types of animal viruses can be propagated. Several types of mammalian cells are registered as continuous or established cell lines and can be purchased from the American Type Culture Collection, Rockville, Maryland.

By utilizing tissue culture systems for cultivating animal viruses and a variety of biochemical methods for analyzing subcellular materials, molecular biologists have expanded research on animal viruses in many directions. Such viruses are being studied extensively from the standpoint of basic mechanisms involved in host–parasite interactions, and as agents that cause important human diseases. Of recent interest is the possible relationship between certain animal viruses and cancer in humans.

Cancer is a general term used to characterize an abnormal state (or group of diseases) that results from unregulated and progressive growth of cells in the body of animals. For reasons that are not clearly understood, the mechanisms that regulate cell division and/or other normal functions are inhibited in cells that produce cancer. Unregulated cells that divide continuously and pile up or accumulate in masses confined to a self-limiting area of the body are called a *tumor.* Self-limiting tumors are referred to as *benign.* On the other hand, *neoplasms* or *malignant tumors* are characterized by *metastasis* or spreading, and grow continuously in a manner that is not self-limiting. Many different kinds of human cancers have been characterized and the classification system for them is complex. One general classification system is based upon the types of cells affected. Cancers that develop primarily in epithelial cells, for example, are called *carcinomas.* When cancers develop in cells of bones and soft tissues, they are called *sarcomas.* Cancer in blood cells, characterized by the unregulated proliferation of white blood cells (leukocytes), is referred to as *leukemia.* Although many chemical substances, such as pollutants and food additives, have been implicated as cancer-producing agents, a great deal of current research is being conducted on viruses that cause tumors in animals, the *oncogenic viruses.*

The first link between viruses and cancer was reported in the early part of the twentieth century. The Rous sarcoma virus is named for an early investigator, Peyton Rous, who described the successful transfer of a chicken sarcoma by cell-free filtrates from tumors of diseased birds to susceptible chickens. In recent years, a number of other viruses have been demonstrated to be oncogenic, among them the SV-40, polyoma, and papilloma viruses. Viruses within these groups will induce the formation of different kinds of tumors when injected subcutaneously into lower animals (e.g., mice, rabbits, or guinea pigs). Furthermore, some papilloma viruses are known to be the causative agents of benign tumors (warts) in humans. Other oncogenic viruses have been implicated, but not confirmed, as agents of human cancer. Most early studies with oncogenic viruses were concerned with the induction of tumors in laboratory animals, but soon after the development of tissue culture techniques for the cultivation of viruses, scientists demonstrated that the oncogenic effect could be reproduced as a viral-induced cell transformation.

Polyoma viruses and the SV-40 virus were among the first to be used in viral-induced transformation experiments. In order to observe virus-induced transformation of mammalian cells in tissue culture, the lethal effects of the virus infection must be prevented. Animal cells in which a particular kind of virus can replicate itself are called *permissive*. Cells that can be infected with a particular virus but in which viral replication will not occur, called *nonpermissive*, will not support a lethal viral infection. Therefore, when nonpermissive cells in tissue cultures are infected with a suspension of oncogenic virions, the cells will undergo transformation and acquire the ability to divide (grow) continuously in an unregulated manner. Furthermore, when a suspension of virus-induced transformed cells is injected into an animal, it will induce tumor formation. Another point to remember about virus-induced transformed cells is that viral genes persist in them in the provirus state and are replicated in synchrony with the host cell's genes. This condition is similar in many aspects to bacterial cells that are lysogenized with a temperate bacteriophage.

The relationship between viruses and cancer in human beings is being investigated by scientists in many countries. Although the specific nature of the research varies from one laboratory to another, the investigators use various model systems in which nonpermissive mammalian cells are transformed by oncogenic viruses, both DNA and RNA. Scientists have learned a great deal about the mechanisms of viral-induced cell transformation by the use of such systems. It is now known that viral-induced transformed cells contain viral genes in an integrated form with host genes, and that viral antigens are associated with the surfaces of transformed cells. In recent years a number of important mechanisms through which certain viruses interact with host cells have also been elucidated. However, a discussion of the theory and/or rationale for such mechanisms and their relationship to cancer is beyond the scope of this book. Yet, we can appreciate the significance of those recent discoveries by noting that the 1975 Nobel Prize in Physiology or Medicine was awarded to David Baltimore, Howard M. Temin, and Renato Dulbecco for generating new information on the molecular mechanisms of virus–host cell interactions as they pertain to viral-induced cancers.

Key Words

aerobe An organism that requires molecular oxygen for growth.

anaerobe An organism that does not require molecular oxygen for growth.

aqueous Watery suspensions; or pertaining to waterlike liquids.

ascospore A dormant stage, or sexual spore, of fungi that belong to the Ascomycetes.

asexual Cellular reproduction in the absence of discrete sexual gametes (opposite mating types).

autotroph An organism that obtains its energy from inorganic elements and its carbon from carbon dioxide.

Bacillus The name of a bacterial genus, as opposed to "bacilli," any rod-shaped bacteria.

bacteriophage A virus that infects bacteria.

binary fission A method of asexual reproduction in which a single parent cell splits or divides transversely into two new cells (daughters or progeny).

biosphere The life-supporting region of the universe.

budding A form of asexual reporduction in which a new cell is formed as a knoblike outgrowth on the parent cell.

capsule A gelatinous material that forms the outermost covering of certain kinds of microorganisms.

chitin A polysaccharide, found as a cell-wall constituent in fungi and in the outer coverings of many invertebrates.

chloroplasts The chlorophyll-containing organelles in photosynthetic procaryotic cells.

chromosome A structure that contains the DNA of all cells.

conjugation The transfer of genetic material through a mating process that requires cell-to-cell contact.

electron microscope A microscope that uses electrons instead of light, magnets instead of glass lenses, and is capable of great magnification.

enucleated   Cells with no discrete nucleus. The genetic material is distributed throughout the cytoplasm in areas called nuclear regions.

etiological   Pertaining to the cause of a disease or abnormal condition.

eucaryon   A cell that has a well-defined nucleus surrounded by a nuclear membrane and has intracellular organelles.

fomite   Any inanimate item (such as a pencil or towel) that, when contaminated, can transport microorganisms passively.

free-living   Capable of living without the aid of a host cell.

gene   A unit of genetic material that codes for a specific trait.

generation time   The time required for a population of any organism to double.

genome   The complete set of genetic material present in an organism.

heterothallic   The ability of some fungi to produce hyphae that contain opposite sexual gametes.

heterotroph   Any organism that requires organic materials for carbon and for energy.

hyphae   Filaments or threadlike structures that form the mycelium of a fungus.

in vitro   Applies to any biological experiment in which organisms are grown or maintained in vessels apart from their natural habitat.

lysogeny   A state in which certain bacterial cells harbor a piece of bacteriophage DNA integrated into the bacterial chromosome as a prophage.

medium   A nutrient-containing substance used for the growth and multiplication of microorganisms.

meiosis   A process that occurs at various points in the life cycle of eucaryotic organisms, reducing the chromosome number by half, from diploid to haploid cells.

mitochondrion   The organelle in eucaryotic organisms in which energy is produced for metabolic processes.

mitosis   A form of nuclear division in eucaryotic organisms characterized by complex chromosome movements and exact chromosome duplication.

morphological   Relating to the form, size, and shape of an organism.

mycelium   A mass of threadlike filaments forming the vegetative structure of a fungus.

pasteurization   The process of heating a substance to kill

certain microorganisms but not changing the quality or
texture of that substance.

pathogenic   The ability to cause disease.

pellicle   A slime layer or film on a liquid surface due to the
growth of microorganisms.

peritoneal cavity   The cavity of the abdomen (in mammals)
that lies between the viscera and the skin.

photosynthesis   Processes in which certain kinds of organisms
utilize solar energy and carbon dioxide to synthesize
carbohydrates.

prophage   A piece of DNA from a bacteriophage (phage) when
integrated into the bacterial chromosome.

pseudomycelia   False mycelia.

pure culture   The growth of microorganisms in which all cells
are of a single type.

saprophytes   Organisms that obtain and utilize preformed
nutrients from dead or decaying organisms.

sexual reproduction   A method of reproduction that requires
the interaction and fusion of opposite sexual gametes.

submicroscopic   Structures too small to be seen with a light
microscope (although some can be seen with an electron
microscope).

symbiotic   The living together of two or more dissimilar
organisms with some degree of constancy.

transduction   The transfer of a gene or of several genes from a
donor bacterial cell to a recipient bacterial cell by an
intermediate agent called a bacteriophage or bacterial virus.

transformation   The process through which a portion of
bacterial chromosome (DNA) is incorporated into a recipient
bacterial cell in the absence of cell-to-cell contact, and
without the aid of a bacterial virus.

vascular system   Tissues concerned with the internal transport
of fluid substances in higher plants and animals.

virus   An infectious agent that contains either RNA or DNA
and is able to alternate between intracellular and extracellular
states, and replicates only within living cells.

Selected Readings

1. Baltimore, D. 1976. Viruses, polymerases, and cancer. *Science*
192:632–636.
2. Brock, T. D., and K. M. Brock. 1973. *Basic Microbiology with
Applications.* Englewood Cliffs, N.J.: Prentice-Hall, Inc.

3. Dulbecco, R. 1976. From the molecular biology of oncogenic DNA viruses to cancer. *Science* 192:437–440.

4. Echlin, P. 1966. The blue-green algae. *Scientific American* 214:75–81.

5. Frobisher, M., R. D. Hinsdill, K. T. Crabtree, and C. R. Goodheart. 1974. *Fundamentals of Microbiology*, 9th ed. Philadelphia: W. B. Saunders Company.

6. Hankins, W. A., and H. J. Hearns. 1970. Direct assessment of viral aerosols on cell cultures. *Appl. Microbiol.* 20:284–285.

7. Hess, E. L. 1970. Origin of molecular biology. *Science* 168:664–669.

8. Hull, R. H., D. C. Delong, and I. S. Johnson. 1966. Tissue culture in virus and cancer research. *BioScience* October:715–719.

9. Lechevalier, H. A., and M. Solotorovsky. 1965. *Three Centuries of Microbiology*. New York: McGraw-Hill Book Company.

10. Margulis, L. 1972. Symbiosis and evolution. *Scientific American* 224:48–57.

11. Perlman, D. 1968. Value of mammalian cell culture as a biochemical tool. *Science* 160:42–46.

12. Stanier, R. Y., E. A. Adelberg, and J. Ingraham. 1976. *The Microbial World*, 4th ed. Englewood Cliffs, N.J.: Prentice-Hall, Inc.

13. Temin, H. M. 1976. The DNA provirus hypothesis. *Science* 192:1075–1080.

14. Whittaker, R. H. 1969. New concepts of kingdoms of organisms. *Science* 163:150–160.

CHAPTER 2

Nutrition and metabolism

- **Essential chemical concepts**
  - *Atoms and molecules*
  - *Chemical bonds*
  - *Acids and bases*
  - *Biological molecules*
- **Requirements for growth**
  - *Water*
  - *Carbon*
  - *Energy*
  - *Nitrogen*
  - *Accessory materials*
- **Nutritional types of organisms**
  - *Autotrophs*
    - *Photoautotophs*
    - *Chemoautotrophs*
  - *Heterotrophs*
    - *Photoorganotrophs*
    - *Chemoorganotrophs*
- **Cellular metabolism**
  - *Enzymes*
  - *Energy-yielding processes*
    - *Photosynthesis*
    - *Fermentation*
    - *Aerobic respiration*
    - *Anaerobic respiration*
  - *Energy-utilizing processes*
    - *Biosynthesis of macromolecules*
    - *Biosynthesis of proteins*
  - *The molecular basis of mutation*
- **Key words**
- **Selected readings**

Those attributes of life described as the ability to grow, to reproduce, to adapt, and to respond to stimuli cannot be expressed by microorganisms unless specific utilizable materials are available in their environment. In general, the types of materials that microorganisms utilize for foodstuffs are numerous and include a variety of substances, which we call *nutrients*. This nutritional versatility among microorganisms is unmatched by members of other biological groups.

Our discussion of nutrition will be considered in relation to the following questions: What are the essential requirements for microbial growth? What is the physiological role (or function) of required nutrients? How are nutritional substances utilized by microorganisms? Answers to these questions will be presented in the form of a survey of current information on these topics.

Since microorganisms are single-celled entities, some insights into their functional abilities can be gained by recognizing the basic significance of the cell to all forms of life. The cell, whether existing as a unicellular microorganism or as the smallest functional unit in complex higher organisms, is composed of atoms, molecules, and macromolecules which interact to sustain life processes. These interactions take place in the form of chemical reactions. Therefore, to understand microbial nutrition and the functional aspects of microorganisms in nature, some knowledge of basic chemical concepts is a prerequisite.

Essential chemical concepts

In this section the chemical background will be developed for subsequent discussions. For students with previous training in chemistry and biology, the material may constitute only a superficial review, but for others it may serve as an introduction to the chemistry of life.

Atoms and molecules

Since living things are composed of matter and involve the reactions of matter, it is imperative that we acquaint ourselves with the basic substances from which all matter is formed—the *chemical elements*. More than 100 chemical elements have been characterized, but each contains only one kind of *atom*. A single atom is composed of several subatomic particles, but only three of them are pertinent to the material to be discussed: the *electron*, the *proton*, and the *neutron*.

A discussion of atomic theory and the structural arrangement of subatomic particles in the atom is beyond the scope of this book. However, it is important to know the electrical charge and the structural location of each major subatomic particle. The electron (e^-) has a negative charge of -1, the proton (p) has a positive charge of $+1$, and the neutron (n) is a neutral particle, with neither a positive nor a negative charge. Structurally, each atom is composed of a nucleus surrounded by electrons. Protons and neutrons are located in the nucleus, around which electrons rotate in "orbitals" or clouds. Each electron is attracted by a proton in the nucleus. In a neutral atom, the positive charge of protons is exactly balanced by the negative charge of its electrons. For example, hydrogen is a neutral atom. It has one proton, no neutrons, and a single orbiting electron. Consequently, the net electrical charge for the hydrogen atom is zero.

Each chemical element is unique and has a definite number of protons in its nucleus. Hydrogen is the simplest element, with only one proton; helium has two protons, and carbon has six protons. Correspondingly, hydrogen has one electron, helium has two electrons, and carbon has six electrons. They are neutral atoms, because each of them contains an equal number of protons and electrons.

To attain stability, atoms undergo electron rearrangement. They accomplish this by gaining, losing, or sharing electrons with other atoms. When rearrangement of electrons between atoms of different elements takes place, the result is a chemical reaction.

Chemical reactions do not occur randomly between elements. They are regulated by the combining capacity or *electron affinity* of each element. The combining capacity of an element is referred to as its *valence*. Each atom has a characteristic spectrum of valence electrons which occupy the outermost position in the cloud around its nucleus. The valence of an element is expressed as small whole numbers (1, 2, etc.). It does not tell us what atomic combinations are possible, but it determines the ratio by which atoms will combine. For example, hydrogen, sodium, and chlorine each have a valence of 1, and carbon has a valance of 4. The valence electrons of each atom may then be indicated by a bond, as follows:

$$
\text{H—} \qquad \text{Na—} \qquad \text{Cl—} \qquad \overset{|}{\underset{|}{-\text{C}-}}
$$

$$
\textbf{HYDROGEN} \quad \textbf{SODIUM} \quad \textbf{CHLORINE} \quad \textbf{CARBON}
$$

Two atoms, each with a valence of 1, will combine with each other in a 1:1 ratio. Two atoms, each with a valence of 2, will combine with each other in a 1:1 ratio. Two atoms of an element that has a valence of 1 will combine with one atom of an element with a valence of 2 in a 2:1 ratio.

Atoms sometimes exist by themselves, but most often they exist in combination with other atoms. When combined, there may be only two atoms in a group or there may be several hundred atoms in a group. Groups of atoms are called *molecules*. Some of the natural gases are molecules that contain only two atoms of the same element: hydrogen (H_2), oxygen (O_2), and nitrogen (N_2). Many other molecules are composed of atoms from different elements. For example, one carbon atom can combine with four hydrogen atoms to form a molecule of methane gas (CH_4), and one sodium atom can combine with one chlorine atom to form a molecule of sodium chloride (NaCl). When two or more atoms of different elements combine in definite proportions, the resulting unit is called a *compound*. Now, it becomes obvious that a compound is also made of molecules. Then a molecule can be defined as the smallest single unit of any compound. Thus, we have a progression:

atoms ⟶ molecules ⟶ compounds

Chemical bonds

Chemical bonds have many functions in biological systems. Basically, they are forces that hold atoms together in molecules, but in some cases they determine the manner in which molecular structures associate in nature. Consequently, the stability and behavior of a molecular structure are related to types of chemical bonds present in that structure. Some chemical bonds form very weak connections with only a few specific atoms and will break easily. Other types of chemical bonds form firm connections between atoms, and structures in which such bonds are found have a high degree of stability. Therefore, the behavior of chemical bonds is characteristic of its type.

In biological systems three types of chemical bonds are important: hydrogen bonds, ionic (electrovalent) bonds, and covalent bonds. *Hydrogen bonds* are relatively weak forces which form connections between atoms in different molecules. Hydrogen bonds always involve an atom of hydrogen bonded to either an atom of oxygen or an atom of nitrogen. This type of bond is made possible by the phenomenon of polarity, which results from the fact that different atoms have different affinities for electrons. Polar molecules are electrically asymmetrical and are oppositely charged at two points (a dipole). The water molecule is polar, and hydrogen bonds play an important role in the association of water molecules in nature (Figure 2.1). Remember that each separate water molecule is neutral. However, a hydrogen atom of one molecule will be attracted to an oxygen atom of another molecule, resulting in the formation of a bridge, the hydrogen bond.

Ionic (electrovalent) bonds are formed when an electron is completely transferred from one atom to another. When any atom loses or gains an electron, it will acquire a positive or negative electrical charge. As a result, charged atoms have been given the name *ions*. A positively charged ion is called a *cation,* and a negatively charged ion is called an *anion*. Oppositely charged ions are attracted to each other by electrostatic forces. Perhaps the concept of ionic bonds can be visualized by considering the manner in which sodium and chlorine atoms interact to form molecules of sodium chloride. Each sodium atom gives an electron to a chlorine atom. After losing a negatively charged electron, the sodium atom becomes a positively charged sodium ion (Na^+).

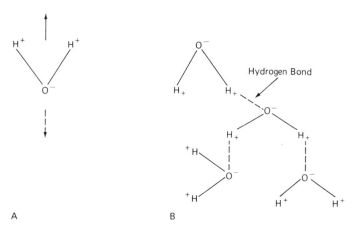

Figure 2.1. Structure of the water molecule: (A) single polarized water molecule; (B) hydrogen bonding among water molecules.

In a similar manner, each chlorine atom gains one electron and becomes a negatively charged chloride ion (Cl^-). Thus, a molecule of sodium chloride (NaCl) is formed from the electrostatic attraction that exists between the sodium ion and the chloride ion, and these ions are held together in the molecule by ionic bonds.

Covalent bonds are formed when electrons are shared between atoms. For example, the formation of the hydrogen molecule (H_2) involves the sharing of one electron from each atom by two positively charged nuclei. This binding of electrons through a single covalent bond gives the hydrogen molecule more stability than existed in either of the separate hydrogen atoms. A carbon atom can form four single covalent bonds by sharing its valence electrons with other atoms. Carbon also has the ability to bind to itself and to yield chains of carbon atoms successively linked by covalent bonds. When this occurs, the remaining valence electrons of the carbon atoms are shared with other atoms. Thus, covalent bonds are extremely versatile and are the strongest type of chemical bonds found in living things.

Acids and bases

When sodium chloride crystals are placed in water, the molecules dissociate into sodium ions and chloride ions. The polar water molecules surround each ion as it dissolves (Figure 2.2). The negative end of the polar water molecule is attracted to the sodium ion, and the positive end is attracted to the chloride ion. Thus, in solution, sodium and chloride ions become attracted to the water dipoles, and their attraction for each other is less than existed in the crystalline state. This is

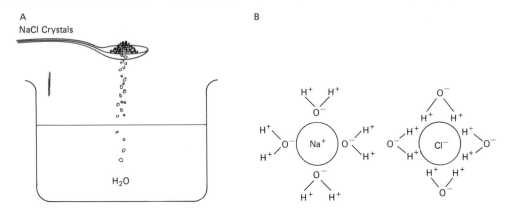

Figure 2.2 (A) Dissociation of sodium chloride in water to form (B) sodium ions (Na$^+$) and chloride ions (Cl$^-$).

a common phenomenon among crystalline salts. They already exist as ions, but separate into individual entities when dissolved in water, a process called *dissociation. Ionization* is a similar process, but differs from dissociation in that ions are actually produced by the reaction of the substance with water. In either case, the resulting solution of ions becomes a conductor of electricity and is the basis for determining if a solution is acidic or basic. For example, when two oppositely charged electrodes from a battery are placed at opposite sides of a solution of ions, the cations will migrate to the negative electrode, the *cathode,* and the anions will migrate to the positive electrode, the *anode.* Thus, the ions will form molecules of gases at the respective electrodes and pass out as bubbles.

Acids are substances that will produce hydrogen ions (H$^+$), and *bases* are substances that will produce hydroxyl ions (OH$^-$) when placed in water. In other words, an acid is a substance that will dissociate to produce a proton, and a base is a substance that combines with, or accepts, a proton. What, then, is an acidic or a basic (alkaline) solution? Since hydrogen ions are associated with acids, the concentration of hydrogen ions present in a solution determines its degree of acidity. Ionic concentrations are usually expressed as the gram-atomic weight of ions present in 1 liter of solution. A gram-atomic weight is the atomic weight of a substance measured in grams. Pure water, at approximately 25°C, is a neutral substance, neither acidic nor basic. Each water molecule (H$_2$O), when ionized, contains a 1:1 ratio of hydrogen ions (H$^+$) to hydroxyl (OH$^-$) ions. One liter of pure water contains 0.0000001 (1×10^{-7}) gram of hydrogen ions and 0.0000001 (1×10^{-7}) gram of hydroxyl ions. Consequently, the product of the ionic concentration is always equal to a constant value (1×10^{-14}). This value is called the *dissociation constant (K)*. It is the ratio of ionized water to nonionized water and can be calculated as follows:

$$K = \frac{[H^+] \times [OH^-]}{[HOH]} = \frac{(1 \times 10^{-7}) \times (1 \times 10^{-7})}{1} = 1 \times 10^{-14}$$

Thus, the dissociation of the water molecule forms the basis for measuring the hydrogen-ion concentration on a pH scale of logarithmic units which ranges from 0 to 14 (Figure 2.3). Therefore, the acidity or alkalinity of any solution can be described in terms of its pH.

In modern laboratories, pH is measured most commonly on a pH meter. Numerous types of pH meters are available, but all are equipped with electrodes sensitive to hydrogen ions when submerged in a solution, and a system for indicating the pH units, either a dial that will reflect the value on a pH scale or an electronic digital printout

	pH Unit	Natural Environment	Representative Microbe
	0		
	1		
	2	Acid Mine Drainage	*Thiobacillus thiooxidans*
Increasing Acidity	3	Spoiled Wines	*Acetobacter* Species
	4	Human Skin	*Corynebacterium acnes*
	5		
	6		
Neutral	7	Pure Water	
	8		
	9	Alkaline Soils	*Azotobacter* Species *Nitrosomonas* Species
Increasing Alkalinity	10	Alkaline Lakes	*Microcystis* Species
	11		
	12		
	13		
	14		

Figure 2.3. The pH scale and representative natural environments where specific microorganisms are commonly found.

of pH units. A solution of pH 7 is neutral. Acidity increases as the numbers approach zero, and the solution becomes more basic as the numbers approach pH 14. Because the logarithmic scale is used, there is a considerable degree of change in hydrogen-ion concentrations between units. A solution of pH 4 is 10 times more acidic than a solution of pH 5, and 100 times more acidic than a solution of pH 6.

It is important to remember that the pH of a solution is not a constant value. The acidity can be increased by adding substances that will combine with or consume hydrogen ions. In laboratory experiments the pH of a solution may be adjusted for a specific purpose by the addition of such substances as hydrochloric acid (HCl) to increase the acidity or sodium hydroxide (NaOH) to increase the alkalinity. In natural environments the pH may be altered by excretory products from biological systems, pollutant waste products from domestic and industrial operations, and the application of commercial fertilizers.

Various microbial types proliferate in environments that encompass most of the pH range. However, most organisms grow optimally between pH 6.5 and 8.5. Molds and yeasts tolerate acidic environments better than other microbial groups, and alkaline environments are tolerated well by blue-green algae.

Under some conditions radical changes in pH may be undesirable for biological systems. In the laboratory a variety of substances can be added to solutions to restrict pH changes within narrow limits. Such substances are called *buffers*. Buffered solutions resist changes in acidity that would otherwise result from the addition of acids or bases. Buffers are also found in nature. For example, human blood (pH 7.3 – 7.5) contains sodium bicarbonate ($NaHCO_3$) as a natural buffer. This substance dissociates into sodium ions (Na^+) and bicarbonate ions (HCO_3^-). When the blood becomes acid, bicarbonate ions serve as proton scavengers. During the process carbonic acid (H_2CO_3) is formed and the acidity of the blood is reduced. An analogous process takes place in natural waters, where carbon dioxide (CO_2) from the atmosphere combines with water (H_2O) to form carbonic acid (H_2CO_3), which dissociates into hydrogen (H^+) and bicarbonate (HCO_3^-) ions. Subsequently, if acids are introduced, the buffering action of bicarbonate acts as a stabilizer by combining with hydrogen ions.

Biological molecules

In general, we refer to organic substances as bioligical molecules, because all of them contain the carbon atom as their main structural element. All carbon-containing substances (except carbon dioxide, carbonates, and bicarbonates) are included in this group. Substances of this nature were once considered only as structural components or products of living things. In fact, many pioneering chemist harbored mystical concepts about organic substances and believed they contained a "vital force" which was beyond the synthetic capabilities of man. However, these beliefs were shattered when Friedrich Wöhler synthesized urea in 1828. Subsequently, interest in the properties and reactions of organic compounds was intensified, and the study of such compounds evolved into a discipline that became known as organic chemistry. It is now a vast field of science, and includes the study of numerous types of synthetic carbon-containing materials, many of which are foreign to biological systems. Knowledge of biological systems has expanded tremendously in recent years, and chemical reactions that occur in living things are currently described more precisely in the science of biochemistry — the chemistry of life.

Since the urea compound has historical significance, its structure can be used as an appropriate model for reviewing valence and for emphasizing some of the unique features of life molecules.

H
 \
  N
 /   \
H   \
       C=O
H   /
 \   /
  N
 /
H

       UREA

In this structure the combining capacity of all elements is indicated by covalent bonds: carbon has four valence electrons (two are bonded to oxygen, and the other two are bonded separately to a nitrogen atom); oxygen has two valence electrons (both are bonded to the carbon atom); each nitrogen has three valence electrons (one is bonded to the carbon atom and one to each hydrogen atom); and each hydrogen atom has one valence electron (bonded only to a nitrogen atom).

Characteristically, six elements are universal constituents of life molecules. They are the four present in the structure of urea (carbon, hydrogen, oxygen, nitrogen), plus phosphorus and sulfur. These six elements, arranged in an infinite variety of combinations, comprise one of the unifying features among all members of the biological world. Several molecules in which these elements serve as structural components are listed in Table 2.1.

Table 2.1.   Universal Elements in Life Molecules

Constituents	Small Molecules	Macromolecules (Polymers)
C, H, N, O, S	Amino acids	Peptides, proteins
C, H, O	Fatty acids	Fats, oils, complex lipids
C, H, O	Sugars	Carbohydrates (starch, glycogen, cellulose)
C, H, O, P, N	Purines and pyrimidines	Components of nucleic acids, nucleotides

During the course of evolution, the elements shown were incorporated into small molecules. Subsequently, the small molecules became building blocks for macromolecules. All macromolecules are *polymers,* large structures composed of subunits *(monomers).* Polymers in which the subunits are identical are called *homopolymers,* and those in which the subunits are different are called *heteropolymers.* The universality of the elements mentioned above will become evident as we describe different kinds of organic structures later in this section.

In many ways cells are analogous to chemical factories in which reactions occur on a miniature scale. Consequently, to understand the functional aspects of microorganisms, it is necessary to have some familiarity with classes of organic compounds and their functional groups.

Perhaps the concept of functional groups can be visualized if we consider the compounds shown in Table 2.2 to be derivatives of methane (CH_4), a structure chosen to represent the class of organic materials called *hydrocarbons.* If one hydrogen atom is removed from the methane molecule, the resulting unit is called a *methyl group* (CH_3—). This methyl group may combine with another atom or other groups to form a variety of structures, each of which can then be characterized by specific reactive components, or sites, which are called *functional groups.* Now, if we consider the methyl group as the base unit, other groups can be connected to form compounds characteristic of a specific class. When a hydroxyl group (—OH) is attached to the methyl group, the class is called *alcohols.* In this specific example, methyl alcohol (CH_3OH) is the product. Similarly, compounds that contain a carboxyl group (—COOH) are called *carboxylic* or *organic acids.* In addition to carboxyl groups, which characterize the class, organic acids may have other groups attached, from which they are given specific names. For example, hydroxy acids always contain a hydroxyl group, keto acids contain a ketone group, and amino acids contain an amino group.

In a similar manner, simple sugars are characterized by having an aldehyde or ketone group, in addition to several hydroxyl groups, in their structure. They are polyhydroxy aldehydes or polyhydroxy ke-

Table 2.2. Classes and Functional Groups of Some Small Compounds

Class	Functional Group		Representative Compound	
	Name	Structure	Name	Structure
Hydrocarbons	Methyl	H | H—C— | H	Methane	H | H—C—H | H
Alcohols	Hydroxyl	—OH	Methyl alcohol	H | H—C—OH | H
Carboxylic acids	Carboxyl	—COOH	Acetic acid	H | H—C—COOH | H
Hydroxy acids	Hydroxyl Carboxyl	—OH —COOH	Lactic acid	H OH | | H—C—C—COOH | | H H
Keto acids	Ketone Carboxyl	C=O —COOH	Pyruvic acid	H O | || H—C—C—COOH | H
Amino acids	Amino Carboxyl	—N(H)(H) —COOH	Alanine	H H H N | | H—C—C—COOH | | H H
Simple sugars (Monosaccharides.	Aldehyde	—C=O (H)	Glucose	H H H OH H H | | | | | | H—C—C—C—C—C—C=O | | | | | OH OH OH H OH
	Ketone	C=O	Fructose	H H H OH O H | | | | || | H—C—C—C—C—C—C—H | | | | | OH OH OH H OH

tones, from which many derivatives may be formed. For example, if a hydroxyl group in glucose is replaced with an amino group, the resulting structure is called *glucosamine*.

It is important to remember that functional groups are distinguishing features among classes of small organic compounds, but larger, more complex organic structures (macromolecules) are not so easily differentiated. However, they may be grouped into at least four major classes on the basis of their inherent subunits and/or chemical properties: carbohydrates, proteins, lipids, and nucleic acids.

The most common macromolecules in the class of *carbohydrates* are cellulose, starch, and glycogen. All are homopolymers in which the subunits are glucose. In each structure the glucose subunits are linked together in a specific manner by glycoside bonds. Such bonds are covalent, but result from the reaction between a carbon atom in one molecule and a hydroxyl group in an adjacent structure. Structures of these macromolecules are shown in Figure 2.4.

Cellulose, the major structural component in cell walls of higher plants, is the most abundant natural source of organic carbon in the world. Its structure is very simple, merely a collection of glucose units linked together by covalent bonds to form a straight chain. Note that the repeating disaccharide unit (two glucose molecules) in cellulose is called *cellobiose.*

Starch is the major storage product or nutritional reservoir in higher plants and in some microorganisms. It is very similar to cellulose, because it also is a chain of glucose subunits. However, starch can be differentiated from cellulose by the manner in which the glucose subunits are linked together, and by having *maltose* as its repeating disaccharide unit.

Glycogen, sometimes called *animal starch,* is a nutritional reservoir in higher animals and in some microorganisms. It, also, is composed of glucose subunits but differs from cellulose and starch in having a branched-chain structure. The branching results from the manner in which the glucose subunits are linked together by glycoside bonds.

In addition to the homopolymers discussed above, many heteropolymers of carbohydrates are found in nature. Lipopolysaccharides are examples of such structures. These macromolecules are composed of a polysaccharide covalently bonded to a lipid. Such materials are found as structural components in the cell walls of some bacteria.

Proteins are complex macromolecules which have many functions in biological systems. Each protein is composed of various combinations of subunits called *amino acids.* Twenty different naturally occurring amino acids have been characterized (Table 2.3). As the name implies, amino acids are compounds that contain at least one amino group and one carboxyl group in their structure. The amino acids in different proteins may be arranged in an infinite number of sequences, but all amino acids in a sequence are connected by peptide bonds, and the resulting structure is called a *polypeptide.* The peptide bond is a specific covalent bridge which results from the reaction between the

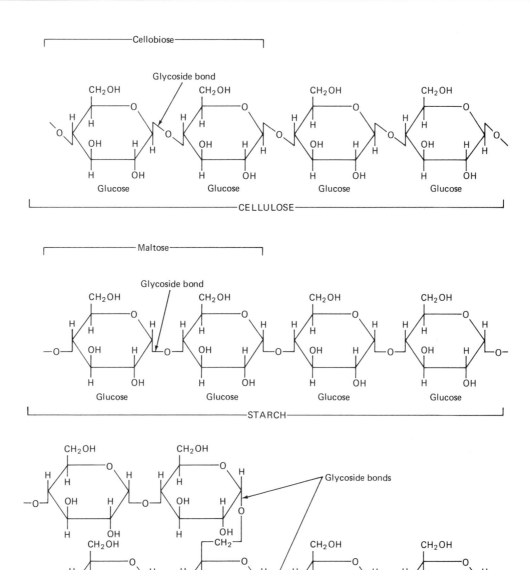

Figure 2.4. Some common carbohydrate macromolecules.

Table 2.3. Natural Amino Acids

1. Glycine	11. Aspargine
2. Alanine	12. Glutamic acid
3. Valine	13. Glutamine
4. Leucine	14. Lysine
5. Isoleucine	15. Arginine
6. Serine	16. Histidine
7. Threonine	17. Phenylalanine
8. Cysteine	18. Tyrosine
9. Methionine	19. Tryptophan
10. Aspartic acid	20. Proline

carboxyl group of one amino acid and the amino group of an adjacent amino acid (Figure 2.5).

Peptide chains have the remarkable ability to arrange themselves into a variety of shapes and configurations. The linear sequence of amino acids in a protein is called its *primary structure*. When polypeptide chains occur as parallel strands, specific amino acids in separate strands may interact through hydrogen bonding or other forces to hold the strands together in a flat, sheetlike manner. Such strands may also appear as coils. These coiled and sheetlike arrangements are referred to as the *secondary structure*. *Tertiary structures* are formed when groups of parallel strands fold or become twisted into ropelike structures. Finally, tertiary structures may aggregate into what is known as the *quaternary structure* of a protein. Consequently, some of the most complex macromolecules found in biological systems are proteins.

Proteins also have the ability to combine with other kinds of compounds, such as carbohydrates and lipids, to form glycoproteins and lipoproteins, respectively.

Lipids are a heterogeneous group of compounds which are only partially soluble or nonsoluble in water, but soluble in fat solvents

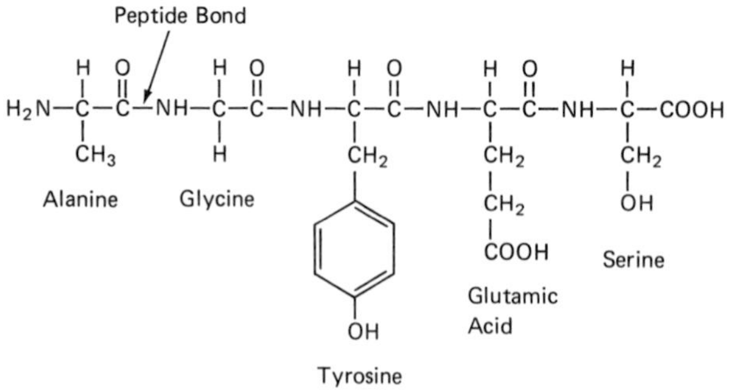

Figure 2.5. Polypeptide chain.

(e.g., alcohol, ether, acetone, chloroform). Thus, the term "lipid" is operational rather than structural.

Lipid macromolecules may be categorized, in a general manner, as simple lipids and complex lipids. *Simple lipids* are structures that contain fatty acids and glycerol. The fatty acids, which occur naturally, may have saturated or unsaturated structures. *Saturated fatty acids* are characterized by having all the valence electrons on carbon atoms filled with hydrogen atoms or some other group. *Unsaturated fatty acids* have adjacent carbon atoms without attached groups. Their structures contain carbon-to-carbon bonds. Typical structures for both types of fatty acids are as follows:

A saturated fatty acid:

$$CH_3—(CH_2)_{14}—COOH$$
PALMITIC ACID

An unsaturated fatty acid:

$$CH_3—(CH_2)_7—\overset{\overset{\displaystyle H}{|}}{C}=\overset{\overset{\displaystyle H}{|}}{C}—(CH_2)_7—COOH$$
OLEIC ACID

Glycerol, a trihydroxy alcohol with the following structure, is the other component in simple lipids:

$$CH_2—OH$$
$$|$$
$$CH—OH$$
$$|$$
$$CH_2—OH$$
GLYCEROL

One, two, or three fatty acids (saturated or unsaturated) may be covalently bonded to glycerol to form a mono-, di-, or triglyceride, respectively. In such structures the carboxyl end of each fatty acid is attached to a hydroxyl group in glycerol through a specific linkage called an *ester bond*. Thus, simple lipids are esters of fatty acids and glycerol. A typical triglyceride structure is shown in Figure 2.6. Structures of this type are common in nature.

Complex lipids include such substances as phospholipids, waxes and sterols. *Phospholipids* have structures which, to a degree, resemble triglycerides. They are glycerol-containing lipids. These com-

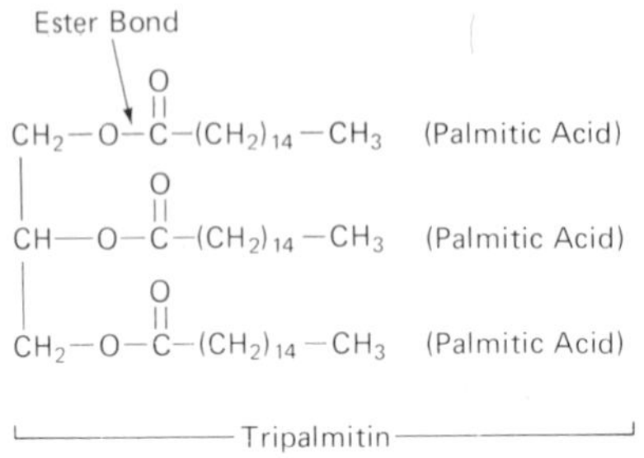

Ester Bond

$$
\begin{array}{l}
\quad\quad\quad O \\
\quad\quad\quad \| \\
CH_2-O-C-(CH_2)_{14}-CH_3 \quad \text{(Palmitic Acid)} \\
| \quad\quad\quad O \\
\quad\quad\quad \| \\
CH-O-C-(CH_2)_{14}-CH_3 \quad \text{(Palmitic Acid)} \\
| \quad\quad\quad O \\
\quad\quad\quad \| \\
CH_2-O-C-(CH_2)_{14}-CH_3 \quad \text{(Palmitic Acid)}
\end{array}
$$

$\llcorner$———————— Tripalmitin ————————$\lrcorner$

Figure 2.6.   Structure of a simple lipid (triglyceride).

pounds are characterized by having two fatty acids esterfied to two of the hydroxyl groups, and phosphoric acid plus one other nitrogen-containing compound (choline, ethanolamine, or serine) attached to the third hydroxyl group in glycerol (Figure 2.7). Names corresponding to the structures are phosphatidyl choline, phosphatidyl ethanolamine, and phosphatidyl serine, respectively.

In addition to the phospholipids, sterols and waxes play important roles in biological systems. The reader should consult one of the basic microbiology books listed at the end of the chapter for information relative to the nature and characteristics of these complex structures.

As previously mentioned in our discussion of carbohydrates and proteins, lipids have the ability to form conjugated structures such as lipoproteins (lipid + protein), and lipopolysaccharides (lipid + carbohydrate). Lipoproteins are important constituents of cell membranes, and lipopolysaccharides are found in the cell walls of many bacteria.

One other compound among the lipid macromolecules deserves special recognition. It is poly-β-hydroxybutyric acid (PHB). It is a homopolymer composed of β-hydroxybutyric acid subunits. The structure of a subunit is:

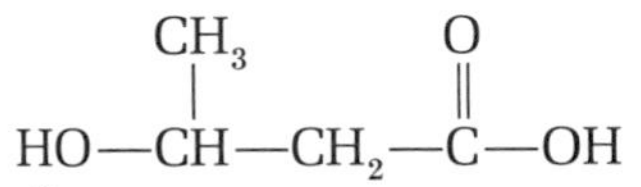

$$
\begin{array}{l}
CH_3 \quad\quad\quad O \\
| \quad\quad\quad\quad\quad \| \\
HO-CH-CH_2-C-OH
\end{array}
$$

β-HYDROXYBUTYRIC ACID

The subunits of this acid are connected by ester linkages, forming long polymers that aggregate into granules within procaryotic organisms. Consequently, this compound is a major reserve nutritional product in many microorganisms.

Nucleic acids are heteropolymers in which the subunits are called *nucleotides*. Each nucleotide has three components: a nitrogen-containing compound, a pentose (five-carbon) sugar, and a phosphate molecule. The nitrogen-containing compound in each nucleotide is either

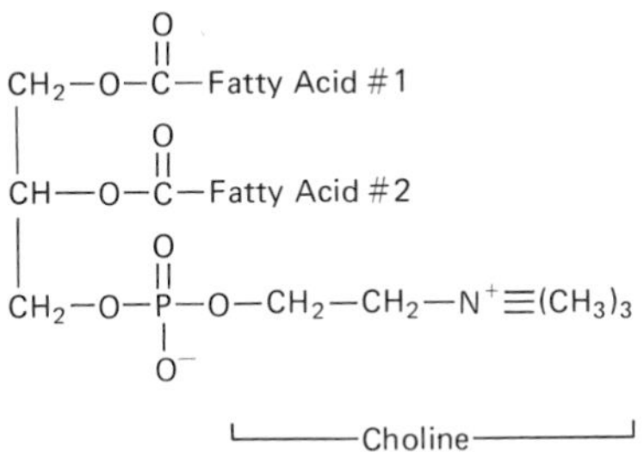

CH₂—O—C—Fatty Acid #1
│
CH—O—C—Fatty Acid #2
│
CH₂—O—P—O—CH₂—CH₂—N⁺≡(CH₃)₃
│
O⁻
└────Choline────┘

Figure 2.7. Structure of a phospholipid (phosphatidyl choline). When choline is replaced with ethanolamine or serine, the resulting phospholipids are called phosphatidyl ethanolamine and phosphatidyl serine.

adenine, guanine, cytosine, thymine, or uracil. The larger structures, which consist of two rings, are called *purines* (adenine and guanine), and the smaller, single-ring structures are called *pyrimidines* (cytosine, thymine, and uracil). Each nucleotide may have either ribose or deoxyribose as the pentose sugar in its structure. However, the phosphate molecule in all nucleotides is always identical. The structures of components commonly found in nucleotides are shown in Figure 2.8.

There are two kinds of nucleic acids: deoxyribonucleic acid (DNA) and ribonucleic acid (RNA). Their names were derived from the type of sugar present in their structure. Deoxyribose is the pentose sugar in DNA, and ribose is the corresponding pentose sugar in RNA. Another difference between the types of nucleic acids is found in their pyrimidine components. Thymine is present only in DNA, and uracil is present only in RNA. As a result of this specificity, nucleic acids contain only four nucleotides.

In either type of nucleic acid, the subunits are linked together in a sequence by a covalent ester bond between the phosphate of one nucleotide and the sugar of the adjacent nucleotide, forming a polynucleotide chain. Both types of nucleic acids usually occur as helical structures. In double-stranded nucleic acids, the polynucleotide chains are held together by hydrogen bonds. A purine in one polynucleotide strand is always bonded to a pyrimidine in the other strand. Because of their structural configuration, purines and pyrimidines are always connected in a characteristic manner. In DNA, adenine is bonded to thymine and guanine to cytosine. For this reason it is common to refer to them as nucleotide pairs, and to describe them in abbreviated form as A-T and C-G. The helical structure for DNA is shown in Figure 2.9. In most organisms DNA is double-stranded and RNA is single-stranded. However, double-stranded RNA and single-stranded DNA do occur in some viruses.

The function of these important macromolecules will be discussed in the section on metabolism.

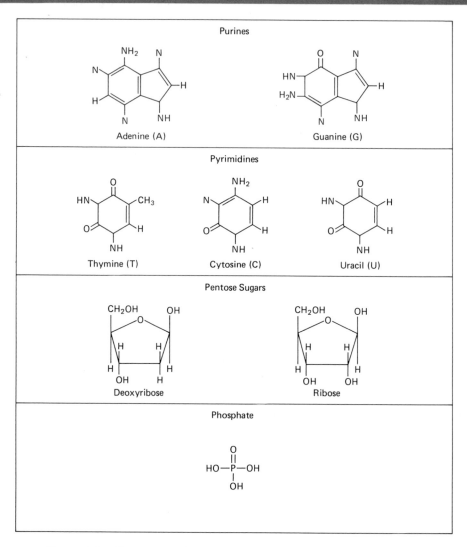

Figure 2.8. Structures of components (subunits) commonly found in nucleic acids (DNA and RNA).

Requirements for growth

The term *growth* may have different meanings when used to characterize a physiological response among different members of the biological world. In multicellular organisms, cellular growth leads to an increase in size of individuals, but in unicellular microorganisms growth is characterized by an increase in the number of individuals.

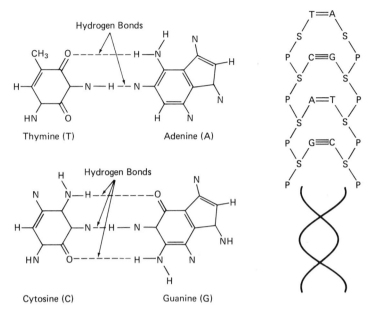

Figure 2.9. Representation of hydrogen bonding between bases in double-stranded DNA, and a schematic drawing of the double-helical molecule. Note that the strands are held together by hydrogen bonds between the complementary bases, and each base is connected to a pentose sugar (S). The sugar is connected to a phosphate (P), which together form the sides of the coils.

Some authorities have further defined growth as an orderly increase of chemical constituents within an organism.

Materials that an organism cannot make (synthesize) but which are absolutely necessary for its life processes are called its *essential nutrients*. Microorganisms as a group can utilize an incredible variety of materials for nutrients. However, the requirements for a given species may be highly specific. Despite the nutritional diversity among microorganisms, all living things have a common set of minimal growth requirements: a source of water, a source of carbon, a source of energy, a source of nitrogen, and traces of accessory materials. When microorganisms are grown under laboratory conditions, these substances must be supplied to the environment by the investigator.

Water

Regardless of an organism's type or its mode of nutrition, water is absolutely essential for cellular function. It is the universal solvent for microorganisms. In the extracellular environment, nutrients are solubilized in water for subsequent transport into the cell. Intracellularly, water serves as the medium in which all chemical reactions take place.

Carbon

Two sources of carbon are utilized by microorganisms: *inorganic* and *organic*. Carbon dioxide (CO_2) is the source of inorganic carbon for eucaryotic plants and some microorganisms. Organic compounds serve as the carbon source for organisms that cannot utilize carbon dioxide. Regardless of source, the element carbon is an absolute requirement for growth and cell function, because it is the major structural element in cell constituents.

Energy

There are several kinds of energy, but in these discussions we will concentrate on types of energy utilized by biological systems. *Energy* is required by all living things for cellular growth, and it is involved in all reactions that occur within cells. In general, three kinds of energy can be utilized by microorganisms: light energy, inorganic chemical energy, and organic chemical energy.

Light (radiant energy) is utilized by *photosynthetic* organisms, organisms that contain chlorophyll or other pigment centers for trapping light. Organisms with such properties are eucaryotic green plants, blue-green algae, and photosynthetic bacteria. These organisms will not grow unless the appropriate wavelengths of light can penetrate their environments.

Inorganic chemicals (substances that do not contain carbon) such as hydrogen (H_2), sulfur (S), and hydrogen sulfide (H_2S) are examples of energy sources for some organisms. Organisms in this category cannot utilize radiant energy and are indifferent to light.

Organic chemical energy is stored in macromolecules carbohydrates, lipids, proteins, etc.) and other carbon-containing compounds. Organisms that require this type of energy cannot utilize inorganic chemical energy or radiant energy, being similar to human beings in this respect.

Nitrogen

Nitrogen sources for microorganisms may be inorganic or organic. Inorganic sources of nitrogen may be materials such as atmospheric nitrogen (N_2), nitrate (NO_3), ammonia (NH_3), or ammonium sulfate ((NH_4)$_2SO_4$. Organic sources may include such materials as amino acids or any organic nitrogen-containing compound. Some microorganisms utilize nitrogen-containing compounds as energy sources, but the element nitrogen is also an important structural component, being a constituent of proteins and nucleic acids. Some microorganisms can utilize a single organic compound to satisfy all their nutritional requirements for carbon, energy, and nitrogen.

Accessory materials

Included among the *accessory materials* are a variety of substances that microorganisms require in small quantities for life processes. These materials can be divided into two categories: inorganic trace elements and organic micronutrients.

The most important *inorganic trace elements* that cells require are listed in Table 2.4. None of the inorganic trace elements are synthesized by cells; all are obtained preformed from their environments. However, the specific requirements, in number, kind, and amount, vary considerably among different types of microorganisms.

Organic micronutrients include such substances as vitamins, amino acids, purines, and pyrimidines. It is noteworthy that some microorganisms require none—all their micronutrients can be synthesized within the cell. In contrast, many other microorganisms are limited in their ability to synthesize such substances and thus have many specific requirements for micronutrients. For example, tryptophan is an essential amino acid for human beings and it is also an essential micronutrient for some bacteria. The bacterial organisms that cause typhoid fever *(Salmonella typhi)*, tetanus *(Clostridium tetani)*, and diphtheria

Table 2.4. **Some Essential Inorganic Trace Elements for Microorganisms**

Element	Biological Function
Magnesium	Cofactor in some enzymes
Iron	Required by all aerobes
*Cobalt	Component of vitamin B_{12}
*Copper	Cofactor for oxidative enzymes
*Zinc	Cofactor for many enzymes
*Molybdenum	Cofactor for a few enzymes
Sodium	Major extracellular cation
Phosphorus	Involved in energy transfer; component of some macromolecules
Sulfur	Constituent of some amino acids (cysteine and methionine)
Chlorine	Major anion (intra- and extracellular)
Potassium	Involved in protein synthesis
Calcium	Component of membranes and spores

*Requirements for these are exceptionally minute. In laboratory cultures, an organism can usually obtain sufficient quantities as contaminants in water or on glassware.

(Corynebacterium diphtheriae) cannot grow unless their surrounding environment contains trace quantities of tryptophan.

Nutritional types of organisms

The type of energy and carbon that a specific organism utilizes is determined by the enzymatic systems inherent to that organism. Microorganisms, as a group, can obtain energy and carbon from many kinds of substances. In general, other kinds of organisms are less diverse, but all living cells can be categorized into nutritional types on the basis of their metabolic systems.

Autotrophs

Autotrophs are organisms with enzymes that enable them to obtain energy and carbon entirely from inorganic materials. Because they are independent of organic nutrients, some members are considered to represent the most primitive forms of cells, those which existed at an early stage of evolution. All autotrophs obtain their carbon from carbon dioxide, but they obtain energy from either of two sources: solar radiations or inorganic chemicals.

Photoautotrophs   *Photoautotrophs* include all the photosynthetic eucaryotic plants, the procaryotic algae, and some types of photosynthetic bacteria. In order to carry out the reactions of photosynthesis, the following requirements are essential: (1) cells must have receptor photopigments, (2) the environment must contain suitable electron donors (reduced compounds), and (3) the environment must contain an appropriate electron acceptor.

In a subsequent section, the general aspects of photosynthesis will be described. The objectives here are to specify some differences in the processes with respect to the requirements listed above. The electron donor for higher eucaryotic organisms is always water, and the reactions always produce gaseous oxygen (O_2). In contrast, the electron donor for photosynthetic bacteria is never water, but an inorganic substance such as hydrogen sulfide (H_2S); and the reactions never produce oxygen, but yield an inorganic chemical such as elemental sulfur (S). Related to the type of electron donors are differences in energies of electromagnetic rays that are absorbed by photopigments: eucaryotic pigments absorb light rays in the range 400 to 700 nm — energy sufficient to split the water molecule. In contrast, procaryotic cells use electron donors that can be split by lower-energy rays. It is also significant to note that all photoautotrophic bacteria are obligate anaerobes.

Their enzyme systems, which mediate the reactions of photosynthesis, will not function in the presence of oxygen.

Chemoautotrophs Bacteria are the only organisms in the *chemoautotroph* group. These types of bacteria represent a select group of organisms, because their enzymatic systems are truly unique. Most are divided into groups on the basis of the kind of inorganic materials they use as energy sources: iron bacteria, iron; sulfur bacteria, sulfur compounds; hydrogen bacteria, molecular hydrogen; and nitrifying bacteria, ammonia and nitrites. Bacteria within each of the groups are equipped with mechanisms that can extract electrons from the respective inorganic materials and use them to generate energy in the form of high-energy molecules. Much of their energy is subsequently used to reduce carbon dioxide in a manner similar to the dark reactions of photosynthesis. However, chemoautotrophs cannot use light energy; their carbon source is always carbon dioxide. These organisms are often the first to occur in environments that do not contain usable nutrients, preceding organisms that have enzyme systems which are less complete. Thus, we may describe this group of bacteria as being the most completely endowed organisms in respect to their enzyme complement for self-sufficiency. The important roles that these bacteria play in the cyclical conversion of material in the environment are discussed in Chapter 6.

Heterotrophs

Organisms in the *heterotroph* group may be *obligate* or *facultative* heterotrophs. Both types require organic compounds for carbon, but the facultative heterotrophs may use radiant energy under some environmental conditions.

Photoorganotrophs Although resembling the autotrophs in respect to energy source, *photoorganotrophs* are facultative heterotrophs. In dark, aerobic environments, they obtain energy and carbon from organic compounds; in light, anaerobic environments, they carry out the reactions of photosynthesis but require an organic carbon source.

Chemoorganotrophs *Chemoorganotrophs* are the most widely distributed organisms in the biosphere. Most bacteria, the fungi, the protozoa, and higher animals (including the human organism) are chemoorganotrophs. All organisms within this group have an absolute requirement for organic compounds as sources of carbon and energy. Consequently, the ultimate survival of each organism within this group is dependent upon the preexistence of autotrophs as sources of organic metabolites. However, the manner in which a specific organism obtains its energy and carbon may vary. Those organisms that obtain their energy and carbon from dead or decaying organic matter are saprophytes, and those which obtain their energy and carbon from another living cell are parasites. Many other types of relationships among chemoorganotrophs will be described in Chapters 7, 8, and 9.

Cellular metabolism

We use the term *metabolism* to designate the functional capacity of cells. This capacity can be described as two separate processes: catabolic reactions and anabolic reactions, both of which occur simultaneously. *Catabolic reactions* of cells are involved in the extraction and/or trapping of energy from inorganic and organic materials. Such reactions provide the cell with raw materials (building blocks) and a supply of energy for utilizing them. *Anabolic reactions* utilize the energy for producing new molecules, building new structures, and for growth, maintenance, and other activities. The result is a functional cell in its environment.

As unicellular entities, microorganisms are responsible for, or involved in, an array of changes that occur in nature. In some instances, microbial activities are undesirable and may cause injury or produce diseases in other organisms. Under other environmental conditions, microorganisms are responsible for the production of highly beneficial substances, such as antibiotics. In both examples, the changes result from the action of substances we call enzymes.

Enzymes

Enzymes are organic catalysts which participate in metabolic reactions as mediators. They do not make reactions possible, but act only as accelerators. Such reactions can occur without the aid of enzymes, but at a much slower rate. In other words, substances must be capable of reacting chemically; otherwise, enzymes cannot exert their effect. Although they take an active part in reactions, enzymes are not used up in the process and are not changed chemically or structurally during the event. They can be used repeatedly and are responsible for the self-perpetuating processes of metabolism. Some important features of enzymes are listed in Table 2.5.

Because of the sensitive nature of enzyme molecules, optimal activity will occur only within a narrow range of many environmental variables; the upper and lower limits for pH and temperature are particularly critical points. In fact, abrupt changes in physical and chemical parameters in an organism's environment may inactivate, or completely destroy many enzymes. Even under optimum environmental

Table 2.5. Some Properties of Enzymes

1. All are composed of proteins.
2. All are denatured (destroyed) by heat.
3. All can be precipitated by a high concentration of salts, such as ammonium sulfate, $(NH_4)_2SO_4$.
4. All are highly specific for structural sites on substrates.
5. Their activity is optimal only within a narrow pH range.
6. Their activity is optimal only within a narrow temperature range.

conditions, enzymes will not function indefinitely. Eventually they will wear out or lose their activity. Consequently, they are being replaced continually through synthetic reactions within the cells.

Those enzymes which remain inside cells and participate in intracellular reactions are called *endoenzymes,* and those which are excreted to the cell's exterior and participate in extracellular reactions are called *exoenzymes.* To function, a single cell must utilize many types of enzymes. All are not present at any given time, but the number and type are determined, to a degree, by the genetic information carried by the cell and by the nutrient composition of the environment surrounding the cell.

Enzymes may be classified in several ways, but our objective is not to present a detailed discussion of enzyme nomenclature. However, consideration must be given to the process of naming enzymes if we are to understand, in a general way, the role enzymes play in metabolic reactions. Prior to the establishment of rules for classifying enzymes, some were given names such as trypsin and pepsin. Those names are still being used, but the current practice is to name enzymes by adding the suffix "-ase" to the name of the substrate (a molecule being acted upon by an enzyme) or to a term that describes the reaction being catalyzed. Some examples of enzymes that have been named in this manner are listed in Table 2.6.

It is important to note that our examples are names for classes of enzymes that will only react with substrates within the class of compounds from which the name was derived. All the reactions in which these enzymes participate are hydrolytic: they add H_2O across the bro-

Table 2.6. Some Hydrolytic Enzymes Named on the Basis of Their Substrate Affinity

Substrate	Enzyme Class	Products
Protein	Proteinases	Polypeptides, amino acids
Carbohydrates	Carbohydrases	Disaccharides, monosaccharides
Simple lipids	Lipases	Glycerol, fatty acids
Deoxyribonucleic acid	Deoxyribonuclease	DNA nucleotides, nucleosides
Ribonucleic acid	Ribonuclease	RNA nucleotides, nucleosides

ken bond. Therefore, they are also called *hydrolases,* because they hydrolyze (split) macromolecules into smaller components (subunits). However, enzymes within a particular class can be further divided and named on the basis of a specific substrate molecule. For example, lactose (commonly called milk sugar) is a disaccharide that can be hydrolyzed into two monosaccharides (galactose and glucose) by the enzyme β-galactosidase (Figure 2.10).

Every enzyme molecule exhibits a high degree of specificity for a particular substrate. For a reaction to be catalyzed, an enzyme molecule and a substrate molecule must have complementary structural configurations called *active sites.* Such complementarity implies a recognition mechanism, because an enzyme and its substrate must match in a manner analogous to that of a lock and key (Figure 2.11).

In the preceding example, only one substrate molecule (substrate A) is recognized by a particular enzyme. Owing to this structural specificity, the enzyme binds to the substrate molecule, causing a chemical change in the substrate. The reaction results in the formation of new products (D and E), with the enzyme being released unchanged and free to react with another substrate molecule.

Utilizing the concept of structural specificity between an enzyme and its substrate, scientists have learned to regulate the growth of microorganisms in many environments. They have developed synthetic molecules that are structurally similar to natural-substrate molecules. When a specific synthetic substrate is added to the organism's environment, which contains a structurally similar natural substrate,

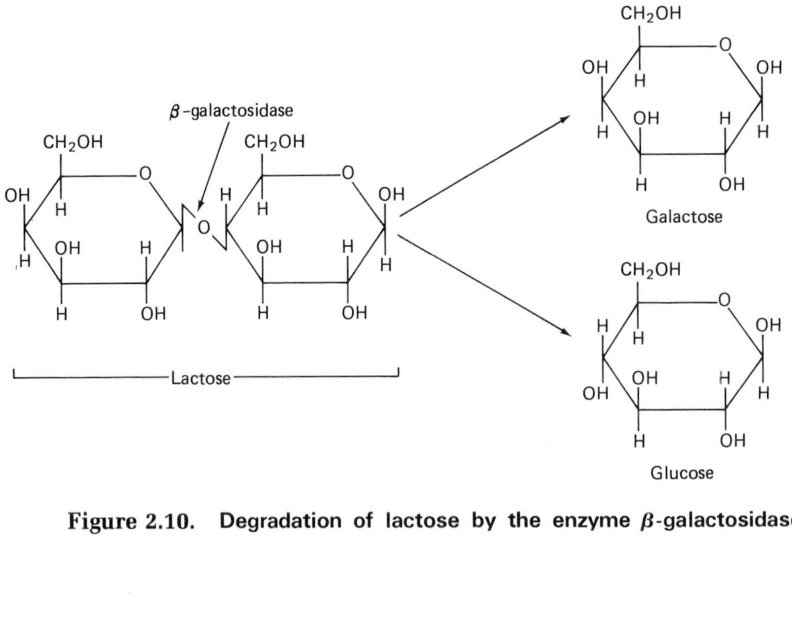

Figure 2.10. Degradation of lactose by the enzyme β-galactosidase.

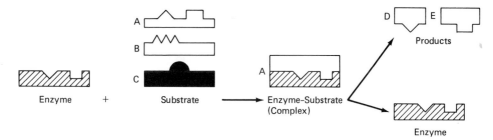

Figure 2.11. Generalized version of enzyme specificity for a substrate molecule.

a particular enzyme can combine with the "false" substrate or with the natural substrate. This process is called *competitive enzyme inhibition.* In such instances the possibility of enzyme inhibition is enhanced when the false substrate is introduced to the organism's environment in large quantities (Figure 2.12).

In the practice of medicine, physicians use sulfa drugs (powerful antimicrobial agents) to control disease-producing bacteria through the process of competitive enzyme inhibition. All sulfa drugs have as part of their structure a sulfanilamide ring (Figure 2.13). The ring structure is similar to para-aminobenzoic acid (PABA), which is an essential metabolite used by many bacteria for the synthesis of folic acid. When sulfa drugs are present in the organism's environment in large quantities, the sulfanilamide ring (false substrate) will combine with the bacterial enzyme and prevent it from reacting with the natural substrate (PABA). Consequently, bacterial growth will be inhibited. As long as the drug is present in a greater quantity than PABA, bacterial growth will be suppressed.

In addition to structural specificity, many enzymes have an absolute requirement for *cofactors,* inorganic ions such as magnesium and iron. Cofactors function primarily to activate enzymes, because some enzymes are present in cells as inactive molecules.

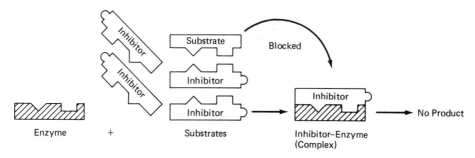

Figure 2.12. Generalized version of competitive enzyme inhibition.

Sulfanilamide (Inhibitor)   PABA (Substrate)

Figure 2.13. **Structures of para-aminobenzoic acid (PABA) and sulfanilamide.**

Many other enzymes require the cooperation of *coenzymes* for activity. Coenzymes function as carrier molecules. They are small organic nonprotein molecules which possess vitamins as an inherent part of their structures. Coenzymes are required by all enzymes that catalyze the removal of an element or group from a substrate molecule. They function as acceptors for the portion removed from the substrate. The dehydrogenases are among this category of enzymes. They catalyze the removal of hydrogens from substrates. The most important coenzymes, which participate in these reactions as hydrogen acceptors, are nicotinamide adenine dinucleotide (NAD), nicotinamide adenine dinucleotide phosphate (NADP), flavin mononucleotide (FMN), and flavin adenine dinucleotide (FAD). It is important to remember that dehydrogenases and their coenzymes differ with respect to specificity. Dehydrogenases, like other enzymes, have structural specificity for substrate molecules. However, coenzymes are not specific for dehydrogenases. A single coenzyme may participate in reactions with several different dehydrogenases, or several dehydrogenases may use the same coenzyme. An example of how coenzymes function is the following:

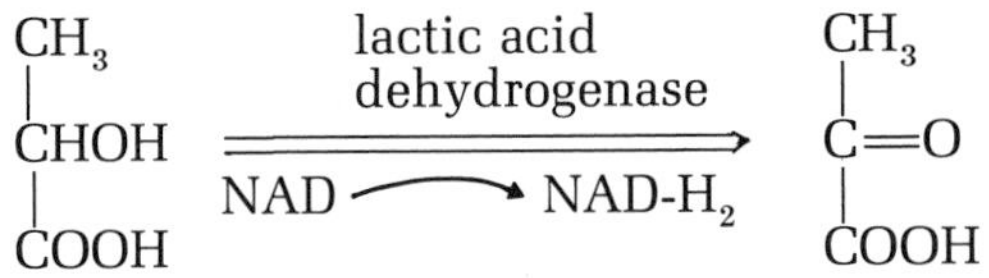

Note that hydrogens are removed from the substrate (lactic acid) and transferred to the coenzyme (NAD), which becomes reduced ($NADH_2$), and the product of the reaction is pyruvic acid.

The concept of enzymatic functions will become clearer as we delve further into our consideration of metabolic activities in microorganisms.

Energy-yielding processes

Previously, the catabolic phase of metabolism was characterized as reactions that provide cells with raw materials and a supply of energy for utilizing them. Those energy-yielding reactions which occur intracellularly are a continuation of the catabolic process. Regardless of energy source or type of organism, these reactions involve the transfer of energy from substrate sources to a form of chemical energy that can be used for cellular work.

Chemical reactions in cells obey the laws of thermodynamics. In this regard, a definite quantity of energy (heat) is released or consumed in every reaction that proceeds to completion. By using the calorie to designate a quantity of heat, energy released from or supplied to various reactions can be compared. A *calorie* is the amount of heat required to raise the temperature of 1 gram of water 1 degree Celsius. Reactions that occur spontaneously and release energy to the surrounding environment are called *exergonic*. Those reactions which do not occur spontaneously but will proceed to completion when energy is supplied to them are called *endergonic*. When exergonic reactions occur in nonbiological systems, the temperature of the surrounding environment always increases. In biological systems, exergonic reactions proceed to completion without a temperature change. This phenomenon results from the fact that enzymes and coenzymes function cooperatively in the transfer of energy to a special compound, adenosine triphosphate (ATP). ATP is universally found in cells, and it serves as the "link" between exergonic and endergonic reactions in all forms of life. The structure of this important compound is shown in Figure 2.14. Note the similarity between this compound and the mononucleotides found in the nucleic acids. ATP is a mononucleotide with three attached phosphate groups. When one phosphate molecule is split off

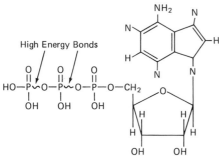

Figure 2.14. Structure of adenosine triphosphate (ATP). After the removal of one phosphate molecule, the resulting compound is adenosine diphosphate (ADP).

enzymatically, approximately 8,000 calories of chemical energy is released, and the resulting compound is adenosine diphosphate (ADP). Now, we can consider one ATP molecule as equivalent to 8,000 calories of chemical energy. On the basis of this approximation, the amount of energy obtained from exergonic reactions in cells can be calculated from the number of ATP molecules formed during a specific reaction or during a series of reactions.

ATP is not the only high-energy compound present in cells. In fact, some phosphorylated compounds have more energy, but ATP is the most important phosphorylated compound because (1) it has the ability to bind to active sites of many enzymes, and (2) it functions as an intermediate among the high-energy compounds by transferring phosphate groups from compounds with higher energy to those with lower energy. ATP is only a short-term energy source. If not used up, the ATP will be hydrolyzed to ADP.

In all cells, energy is transferred through a series of oxidation and reduction reactions. Substances become oxidized when hydrogens or electrons are removed, and reduced when hydrogens or electrons are accepted. Oxidation and reduction reactions always occur simultaneously and are considered as coupled reactions. Whenever a substrate or element is oxidized, another compound or element is reduced. In a similar manner, ATP is formed from ADP through coupled reactions during energy-yielding processes in all organisms.

It is important that we remain cognizant of energy relationships, because the survival of all living things (including the human organism) is intimately related to the flow of energy through the biosphere. Since the sun is the ultimate source of all energy, our discussion of specific energy-yielding mechanisms will proceed by reviewing the general aspects of photosynthesis, fermentation, aerobic respiration, and anaerobic respiration.

Photosynthesis In terms of magnitude and importance, *photosynthesis* is one of the major events in the biosphere. During the process of photosynthesis, radiant (light) energy is trapped and subsequently converted into carbohydrates by specific kinds of organisms. The only organisms in which the reactions of photosynthesis can occur are the higher green plants, eucaryotic algae, procaryotic algae, and some bacteria. All photosynthetic organisms have specialized pigment centers that serve as receptors for light energy from specific regions in the electromagnetic spectrum (Figure 2.15). Higher plants contain chlorophylls and other pigments that absorb light in the range 400 to 700 nm, whereas the bacteriochlorophylls absorbs lower-energy radiations in the range 700 to 950 nm. Some photosynthetic organisms have accessory pigments which contribute to the photosynthetic process.

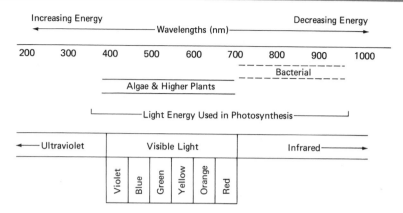

Figure 2.15. Portion of the electromagnetic spectrum. (Not drawn to scale.)

For details relative to the characterization of photopigments, the reader should consult one of the references at the end of this chapter. In this discussion we shall present a general view of the photosynthetic process which is divided into two parts: a light phase and a dark phase.

During the *light phase* of photosynthesis, radiant energy is first absorbed by the receptor pigments. Energy generated in the absorption process forces electrons in the pigments through a series of complex exergonic reactions that result in the formation of ATP. Simultaneously, water or other reduced substances serve as electron donors, and a reduced coenzyme ($NADH_2$ or $NADPH_2$) is formed as a second product.

During the *dark phase* of photosynthesis, ATP and the reduced coenzyme ($NADH_2$ or $NADPH_2$) are used to incorporate (fix) carbon dioxide (CO_2) into a carbohydrate molecule (ribulose 1,5-diphosphate). The carbohydrate molecule can then be further degraded as a source of energy through reactions that are common to most organisms. Thus, the overall photosynthetic process transforms radiant energy into chemical energy and produces carbohydrates for subsequent use by nonphotosynthetic organisms.

Collectively, the mixed group of photosynthetic organisms (higher green plants, eucaryotic algae, blue-green algae, and photosynthetic bacteria) may be considered as power generators for ecosystems, because reactions mediated by them are vital to the movements of energy and materials within the biosphere (see Chapter 6).

Fermentation   The term *fermentation* is applied to those energy-yielding reactions which organisms use to break down carbohydrates in the absence of air. Since oxygen is not present, the environment is called anaerobic, and organisms that have enzymes which function only in such environments are called *obligate anaerobes.* In contrast, organisms that live only in the presence of oxygen are called *aerobes.* However, many microorganisms are facultative and can function in the presence or absence of oxygen. Consequently, facultative microorganisms are highly versatile and obviously have an ecological advantage over those types that can function only in aerobic or anaerobic environments.

Regardless of an organism's ability to live in the presence or absence of oxygen, the major reactions that all organisms use to extract energy from carbohydrates are similar. The sequence of reactions that organisms use to obtain energy (ATP) from glucose under anaerobic conditions is referred to as *glycolysis* or the *glycolytic pathway.* Changes that occur in the glycolytic pathway are believed to represent types of metabolic reactions that took place in primitive microorganisms, because primordial environments were probably anaerobic. Furthermore, these reactions are not only present in microorganisms, but occur universally in the cells of all living things. Although some organisms degrade glucose through different reactions, the variations appear to represent modifications in the glycolytic pathway that occurred during their evolutionary development.

A generalized diagram of the glycolytic pathway is shown in Figure 2.16. Although the sequence involves several reactions, glycolysis begins with a highly reduced (high-energy) six-carbon substrate (glucose) and ends with the formation of a highly oxidized (low-energy) three-carbon substance (pyruvic acid). When considering the energy yields from glycolysis, it is important to note a very significant reaction in which a six-carbon fructose 1,6-diphosphate is split into two three-carbon units. One of the three-carbon substances (dihydroxyacetone phosphate) is not pertinent to our discussion. Therefore, all subsequent reactions with the other three-carbon substance (glyceraldehyde 3-phosphate) must occur twice in order to degrade glucose to pyruvic acid. In terms of energy we will consider two pyruvic acid molecules to be equivalent to one glucose molecule. Therefore, for each molecule of glucose that is degraded to pyruvic acid in the process, there is a net gain of two ATP molecules.

The end products of glucose fermentation may vary and are determined by the type of organism and environmental conditions. For example, when glucose is fermented by a facultative yeast, *Saccharomyces cerevisiae,* ethyl alcohol is the end product and the gas carbon dioxide is released (Table 2.7). When glucose is fermented in animal cells, lactic acid is the final product. Concomitantly with the formation of fermentation end products from pyruvic acid in either case, $NADH_2$ is oxidized to NAD. Thus, the cycle continues.

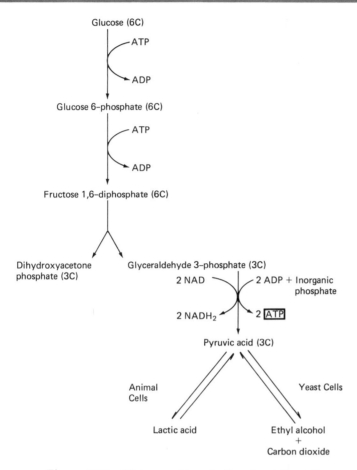

Figure 2.16. Major reactions in the glycolytic pathway of animal cells and yeast cells grown anaerobically. (Note that some intermediate reactions are not shown.) The number of carbon atoms per molecule is shown in parentheses. Enzymes participate in each reaction in the pathway.

Regardless of substrate or kind of organism, all fermentations produce some end products that are incompletely oxidized. It is important to remember that in fermentation, the substrate is the electron donor, and its derivatives (incompletely oxidized end products) always serve as electron acceptors. In other words, the end products are compounds with a great deal of energy remaining in them. For this reason, fermentation is an inefficient energy-yielding process.

Aerobic respiration   It is important to remember that the glycolytic pathway is a generalized sequence of reactions that are utilized

Table 2.7. Important Fermentation Products

Substrate	Microorganism	Product
Gluocse	Yeast *(Saccharomyces cerevisiae)*	Ethyl alcohol and carbon dioxide
Glucose	Bacteria *(Lactobacillus lactis)*	Lactic acid
Glucose	Bacteria *(Clostridium propionicum)*	Propionic acid
Glucose	Bacteria *(Escherichia coli)*	Mixed acids

by organisms living in either aerobic or anaerobic environments. Therefore, glycolysis is an integral part of aerobic respiration. In addition to glycolysis, organisms that obtain energy through aerobic respiration have enzymes that function in the *Krebs* or *citric acid cycle,* and in the electron transport system. These combined enzymatic processes are shown in Figure 2.17. The Krebs cycle and the electron transport system always function coordinately and enable aerobic organisms to oxidize pyruvic acid to carbon dioxide and water. Energy produced during these reactions is conserved in ATP molecules.

The initial reaction of the Krebs cycle involves the oxidation of pyruvic acid through an intermediate reaction in which carbon dioxide is released, and a two-carbon substance (acetyl-CoA) is formed. Then acetyl-CoA combines with a four-carbon substance (oxaloacetic acid) to form citric acid, which is a six-carbon compound. During the sequence of reactions within the Krebs cycle, carbon dioxide is released from two more compounds. The resulting product is the four-carbon oxaloacetic acid, which can then combine with another acetyl-CoA to repeat the sequence. Thus, the Krebs cycle can be visualized as a wheel that must make two complete revolutions to completely oxidize a molecule of glucose to carbon dioxide and water.

Simultaneously, hydrogens are removed from Krebs-cycle intermediates by dehydrogenases and transferred to coenzymes (NAD and FAD). Then the reduced coenzymes ($NADH_2$ and $FADH_2$) transport each pair of hydrogen atoms with their accompanying electrons ($2H^+ + 2e^-$) to the electron transport system. The electrons are oxidized through several cytochrome substances and transported to the final electron acceptor (molecular oxygen)—an ATP-generating process called *oxidative phosporylation.* The hydrogen atoms are not passed through the cytochromes, but are solubilized and ultimately combine with the electron acceptor (molecular oxygen) to form a water molecule.

Significant points to remember about energy production during aerobic respiration are as follows: (1) the sequential passage of a pair of electrons from $NADH_2$ through the cytochrome system to molecular oxygen produces three ATP molecules, (2) the sequential passage of a pair of electrons from $FADH_2$ through the cytochromes to molecular

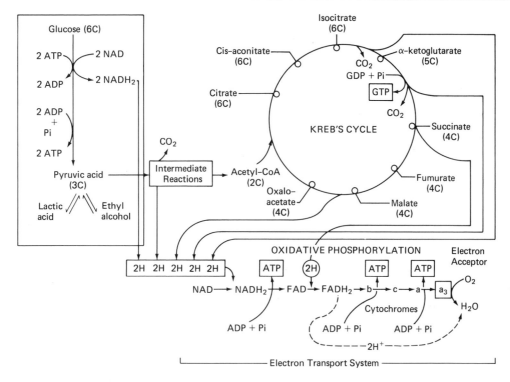

Figure 2.17. Abbreviated version of the combined metabolic reactions that occur in the breakdown of glucose in aerobic environments. Anaerobes use reactions in the upper left only. Pi, inorganic phosphate.

oxygen produces two ATP molecules, and (3) during each revolution of Krebs cycle, one ATP equivalence is produced in the form of quanosine triphosphate (GTP). Thus, the energy produced from the breakdown of one glucose molecule during aerobic respiration can be approximated by calculating the net high-energy molecules generated during the process. As a result of the sequential group of oxidative reactions in Krebs cycles, each glucose molecule is completely degraded to carbon dioxide (released by decarboxylating intermediate compounds) and water, with the concomitant production of 38 ATP molecules.

We can conclude this section by emphasizing the fact that aerobic respiration is a highly efficient energy-yielding process when compared to fermentation, which yields a net of two ATP molecules for each glucose unit that is degraded. Although glucose has been used as the substrate in these discussions, other substrates, such as proteins, fats, and a variety of organic substances, can be processed through intermediates such as pyruvic acid and acetyl-CoA. In addition, other materials can be converted into some Krebs-cycle intermediates. Consequently, cells are extremely versatile in their ability to obtain energy from environmental resources.

Anaerobic Respiration Organisms with the capability to utilize this mode of respiration are either facultative or obligate anaerobes. Facultative organisms carry out the reactions of anaerobic respiration only when living in anaerobic environments. Anaerobic respiration differs from fermentation with respect to types of substances that serve as final electron acceptors. During the process, electrons are removed from substrates (electron donors), cycled through reactions that generate energy, and finally transferred to an external electron acceptor such as nitrate, sulfate, and carbon dioxide. The mechanisms of anaerobic respiration are complex and inefficient in terms of energy yields. However, the reactions are extremely important ecologically.

Many of the organisms that carry out the processes of anaerobic respiration mediate reactions that are critical to the cyclical conversion of essential materials in nature (see Chapter 6). Some examples are as follows: (1) In certain soils, *Clostridium* species and some *Thiobacillus* species reduce nitrates to nitrites, nitrous oxide, or gaseous nitrogen through reactions that are collectively referred to as denitrification, (2); also, in certain soils and aquatic habitats, *Desulfovibrio* species play important ecological roles by converting sulfates to gaseous hydrogen sulfide; and (3) *Methanobacterium,* an inhabitant of the rumen, contributes significantly to the nutrition of ruminants by converting carbon dioxide to methane (see Chapter 5).

Energy-utilizing processes

Prior to our discussion of the various ways in which cells utilize energy, we shall review some features of ATP. It is the cell's major high-energy intermediate through which energy is transferred to molecules of lower energy, and it is also the intermediate that connects degradative processes to biosynthetic processes. Actually, ATP does not participate in every energy-requiring reaction in cells, because its phosphate bond energy can be transferred to other phosphate compounds, namely the nucleotides of DNA and RNA. Such high-energy compounds are recognized as deoxyribonucleotide phosphates (dATP, dGTP, dTTP, dCTP) and ribonucleotide triphosphates (ATP, GTP, UTP, CTP). In terms of energy transfer, each of those eight nucleotides functions in cells as an ATP equivalent.

All cells utilize energy for the production of new molecules, for growth and maintenance, and for performing a variety of other activities. First, cells utilize energy to facilitate the transport of certain nutrients into the cell through processes that are called *active transport.* During active transport, enzymes called *permeases* catalyze energy-dependent reactions that transfer molecules against concentration gradients. Second, cells utilize energy in the performance of mechanical work, such as flagella movement. The discussion to follow will focus on the biosynthetic phase of metabolism and emphasize some aspects of energy utilization common to many kinds of cells.

Biosynthesis of macromolecules The major classes of macromolecules (polysaccharides, proteins, lipids, and nucleic acids) are synthesized within cells from small molecules (building blocks or subunits) and then assembled into cellular components. Such reactions are complex and, in addition to energy, require the participation of an enormous number of enzymes.

The manner in which cells synthesize macromolecules is complex, but some general features can be noted. First, each monomer or subunit of the polymer must be activated by ATP or by one of the equivalent high-energy nucleotides. Second, all subunits must be catalytically assembled at its designated site by specific enzymes. Enzymes that catalyze the formation of polymers are called *polymerases.* Perhaps these reactions can be visualized if we examine the first few steps that lead to the biosynthesis of a polysaccharide (structures for several were shown in Figure 2.4). Note that the glucose molecule is the subunit from which each homopolymer was assembled. In the biosynthesis of such structures, the initial reaction is to add a phosphate molecule to the glucose residue, thus raising its energy content. Next, the phosphorylated glucose molecule undergoes some intermediate reactions and then interacts with uridine triphosphate (UTP) to form uridine diphosphoglucose (UDP-glucose). This structure is a carrier molecule and its function is to transport a single glucose molecule to the end of the polymer, where it can be enzymatically attached by a specific polymerase. Through the process the polymer is increased in length by one glucose unit, and UDP is released. The complete process must be repeated for each glucose subunit that is added to the chain.

The initial reactions that occur in the biosynthesis of a simple lipid are similar, but more complex, because macromolecules of simple lipids are heteropolymers. Note the number of different subunits that are present in phosphatidyl choline (see Figure 2.7). That phospholipid macromolecule contains one glycerol molecule, two fatty acid molecules, one phosphate molecule, and one choline molecule. The first reactions that require energy in the biosynthesis of phosphatidyl choline are the activation of all building blocks (subunits). In activated form, the subunits are: 2 (fatty acyl-CoA); glycerol-1-phosphate; and cytidine diphosphate choline (CDP-choline). The latter, a carrier molecule, functions to transport a molecule of choline to its appropriate attachment site on the growing phospholipid macromolecule. It is important to remember that a specific enzyme catalyzes the attachment of each activated subunit, and each carrier molecule is released.

The above is an oversimplification of the biosynthesis of a polysaccharide and of a phospholipid, but the abbreviated examples will suffice to demonstrate the utilization of energy by cells in the synthesis of two different classes of macromolecules. However, one should remember that biosynthesis of a particular macromolecule may proceed from pyruvic acid or from Krebs-cycle intermediates rather than activated building blocks.

Biosynthesis of proteins We shall now describe the biosynthesis of proteins with greater detail. Proteins are macromolecules that contain amino acid subunits. Since enzymes are proteins, it is important to understand how information in DNA is transmitted into proteins and ultimately to cellular functions. The information required to direct the synthesis of proteins is encoded within DNA (see Figure 2.9). A sequence of three nucleotides within each DNA strand is the *genetic code* or *gene* for a specific amino acid (Figure 2.18). Consequently, the sequence of bases on the polynucleotide chain will determine the sequence of amino acids in a particular protein. It is important to recognize that DNA contains the genetic code, but RNA is the carrier of the information. The mechanisms through which the encoded information in DNA directs the polymerization of amino acids into proteins occur in two steps: transcription and translation.

One strand of DNA serves as a template from which a single strand of RNA is polymerized. Since RNA is complementary to a DNA strand, a segment of DNA consisting of TCG GAT ACT would result in the formation of a complementary RNA sequence of AGC CUA UGA. This process is called *transcription*, because each message in DNA has been transcribed as a result of base pairing to RNA. Now, let us review this process by taking the first code in the DNA segment (TCG) and noting the complementary code in RNA after transcription (AGC). Assume that AGC codes for the amino acid serine; then it was actually determined by the triplet code encoded in the DNA molecule. In this manner any sequence of triplets on a DNA strand can be transcribed to RNA by specific enzymatic processes. This specific RNA is called *messenger RNA* (mRNA), because it carries information from DNA to the site of protein synthesis in the cell.

The next step in transmitting the information from DNA to a specific protein is called *translation*. We shall continue with mRNA and the codon for serine (AGC). The mRNA migrates to a site in the cyto-

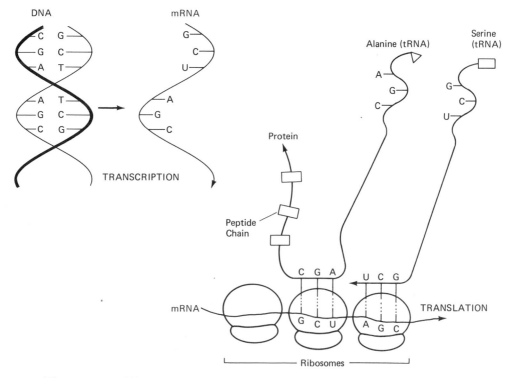

Figure 2.18. Abbreviated version of protein synthesis. Messenger RNA (mRNA) forms complementary to the "heavy strand" in the helical DNA molecule. Then, mRNA migrates to the ribosomes. Subsequently, each specific amino acid is carried to the ribosomes by a specific RNA (tRNA). The code recognizes the incoming complementary site on the tRNA, and the message is translated into a specific amino acid. As the ribosomes rotate in a counterclockwise manner, a peptide chain is formed. Actually, the process involves many complex reactions which require energy and the participation of a number of enzymes.

plasm and becomes associated with some particulate cytoplasmic matter called *ribosomes*. The ribosomes function as an assembly line for processing the assemblage of amino acids into peptides. Also, existing in the cytoplasm is another type of RNA called *transfer RNA* (tRNA). Energy is required to activate each amino acid prior to its enzymatic attachment to a specific tRNA. After attachment the tRNA becomes the carrier and transports each amino acid to the ribosomes. Consequently, tRNA carrying serine has a specific site for the codon AGC. At this point the serine molecule is joined to the preceding amino acid by a peptide bond; tRNA is released, and the process can be repeated in a similar manner with successive amino acids. Thus, a single mRNA can carry codes for many amino acids, and they can be processed into a sequence in a particular protein. Perhaps this oversimplification of protein synthesis will help to visualize the general concept of how the functional activities of cells are determined by genetic information coded within the DNA of their nucleus or nuclear region.

The molecular basis of mutation

Since the elucidation of the double helical structure of DNA in 1953, much has been learned about the characteristics of this unique heteropolymer. In living cells, DNA is a stable macromolecule, and the amount present in all cells of a particular species is relatively constant. The chemical basis for life is now known to be contained within DNA, because the determinants for what cells are and how they function is encoded within its unique structure. Consequently, when a trinucleotide sequence (codon) in DNA is altered, a modified message will be transcribed to mRNA. An altered codon is referred to as a *mutation*. A sequence of nucleotides within a gene can be altered in a number of different ways (Figure 2.19): (1) deletions—one of several bases are removed; (2) insertion—one or several bases are added; and (3) base-pair substitution—a base pair such as (AT) at a specific cite is exchanged with a different base pair (GC). Because the genetic code is said to be *degenerate*, mutations do not always produce an error in the sequence of amino acids that are being assembled into a growing peptide chain. Degeneracy is the phenomenon that accounts for the fact

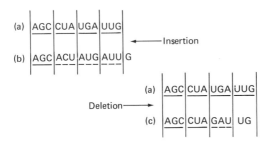

Figure 2.19. Segment of mRNA showing the manner in which a point mutation may occur. The message or code reads in triplets from left to right and is indicated by a solid line for normal reading and by a broken line for misreading after a point mutation has occurred. (A) Normal; (B) mutation resulting from insertion of A in the fourth position; (C) mutation resulting from the deletion of U in the fifth position of the normal mRNA.

that a single amino acid can be coded for by several different, but closely related, codons. For example, UUU and UUC codes for phenylalnine; AGU and AGC codes for serine; and any one of six different triplets can code for leucine. From the preceding we can readily see that if a single nucleotide change in DNA caused UUU to become UUC, phenylalanine would still be coded for and assembled into the peptide chain. An alteration of this nature is referred to as a *silent mutation,* because functions are not altered by the type of change that occurred in DNA. Thus, as a consequence of degeneracy, silent mutations may occur frequently. Note that a silent mutation could also result from the interchange of UUU and UUC. However, alterations in single nucleotides (point mutations) occur more frequently.

Mutations occur in all cells, at a low frequency (one mutation per 10^8 cells), and should be regarded as a normal event. Because the small ratio of normal mutants (in this case one cell) to parent or wild type $(1 \times 10^8$ cells), they are seldom detected. However, certain substances are known for their ability to enhance the mutation frequency in microorganisms. Such substances are called *mutagenic agents.* Ultraviolet light (especially wavelengths of 260 nm) and x-rays are powerful physical mutagenic agents. Known chemical mutagens include nitrous acid, 5-bromouracil, and 2-aminopurine. The latter two substances are called *base analogs,* because 5-bromouracil can be substituted for thymine, and 2-aminopurine can be substituted for adenine. After such substitutions in DNA, the analogs may cause false pairing of nucleotides during subsequent replications.

Scientists have learned to modify the environment of cells in a manner to select mutations of a particular kind. One convenient method for the selection of mutants that are resistant to an antibiotic such as streptomycin is to cultivate the bacterium on a medium that contains streptomycin in concentrations that range from very low to high. Resistant cells will grow in the higher concentrations, whereas sensitive cells will be suppressed. In experiments of this type, the antibiotic does not induce mutations; it functions only as an environmental selecting agent by causing the environment to become more favorable to the growth of mutants that occur normally but at low frequency. Currently, one of the major problems that physicians must deal with is the management of infections caused by antibiotic-resistant microorganisms. The widespread and sometimes indiscriminate use of antibiotics as prophylaxis is presumed to be partially responsible for creating selective pressures that favor the selection of resistant strains. The problem not only occurs in therapeutic prophylaxis but is severe in the hospital environment, where the atmosphere and surfaces are laden with traces of antimicrobial agents.

Key Words

acid Any substance capable of producing or liberating hydrogen ions (H^+) when placed in a solution.

anion A negatively charged atom.

autotroph An organism that obtains its energy and carbon entirely from inorganic materials.

base Any substance capable of producing or liberating hydroxyl ions (OH^-) when placed in a solution.

calorie That quantity of heat required to raise the temperature of 1 gram of water 1 degree Celsius.

cation A positively charged atom.

chemoautotroph Bacteria that obtain their energy from the oxidation of inorganic compounds and their carbon from carbon dioxide.

coenzyme A compound that functions coordinately with those kinds of enzymes that catalytically remove an element or chemical group from a substrate, and may be considered a carrier molecule.

covalent bonds Forces that hold molecules tightly together and result from the sharing of one or more electrons between two atoms.

deoxyribonucleic acid (DNA) A heteropolymer composed of nucleotide subunits in which the sugar deoxyribose is a component. DNA is the carrier of the hereditary material (genetic code) in all cells.

endoenzyme An organic catalyst that is synthesized within cells and participates in intracellular reactions.

enzyme An organic catalyst that participates in metabolic reactions as mediator.

exoenzyme An organic catalyst, synthesized within cells, but excreted to the cell's exterior, where it participates in extracellular reactions.

heteropolymer A macromolecule in which the subunits (monomers) are different.

heterotroph An organism that requires organic materials as a source of energy and carbon.

homopolymer A macromolecule in which the subunits (monomers) are identical.

hydrogen bond A relatively weak force that forms connections between atoms in different molecules and always connects an atom of hydrogen to either an atom of oxygen or an atom of nitrogen.

ion   Any atom that has an electrical charge.

ionic (electrostatic) bond   The force that holds oppositely charged atoms together and results from the complete transfer of an electron from one of the atoms involved to the other.

macromolecule   A large structure that is composed of smaller subunits (monomers).

molecule   The smallest single unit of any compound.

photoautotroph   An organism that obtains its energy from solar radiations and obtains its carbon from carbon dioxide.

photoorganotroph   Bacteria in this category are facultative heterotrophs and are capable of utilizing either radiant energy or energy from organic sources but have an absolute requirement for organic carbon.

poly-β-hydroxybutyric acid (PHB)   An intracellular nutritional source (storage product) for many procaryotic organisms; a homopolymer composed of β-hydroxybutyric acid subunits.

ribonucleic acid (RNA)   A heteropolymer composed of nucleotide subunits in which the sugar ribose is a component, is the carrier of the hereditary material (genetic code), and plays a major role in protein synthesis.

ribosomes   Intracellular granules or particles that are composed of protein and RNA and function in protein synthesis.

valence   The combining capacity or electron affinity of any atom.

Selected Readings

1. Brock, T. D. 1974. *Biology of Microorganisms,* 2nd ed. Englewood Cliffs, N. J.: Prentice-Hall, Inc.
2. Frosbisher, M., R. D. Hinsdill, K. T. Crabtree, and C. R. Goodheart. 1974. *Fundamentals of Microbiology.* Philadelphia: W. B. Saunders Company.
3. Scientific American Offprint, "The Living Cell." 1965. San Francisco: W. H. Freeman and Company.
4. Watson, J. D. 1976. *Molecular Biology of the Gene,* 3rd ed. Menlo Park, Calif.: W. A. Benjamin, Inc.

CHAPTER 3

Diversity of microbial habitats

- **Environmental selecting factors**
    Physical
        Solar radiation (temperature and light)
        Osmotic pressure
        Hydrostatic pressure
    Chemical
        pH (acidity and alkalinity)
        Available gases (O_2 and CO_2)
    Biological
- **Types of microbial habitats**
    Atmospheric environments
        Outdoor air
        Indoor air
    Aquatic environments
        Fresh water
        Salt water
    Terrestrial environments
    Biological environments
    Microenvironments
- **Modes of dispersal**
- **Key words**
- **Selected readings**

Microorganisms are ubiquitous in nature and are intimately associated with all members of the biological world (excluding those maintained in artificially constructed germ-free environments). They proliferate in extreme habitats where environmental pressures are too hostile for the growth and proliferation of other types of organisms. In extreme environments microorganisms may be considered as mavericks – the first arrivals and longest survivors.

The habitat of any organism may be defined as the location in the environment where it is found. In this respect, the term is used to designate a physical position, not a function. The term *niche* is used to specify the functional role of an organism in its habitat. In other words, an organism may have several habitats and only a few niches. However, the terms are still used interchangeably by some ecologists. In natural habitats, microorganisms are an integral part of the *biotic* (living) community which interacts dynamically with *abiotic* (nonliving) components to form *ecosystems* (Figure 3.1).

Within an ecosystem the biotic components can be described in terms of organizational levels. The lowest (simplest) level is the cell, and the highest (most complex) level is the community.

Actually, an infinite number of ecosystems can be found in the *biosphere* (the life-supporting region of the world). Ecosystems are often characterized arbitrarily, such as a pond, a forest, a lake, the pouch of a rumen, and the armpits of human beings. Since microorganisms play an integral role in all ecosystems, they must also be viewed on a microscale as entities that occupy microenvironments.

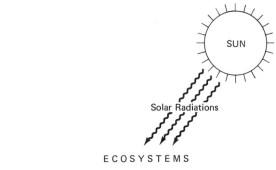

ECOSYSTEMS

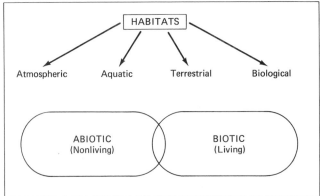

Figure 3.1. Generalized version of interactions between abiotic and biotic components in ecosystems. Existing within the biotic components is the following organizational structure:
unicellular units ⟶ multicellular units ⟶ populations ⟶ communities

Environmental selecting factors

The distribution and abundance of organisms in the biosphere are controlled by factors common to specific environments. In general, the distribution of macroorganisms is greatly influenced by climatic and topographic factors, but microorganisms are less sensitive to such gross changes. Characteristics of microorganisms that enable them to inhabit diverse environments include short generation time (and thus, large populations), ability to synthesize an array of enzymes, and their nutritional versatility. Macroorganisms lacking these properties are less successful in their ability to adapt. In this section we shall discuss representative kinds of physical, chemical, and biological factors that selectively influence the proliferation and distribution of microorganisms in nature.

Physical

Temperature, light, osmotic pressure, and hydrostatic pressure are some of the important environmental factors that influence growth and survival of microorganisms in nature. Although versatile in their ability to colonize new environments, microorganisms are highly sensitive to radical changes in physical conditions. Metabolically active microorganisms can tolerate physical changes only within narrow limits, outside which environmental changes are often destructive.

Owing to this ubiquity of microorganisms in nature, the application of certain principles offers practical benefits. Often we manipulate physical conditions of specific environments to control undesirable types. For example, when a contaminated material is sterilized in an autoclave, its physical condition is changed to a state in which all forms of life are completely destroyed. Thus, sterilization of the materials results from the interaction of physical factors — heat and pressure.

Solar radiation (temperature and light) On a macroscale, temperatures are governed by climatic zones which are partially determined by the amount of sun light (solar radiation) that reaches the biosphere. Although this solar energy is the ultimate powerhouse for all forms of life, much of it is lost in space before it reaches the earth's surface. Approximately 30 percent is scattered by constituents in the atmosphere, and about 20 percent is absorbed on its path through the biosphere. The remaining 50 percent reaches the ground and oceans, but a large portion of this is reradiated into space as heat, and only a small fraction enters the biological world through photosynthetic processes.

Although insensitive to gross climatic changes, growth rates of microorgansism in nature are influenced by temperatures of microenvironments. The thermal environment of hot springs in Yellowstone National Park is highly selective for *Thermus aquaticus,* a non-spore-forming bacterium with optimum growth temperatures of 70 to 75° C, and for *Bacillus stearothermophilus,* a spore-forming bacterium with optimum growth temperatures of 55 to 60° C. Similarly, the gastrointestinal tract of human beings (37° C) is the natural habitat for *Escherichia coli,* a non-spore-forming bacterium with an optimum temperature of 35 to 39° C. *Rhodotorula infirmoniniata,* a yeast with an optimum temperature of 14 to 18° C, proliferates abundantly in marine and freshwater environments. These examples demonstrate the diversity of temperature ranges in natural environments where microorganisms live and grow. However, cardinal temperatures (minimum, optimum, and maximum) are characteristic for a given species. For this reason, microorganisms have been categorized into physiological types on the basis of temperature relationships (Figure 3.2). Organisms that grow optimally in environments with temperatures between 45 and 60° C are called *thermophiles.* Those species that are common inhabitants of warm-blooded animals have optimal growth ranges between 25 and 40° C, and are called *mesophiles.* Cold-loving species grow best at temperatures between 0 and 20° C are called *psychrophiles.*

All metabolic processes result from enzymatic functions, and all enzymes have an optimum temperature range for maximum activity. It is important for us to recognize that optimum temperature for growth of a given species may not be the same as optimum temperature for a specific enzymatic process. It is also known that some mesophilic bacteria mutate to the psychrophilic state by synthesizing temperature-sensitive enzymes. Olsen and Metcalf of the University of Michigan used ultraviolet irradiation to produce psychrophilic mutants from mesophilic organisms, and observed a shift in minimum and maximum temperature from 11 to 44° C to 0 and 32° C, respectively. They also demonstrated that the gene marker for this characteristic could be transferred by a transducing phage. If mutations of this type occur in natural habitats, this mechanism could serve as an aid to microorganisms when colonizing new environments. It is important to remember that environmental temperatures do not influence growth and survival of a given species in a constant manner, because an organism's response to a given temperature may be altered by changes in other environmental factors.

In the biosphere, light and energy are inseparable phenomena; both originate from an ultimate source (the sun) and are emitted in the

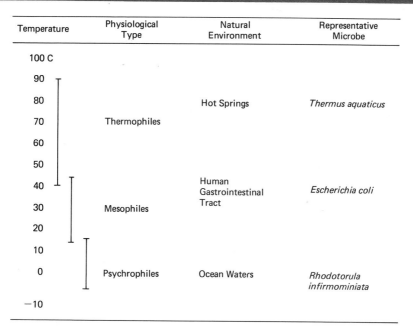

Temperature	Physiological Type	Natural Environment	Representative Microbe
100 C			
90			
80		Hot Springs	*Thermus aquaticus*
70	Thermophiles		
60			
50			
40		Human Gastrointestinal Tract	*Escherichia coli*
30	Mesophiles		
20			
10			
0	Psychrophiles	Ocean Waters	*Rhodotorula infirmominiata*
−10			

Figure 3.2. Temperature range and representative natural environments where specific microorganisms are found.

form of electromagnetic waves or solar radiations (see Figure 2.15). The electromagnetic spectrum is divided into several regions on the basis of wavelengths of light, and extends from the shorter cosmic rays on one end of the spectrum to the longer radio waves on the other. Energies associated with the various regions of this spectrum are referred to as *radiations*. High energies are associated with short waves and low energies with long waves. Among these radiations, the visible portion of the spectrum (400 to 800 nm) lies between the ultraviolet and infrared regions. However, radiations from several different regions interact with biological systems in a variety of ways.

Solar radiations enter the biosphere only through the process of photosynthesis. Higher plants contain photosynthetic pigments which absorb light in the range 400 to 700 nm, whereas photosynthetic bacteria utilize lower-energy radiation in the range 700 to 950 nm. Since many photosynthetic organisms live and function in aquatic environments, their metabolic activities are influenced considerably by the availability of light and the depth of its penetration.

In contrast to light energies which are absorbed by photosynthetic pigments, radiations from other areas of the spectrum function as selective agents in nature. For example, ultraviolet rays are extremely mutagenic, because rays in the region between 260 and 270 nm are absorbed by DNA. Consequently, lethal mutations often result from this event. Likewise, x-rays and gamma rays are destructive to biological systems, because they penetrate cells and produce damaging reactions within the cytoplasm.

Osmotic pressure To avoid ambiguity in our discussion of relationships between osmotic pressure and microbial growth in natural environments, it is imperative that we understand the basic concepts of osmosis. When two solutions of different solute concentrations are separated by a semipermeable membrane, the diffusion of the liquid across the membrane into the area of highest solute concentration is called *osmosis*. The unit membrane surrounds the cytoplasmic constituents of all cells in both procaryotic and eucaryotic organisms, and is the semipermeable boundary that separates the intracellular constituents from the external environment.

As mentioned in an earlier section, all ecosystems contain organisms that interact with other components of the environment in a dynamic way. During these processes, the unit membrane with its associated enzymes selectively regulates the entry of materials from the environment, and the exit of constituents from the cell's interior. In all aqueous environments, water tends to flow in a direction to equalize the solute concentration on either side of the plasma membrane. If the solute concentration of the external environment is more dilute than the solute concentration of the cytoplasmic materials, water will flow into the cell. Concurrently, an intracellular tension (osmotic pressure) develops in an attempt to counterbalance the inward flow of water. During this process, animal cells and cell-wall-less forms swell and may eventually burst. In plantlike cells the plasma membrane is protected from rupture by rigid cell walls. In this condition the environmental solute is *hypotonic* to the cell. If the solute concentration of the external environment is higher than the solute concentration of cytoplasmic materials, water will flow outward from the cell's interior into the environment, and the cell will become dehydrated through the process of plasmolysis. In this condition the environmental solute is *hypertonic* to the cell. When the solute concentration of the external environment and the solute concentration of the cytoplasmic constituents are equal, the environmental solute is said to be *osmotically protective* or *isotonic* to the cell. These processes are shown in Figure 3.3. Certain types of microorganisms proliferate optimally in environments where each of the described conditions prevail. For example, halophilic organisms grow optimally in hypertonic environments such as oceans and saltwater lakes. Similarly, saccharophilic organisms thrive in environments with high concentrations of sugars. In general, we refer to orga-

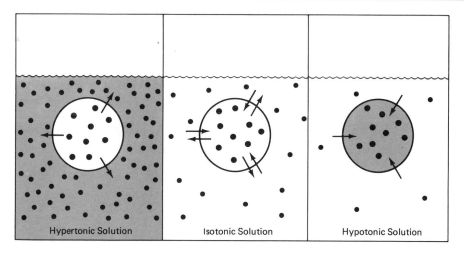

Figure 3.3. Schematic views of osmotic relationships. Arrows indicate the direction of water flow across the semipermeable cell membrane.

nisms with enzyme systems that function optimally in environments with high osmotic pressures as *osmophiles*.

It is significant to note that hypertonic environments are unfavorable to most of the microbial groups that we encounter in nature. Ancient civilizations recognized this fact and treated their meats and vegetables with salt brines to prevent spoilage. In modern food-processing establishments, salt and sugar contents of many products are increased deliberately to reduce spoilage from microbial action. The preservative action of hypertonic environments results from two processes: (1) plasmolysis of the organisms, and (2) dehydration of the product, yielding a moisture content too low to support microbial life.

Hydrostatic pressure　Microorganisms adapted to growth in environments of high hydrostatic pressure have rather restricted ecological niches—the great depths of the oceans. Many species among the various microbial groups have evolved with mechanisms for enduring high pressures; types that proliferate optimally in such environments are called *barophiles*. Hydrostatic pressure of seas is primarily a function of depth. For each 10-meter increase in depth, the hydrostatic pressure increases approximately 1 atmosphere (15 pounds per square inch). Several investigators have isolated organisms from depths of 1,000 to 10,000 meters, which would correspond to pressures of 100 to 1,000 atmospheres. However, very little information is available on the physiology of barophilic microorganisms. This is due, in part to the difficulty of growing microorganisms in the laboratory under high-pressure conditions. Researchers were therefore interested in the extent to which changes in temperature and pressure influence the survivability of barophilic organisms during the transition from natural high-pressure environments to simulated *in vitro* high-pressure systems, and ZoBell and Morita showed that terrestrial organisms were more sensitive to *in vitro* high pressures than were marine organisms. In their experiments, marine species grew well at 600 atmospheres, but terrestrial species were inhibited by 300 atmospheres. Similarly, other investigators showed that protein synthesis in both cell-free extracts and whole cells of *Escherichia coli* were progressively inhibited by *in vitro* pressures of 200 to 680 atmospheres.

Chemical

The intent of this section is to focus on those chemical factors that may function as selective agents in natural environments. In view of the nutritional versatility of microorganisms, a single material in any environment may satisfy more than one physiological requirement. In other words, a single factor may be bi- or polyfunctional, but the purpose here is to deemphasize their nutritional aspects. However, we must remember that physical and chemical components do not act independently, but function in a dynamic state; and one factor may alter an organism's response to other factors in the environment. For example, light, temperature, and oxygen function cooperatively to create a favorable environment for some aquatic organisms. However, we have grouped these factors into artificial categories for clarity of discussion and will emphasize the effects of pH (acidity and alkalinity) and available gases (O_2 and CO_2) on microorganisms in nature. Water is a chemical factor that has a great influence on microbial growth in nature, but because of its ability to function as a solvent in all biological systems, it was discussed in Chapter 2. Other aspects of water will be discussed in relation to aquatic habitats.

pH (acidity and alkalinity)   In our discussion of metabolism (Chapter 2), enzymes were characterized as being pH-dependent. In other words, they function optimally within a narrow pH range. Consequently, environments with a specific pH range are selective for organisms with enzymatic systems that are compatible with that range. As the acidity increases (lower pH values), the environment becomes highly selective for organisms that grow optimally in acid environments. For example, *Thiobacillus thiooxidans,* a bacterium that oxidizes sulfur compounds, grows optimally in environments where the pH is below 3. In contrast, organisms that grow optimally in alkaline environments would be inhibited by low pH values (see Figure 2.3).

Available gases (O_2 and CO_2) The atmospheric phase of the biosphere includes the natural gases (nitrogen, 79 per cent; oxygen, 19 per cent; carbon dioxide, 0.03 per cent; and minute quantities of inert gases), plus an array of gaseous materials (pollutants) which result from human-related activities in modern societies. Do these pollutant gases affect microbial systems in nature? Several pollutants are harmful to higher organisms, but little is known about their toxicity, if any, to microbial systems. However, we do know that microbial populations in specific environments are related directly to the availability of oxygen and carbon dioxide.

Since both of these gases are required by specific groups of organisms, they have a direct influence on types of organisms that will proliferate in a given environment. Obligate aerobes require oxygen for growth, without which their enzymatic systems will not function. In contrast, many obligate anaerobes are equipped with certain enzymes that function only in the absence of molecular oxygen, but other anaerobes can tolerate this gas in varying concentrations. The latter group are referred to as facultative organisms, and will grow in the presence or absence of oxygen and obviously have a wider range of natural habitats.

Carbon dioxide functions cooperatively with oxygen in biological systems. Therefore, microbial communities are influenced directly or indirectly by the availability of this gas. Its most critical function is in the process of photosynthesis (see Chapter 2).

Biological

The biological factors are properties and/or characteristics of living cells that influence the proliferation of microorganisms in or on host cells. These factors could easily be called chemical or physical, but for clarity we refer to them as biological since they are inherent to living cells. The biological factors to be discussed can be divided into two subgroups: (1) those which selectively favor colonization, and (2) those which selectively inhibit colonization (e.g., defense mechanisms).

Tissues of living cells exhibit optimum physical (e.g., temperature, pH) and chemical (nutrients) environments for rapid colonization by heterotrophs. In addition, the environments of living cells (e.g., microbes, plants, insects, human beings and other animals) are the only niches for obligate parasites. They have deficiencies that other living cells must satisfy and cannot function as free-living entities. In this regard, the survival of obligate parasites is dependent on access to a susceptible host, and a means for direct transmission and dispersal among host organisms.

In all environments, microbes compete for space and nutrients.

Therefore, prior arrivals may prevent subsequent colonization of other types. The space may be physically occupied, essential nutrients may be utilized, and toxic metabolic products (e.g., acids, enzymes, antibiotics) may be excreted (see Chapter 4).

The second group of factors selectively inhibit colonization by acting against invading microbes. In other words, they may be considered as barriers to colonization, and are both physical and chemical in nature. Some living cells are not colonized by microorganisms because of natural incompatibility, the phenomenon we call *natural resistance*. For this reason, many microbial diseases of animals are not transmissible to human beings.

Some structures of living cells (skin on animals and bark on trees) physically protect them from invasion by microorganisms. In other words, the outermost cover on living cells is a physical barrier. Also, animals are equipped with an immune system that selectively protects them from invading microorganisms.

Types of microbial habitats

We are now cognizant of critical factors in the biosphere that may influence microbial growth in specific environments. Also, we recognize that a given species tolerates environmental changes only within a narrow range, outside which drastic changes are often inhibitory and may be destructive. However, representatives from among the various microbial groups proliferate and survive in environments that are widely diverse. This diversity is exemplified by thermophiles in the hot springs of Yellowstone National Park, psychrophiles in the dry valleys of Antarctica, and mesophiles in the gastrointestinal tract of human beings.

If we are to avoid ambiguity in the discussion to follow, we must understand the proper use of several terms. First, we must differentiate between survival and growth. *Survival* denotes an organism's ability to remain viable in a state of dormancy or in a state of reduced metabolic activity, whereas *growth* refers to the ability of organisms to proliferate (increase in numbers or size). Second, we must differentiate between allochthonous and autochthonous floras. *Allochthonous* refers to organisms that are foreign or transient in a given environment, whereas *autochthonous* organisms are native inhabitants.

Atmospheric environment

Unlike other environments, the atmosphere contains only allochthonous microorganisms. Many species, representing the various microbial groups, have been isolated from the atmosphere, but they always reflect types present in other environments. Often they are introduced into the atmosphere from human activities (such as industrial operations). Then they become associated with particulate matter and are transported aerodynamically by wind currents.

Outdoor air In urban centers many types of microbes are constantly being expelled into the air. For this reason the air over cities contains a higher concentration of microorganisms than does the air over grasslands or forest areas. In all cases, though, airborne microbes are influenced by climatic conditions. Precipitation (rain, hail, and snow) exerts a cleansing affect on the atmosphere by washing out microbes. Similarly, humidity and suspended matter influence the survival of airborne organisms. Low humidities may enhance destruction of delicate species which cannot withstand dehydration. Likewise, in the absence of suspended matter (fog and smog) on clear days, the destruction of many microbial types may be hastened by increased exposure to ultraviolet radiation.

Aerosols from sewage-treatment plants and fly ash from solid-waste incinerators often contain viable microorganisms. Consequently, organisms emitted from such sources are disseminated in the environment by air movements. Aerosols of raw sewage and droplet nuclei from trickling filters are potential vehicles for contaminating the atmosphere with infectious agents.

Indoor air The microbial content of indoor air has been investigated primarily from the standpoint of controlling the spread of infectious agents in hospital environments. These environments contain a diversified group of pathogenic (disease-producing) and nonpathogenic organisms in a dynamic state. Sensitive strains are destroyed readily by cleaning and disinfecting agents, while others may mutate, acquire resistence to environmental agents, or survive on inanimate surfaces for long periods. Contaminated instruments, clothing, bedding, floors, and furniture represent inanimate sources in the hospital environment from which organisms are made airborne as a result of normal activities associated with patient care. Likewise, patients and personnel contribute to the airborne flora when coughing and sneezing. During these processes, large numbers of bacteria are expelled into the environment in atomized droplets. Subsequently, these complex groups of airborne organisms may be distributed throughout the hospital in air from the ventilation system. Modern hospitals often contain elaborate air-filtration systems through which the flow is unidirectional. Air enters patients' rooms through filters and is exhausted at floor-level outlets, thus greatly reducing turbulance in the vicinity of patients.

Table 3.1. Bacterial Cells Collected from Air Near Laundry Chutes in a Hospital

Chute	Air Samples (5 ft³)			Particles Settling (1 ft³ /min)		
	Before	Pulling Laundry	After*	Before	Pulling Laundry	After*
A	26	260	85	33	480	63
C	45	1500	27	33	140	87
D	77	160	34	15	330	27
E	47	330	54	–	180	54
F	75	700	60	110	420	45

*Taken 5 minutes after laundry pulling.
Source: Proceedings of National Conference on Institutionally Acquired Infections, School of Public Health, University of Minnesota, September 4– 6, 1963; *U.S. Public Health Service Publication No. 1188.*

However, during waste collection and laundry changes, many organisms can be thrust into the air (Table 3.1).

Aquatic environments

Water, in many forms, is an essential component of all ecosystems. In a broad sense, all human activities (e.g., agriculture, urban centers, technology) are directly related to the availability of water.

More than 70 per cent of the earth's surface is occupied by water, confined primarily to the world's oceans. The remainder is distributed among seas, lakes, rivers, and surface and subsurface streams. These bodies of water may be classified in several ways, but fresh- and salt-water types are specified in all classification systems. *Salinity,* high salt concentration, is the major criterion used to distinguish the two systems. Ocean waters and a few inland lakes contain high concentrations of salts and mineral ions, among which sodium and chlorine are the major components. This accounts for their high salinity. In contrast, these ions are consistently low in freshwater systems. Regardless of classification, all aquatic systems are dynamic, and water from both types is moved in the biosphere through the hydrologic cycle.

Fresh water Our vast urban centers and industrial facilities are linked primarily to bodies of fresh water, which to a great extent is a fixed quantity. In recent years we have experienced a very rapid population increase and a simultaneous expansion in our technology to meet our modern living preferences. As a result, our rivers and lakes collect tremendous quantities of complex materials, many of which are toxic and nonbiodegradable. Consequently, the supply of nonpolluted fresh water is being reduced. Water is considered polluted only after the introduction of infectious agents, detergents, pesticides, radioactive compounds, toxic materials, and other such items in quantities that will unfavorably affect aquatic organisms or render the water unsafe for human use. Even when toxicity is not suspected, organically rich effluents (agricultural, domestic, and industrial) enhance the rapidity of *eutrophication,* the natural aging process of lakes. Lake Erie is an example of a freshwater system that is rapidly dying as a result of such processes. Among the Great Lakes, Erie is the most shallow and is surrounded by many urban centers and industrial facilities, from which it collects an array of pollutant materials. Today, this is one of the most dramatic examples of human's destruction of an aquatic system, and according to some environmentalists, this dying lake may have passed the reversible stage.

Deep freshwater lakes are characterized by a definite cycle of events that occur during the year. During the winter season, ice at $0°$ C floats because it is lighter than water. In the liquid state, water reaches its maximum density at $4°$ C. Consequently, the subsurface liquid phase is relatively static. In the spring, water warms and there is a thorough mixing by wind currents when temperature equilibrium is reached, at $4°$ C. As the temperature continues to rise, a definite stratification develops (Figure 3.4). The upper layer (epilimnion) becomes uniformly warm and the lower layer (hypolimnion) becomes uniformly cold. A middle layer (thermocline) separates the warm upper region from the cold lower region. This narrow demarcation zone is characterized by a rapid temperature decline. In the fall season, water cools, and there is a thorough mixing by wind currents when temperatures cool to $4°$ C. As a result of these cyclical events, the entire contents of lakes (microorganisms, pollutants, and nutrients) undergo a thorough mixing *(overturn)* twice each year when temperature equilibrium is reached in the entire body of water. Remember that mixing occurs in both the epilimnion and the hypolimnion during the summer but that the two layers are separated by the thermocline. Regardless of climatic conditions, freshwater temperatures seldom rise above $30°$ C in the summer.

Lakes are often classified on the basis of their organic composition. Eutrophic lakes contain large quantities of organically rich nutrients which promote the growth of microorganisms. In contrast, oligotrophic lakes contain few organic nutrients and have relatively low microbial populations. Under natural conditions, either type of lake can remain relatively stable for thousands of years, but the balance is upset when an overabundance of nutrients or toxic materials is introduced. When considering environmental quality, waters rich in organic materials contribute to other environmental problems. Some species of blue-green algae grow abundantly in such waters and produce undesirable algal blooms and toxic products. Major fish kills and livestock poisoning have resulted from such conditions. Other phytoplanktonic species grow abundantly and deplete waters of dissolved oxygen, which results in the asphyxiation of fish and the cessation of activities by aerobic

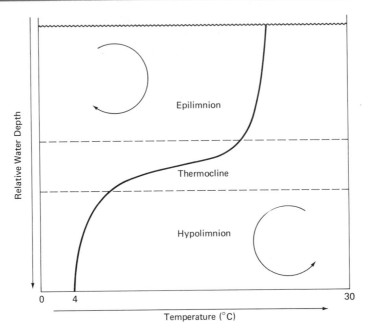

Figure 3.4. Schematic view of temperature stratification in deep waters during the summer. Arrows indicate separate mixing of the upper and lower layers.

microbes. After many years (probably thousands) of these accelerated processes, sludge accumulates, aquatic life ceases, and swamp conditions slowly develop.

Salt water   Aquatic systems that contain high concentrations of dissolved salts are vast and, in terms of volume, represent the major portion of water in the biosphere. Specifically, these environments include all oceanic waters and some inland lakes. In these environments, most chemical elements are present to some degree as dissolved inorganic salts. However, only a few elements are considered to be major constituents of seawater, and they are universally present in relatively constant proportions (Table 3.2). The predominant ions are sodium and chlorine, which accounts for the high salinity.

Seawaters are affected by temperature changes to a lesser degree than are freshwater lakes. Stratification of temperatures does occur, but the waters do not overturn in a comparable manner. Constant mixing takes place in the constant-temperature layer near the surface (epilimnion), which may range from a few feet to several hundred feet in depth. Below this region of constant mixing, there is a zone of rapid

Table 3.2.   Major Ions in Ocean Waters

Ion	Grams/Kilogram of water	Percent of Total Solids
Chloride (Cl^-)	19.40	55.20
Sodium (Na^+)	10.70	30.40
Magnesium (Mg^{2+})	1.30	3.70
Calcium (Ca^{2+})	0.40	1.16
Potassium (K^+)	0.39	1.10
Bromide (Br^-)	0.06	0.19
Bicarbonate (HCO_3^-)	0.14	0.35
Sulfate (SO_4^{2-})	2.70	7.70

temperature decrease which merges smoothly with the colder bottom layer.

Because of its high concentration of salts, seawater has a freezing point of $-1.9°$ C, and surface temperatures seldom rise above 30° C in summer. In the great depths of the ocean, temperatures are relatively uniform (-2.0 to 2.0° C). Consequently, the environmental conditions near the ocean bottoms are highly favorable to the growth of psychrophilic organisms.

The presence of dissolved and suspended materials in natural waters results in considerable variation in the transmission of specific light waves. The portion of oceanic waters that receives adequate light for photosynthetic processes is called the *photic zone*. This region seldom extends to depths greater than 20 meters. The area of continual darkness is referred to as the *aphotic zone*.

Representative species from among all microbial groups inhabit fresh- and saltwater environments, and many of their activities are vital to the maintenance of a balanced ecosystem. Except for a few shallow lakes and streams, environmental factors are more constant in aquatic habitats than in other ecosystems of the biosphere. Changes do occur, but under natural conditions they are relatively slow and extend over a period of many years. Aquatic ecosystems are complex, and microbial activities are regulated to a great extent by the solubility of gases and the distribution of light.

The photic zone is the most productive from the standpoint of photosynthetic processes. It is easy for us to view these processes only in terms of higher plants and some species of planktonic algae. Although unrecognized by many as producers, photosynthetic bacteria contribute significantly to the gross productivity in aquatic systems. Photosynthesis is the propelling force for all organisms except the chemolithotrophic bacteria. Therefore, the chemosynthesis that takes place in the aphotic zone of oceanic environments must also be included as a mechanism of gross productivity.

Heterotrophic organisms are the most predominant in both the photic and aphotic zones of aquatic systems. However, in the deoxygenated regions of aquatic environments, anaerobic bacteria mediate many vital processes.

Terrestrial environments

The loose, weathered, consolidated covering of the earth's crust that we call *soil* is dynamic and contains a complex mixture of macro- and microorganisms. Soil is formed initially from the disintegration and decomposition of rocks by physical, chemical, and biological processes. Soils do not reproduce themselves but, as a result of organismic activities, evolve through stages (young, mature, and old).

Igneous rocks are the starting materials from which soils develop, because they are the source of the major inorganic elements that we find in soils. For this reason, our perspective of soil might become more vivid if we consider the manner in which igneous rocks are formed. Often they are referred to as *volcanic* rocks, because they are formed when magma (the molten material within the earth) consolidates on the earth's surface. This process will become clear as we acquaint ourselves with the events that take place preceding, during, and following an active volcano.

All regions of the earth, at some time in the past, have been the

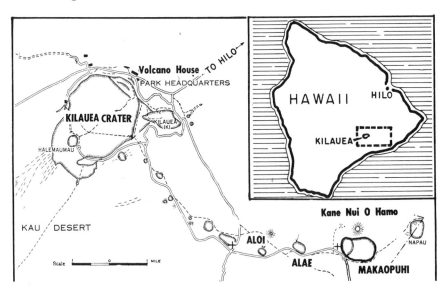

Figure 3.5.   Map showing the location of active volcanoes in Hawaii. Eruptions occurred at Aloi and at Alae in 1974. [Courtesy of *The Honolulu Advertiser*]

site of volcanic activity. At present, active volcanoes are common only in certain regions or belts of the earth, one of which encircles the Pacific Ocean. In such regions magma is more agile than in nonvolcanic belts. Volcanic eruptions occur when gases within the earth build up enough pressure to force magma from the earth's interior onto its surface. The specific location in Hawaii of some recent volcanic eruptions are shown in Figure 3.5. During one of the violent eruptions, magma was ejected explosively from the interior of the earth (Figure 3.6). When gas pressure is less, magma oozes quietly through cracks onto the earth's surface (Figure 3.7). Immediately after an eruption, the temperature of molten material on the earth's surface are between 1000 and 1200° C. Upon cooling, it becomes consolidated into igneous rocks of different sizes and shapes. Such rocks are the basic starting materials from which soil evolves.

Soil formation continues as rocks undergo *weathering,* which is a combination of physical, chemical, and biological processes. Any attempt to enumerate specific reactions or types of processes that occur during weathering would be futile, because the weathering of rocks is influenced by all kinds of interactions that take place in ecosystems.

Figure 3.6. **Active volcanic eruption at Aloi. [Courtesy of** *The Honolulu Advertiser.*]

Figure 3.7. Less-active volcanic eruption in which the molten material slowly oozes onto the earth's surface. In this instance the molten material emerges through a paved highway. [**Courtesy of** *The Honolulu Advertiser.*]

From this brief and simplified overview of soil formation, we can visualize that a very young soil is highly similar in composition to its parent rock. At the other extreme, there is a great deal of dissimilarity between an old soil and its parent rock. Thus, in addition to organismic activity, we must remember that time is an important factor in the evolution of soils.

During eons of evolution, the loose material on the earth's surface was disturbed by climatic changes and transported for considerable distances by wind, water, and glacial ice. As a result of the differential deposition of these materials, we now have topographical regions composed of characteristic soil types. The soil proper is composed of mineral elements and organic matter. The mineral portion is extremely variable with respect to chemical elements and particle size. On a weight basis, organic matter represents but a small proportion of the earth's crust, and it is distributed unevenly among the mineral constituents. As a result we find a graduated series of soil types in the biosphere, ranging from mineral soils at one extreme to organic soils on the other.

Soils are classified on the basis of physical properties and particle

size, the latter ranging from macroscopic in gravel and sand to microscopic in clay. Soils composed of the same range of particle size and of
similar physical properties constitute a single soil class. Soils also have
characteristic features with respect to the presence of water and gases
that occupy interstitial crevices or pore spaces. These spaces are influenced by the composition and particle size of the organic component.
Consequently, the water-holding capacity of soils is determined to a
great extent by particle size. Pore spaces are occupied by water, air,
or both. Therefore, when soil water increases, the amount of air decreases. Thus, poorly drained waterlogged soils are deficient in air,
which results in a progression toward an anaerobic environment.

The soil is a dynamic ecosystem in which microorganisms are in-
terwoven among the physical and biological components. They are associated with root hairs of higher plants, the macroscopic fauna
(worms, insects, etc.), soil crumbs, and interstitial water. Thus, the
microflora of soils are complex and may include representative members of any microbial group. In general, algae are present to a lesser
degree than are fungi, bacteria, and protozoans. However, the magnitude of populations and specific types are related to soil types and environmental factors such as temperature, moisture, and the availability
of nutrients. Unlike aquatic environments, the soil represents a variable habitat for microorganisms. Consequently, microbial populations
in soils are also variable. Each soil environment selects microbial
types that can adapt physiologically to a given set of environmental
conditions. In other words, microbial populations in the soil shift readily in response to environmental changes. For example, in regions
where rainfall is heavy during the spring, populations of green and
blue-green algae increase proportionately. In a similar manner, populations of the same groups decrease as the moisture evaporates, and
the soil becomes drier. These algal groups form resistant cysts that
enable them to survive until moisture increases or until they are
transported to more favorable environments by erosion forces. Under
conditions of increased soil water, bacterial populations respond differently: types that thrive under aerobic conditions decrease, and populations of anaerobes increase. The reverse occurs when the moisture
evaporates and the air in pore spaces increases.

In addition to the role of microorganisms in natural ecosystems, a
few pathogenic species are always present in the soil. For example,
Clostridium tetani and *Clostridium botulinum* proliferate optimally in
the soil under anaerobic conditions. The former is the causative agent
of tetanus and usually enters the body through open or puncture
wounds. The latter is a causative agent of botulism. This is a type of
food poisoning that results from eating unheated or partially heated
packaged or canned foods which contain toxins that were produced in

Table 3.3.   Frequency of Foodborne Botulism in the United States (1971 – 1972)

Year/State	Food Product	Place Processed	Toxin Type	Cases	Deaths
1971					
Alaska	Frozen whitefish	Home	E	2	0
Washington	Vegetables (suspected)	Home (suspected)	Unknown	2	2
New York	Antipasto	Home	A	2	0
	Vichyssoise soup	Commercial	A	2	1
Pennsylvania	Peppers	Home	B	3	1
California	Unknown	Unknown	Unknown	2	0
	Chili peppers	Home	A	5	1
	Bean paste	Home	B	2	0
	Celery	Home	Unknown	2	0
Maryland	Unknown	Unknown	A	1	1
				23	6
1972					
Ohio	Peppers	Home	Unknown	4	2
California	Unknown*	Unknown*	A	18†	0
Colorado	Peppers	Home	A	1	1
Oklahoma	Vegetables	Home	Unknown	1	1
				24	4

*No food item could be definitely incriminated.
†Persons hospitalized, 6; not hospitalized, 12.
Source: Morbidity and Mortality Weekly Report 22 (No. 7): February 17, 1973.

the anaerobic containers by cells of *C. botulinum* that survived the food-processing operations. Foodborne botulism is not uncommon; its frequency and severity in the United States are shown in Table 3.3.

It is important to remember that many types of microorganisms live in the soil; some of them play a vital role in natural ecosystems, but a few types must be controlled in order to safeguard and promote the health of the general public.

Biological environments

Unlike other natural habitats, biological environments are sub-units of larger ecosystems. With the exception of germ-free organisms, microbes live on surfaces and within tissues of all other organismic units. Consequently, physical and chemical factors (e.g., temperature, pH, nutrients) that influence microbial growth are governed by the conditions of the host organism. Thus, biological environments are highly selective for fastidious intracellular parasites and free-living heterotrophic species. In these habitats, the microflora is complex, and there is a great deal of diversity among the various physiological types.

The human body is one example of a biological environment. An intimate relationship with microorganisms begins at the time of birth when the infant becomes exposed to the autochthonous microflora of the birth canal, and it continues throughout life. Microbes have established a variety of ecological niches in and on the human organism.

Many phenomena occur in nature as a result of the association between microorganisms and higher forms of life. Symbiotic nitrogen fixation is a process that results from the metabolic activities of specific bacteria that live in the root nodules of leguminous plants. Bioluminescence in some fishes is possible only because specific bacteria live in their systems. Termites eat wood, and ruminants eat grass, only because they harbor specific microorganisms in their digestive systems which perform the vital function of cellulose digestion.

Many intermicrobial associations also exist in nature. In addition to all viruses that are obligate parasites, protozoans parasitize other protozoans; bacteria parasitize other bacteria; fungi parasitize other fungi; and algae parasitize other algae (most predominant among the blue-green group).

Microenvironments

As the name implies, *microenviroments* are self-contained ecosystems on a miniature scale. An infinite number of microhabitats exist within every macroenvironment. However, microhabitats are unique, because they often exhibit characteristics that are not observable on a macroscale. Some examples of microhabitats are crevices in soil, grains of sand, leaf litter, spaces under rocks, holes in trees, insect eggs, carcasses, and dung. Of the vast number of microenvironments that have been studied, those found associated with the human organism or with microbes that cause human diseases are probably the best characterized.

Although microhabitats are always a subunit of a large ecosystem, microorganisms in them may actually be living under very different physical and chemical conditions. For example, the human organism is an obligate aerobe, but an anaerobic bacterium, *Corynebacterium acnes,* can be isolated from the human skin. Thus, the microhabitat of the bacterium is obviously different from that of the human organism. Microhabitats of obligate intracellular parasites can often be extremely different from that of the host organism.

Modes of dispersal

By utilizing effective dispersal mechanisms, microorganisms have become the most widely distributed kinds of living things. Those processes which contribute to the spread of microorganisms also contribute to their escape from hostile environments. Without provisions for escape, some species would become extinct within a short period. Therefore, survival of a species is related to its ability to be dispersed efficiently, and ultimately to its successful colonization in a conducive habitat. In this regard, dispersal and colonization are important facets of competition. *Dispersal* refers to the spread or distribution of organisms from one place to another, and *colonization* refers to the establishment of residence in a specific site with subsequent reproduction at that locale.

Microorganisms may be dispersed by active or passive means. *Active dispersal* occurs when cells or their reproductive structures move from one place to another in response to inherent physiological processes, and *passive dispersal* occurs when distribution is by means of mechanisms that are not controlled by the cells.

Regardless of the manner in which dispersal occurs, many structural and physiological characteristics enhance the ability of microbes to be dispersed in ecosystems. Such characteristics are often referred to as *mechanisms of dispersal* (Table 3.4). No single type of organism possess all the characteristics listed, but each characteristic or property contributes to the survivability of one or more species.

Forms of active dispersal include all types of motility, outward extensions (growth) of filamentous organisms into surrounding niches, and aerial discharge of spores (common among some fungi). Active dispersal of some species may also be enhanced by their ability to respond to various environmental stimuli, called *taxes* (which include phototaxes, chemotaxes, and aerotaxes). These responses may be positive

Table 3.4.   Some Physiological and Structural Characteristics That Enhance Dispersal of Microorganisms

Characteristic	Function
Rapid growth	
Colonial forms	The large number of cells increases the probability of survival
Filamentous forms	Extend the range of usable resources by penetrating adjacent niches
Small size and light weight	Facilitate aerial transport and movement through soil crevices
Buoyancy	Facilitate transport by floating on water
Organs of locomotion (flagella and cilia)	Provide active motility
Response to taxes (chemotaxis, phototaxes, etc.)	Stimulate directional movements (positively or negatively)
Spores and cysts	Dormant stages that enhance survival under adverse conditions
Ability to live as intracellular parasites	Enhance transport by host, provide protection against adverse conditions

(movement toward the source of stimulus) or negative (movement away from the source of stimulus). Such movements may, under certain conditions, facilitate the escape of organisms from hostile environments.

Successful dispersal alone is not adequate to provide for the continual survival of a species. In this regard, colonization and proliferation in a favorable habitat is the ultimate "goal" of dispersal. Because ecosystems are heterogeneous in structure, they often contain numerous habitats that are similar in terms of environmental parameters. Thus, cells adapted to the microenvironment of one habitat can thrive optimally in other habitats. However, all compatible habitats in ecosystems are not readily accessible to colonization by a given species. Consequently, species residing in one habitat are often separated by gaps or discontinuities from habitats that have similar environments. Such conditions occur when ecological niches are not overlapping or adjacent.

Since the niche is a natural reservoir from which microorganisms are dispersed, it is important for us to consider the manner in which all habitats are connected in ecosystems. Connections or linkages among habitats result from the fact that ecosystems are dynamic. Thus, passive dispersal of microorganisms is made possible through interactions among the meterological, geological, and biological components of the biosphere, which facilitate dispersal by functioning as vehicles and vectors for transporting microorganisms. The conditions are similar to the passive transport of human beings by modern vehicles during global travel: on land by motor vehicles; on water by seagoing vessels; and through the air by airplanes and spaceships. The manner in which each meterological, geological, or biological component contributes to the dispersal of microorganisms is depicted in Figure 3.8.

The meterological components include the transport of airborne microorganisms by wind currents and precipitation (rain, hail, or snow). Microorganisms in the form of vegetable cells, cysts, and spores can be transported for considerable distances by wind currents. However, during precipitation some microorganisms are removed from the atmospheric habitat and transported to terrestrial, aquatic, or biological habitats. Since airborne microbes behave as particles, they tend to settle or fall out of the air by gravitational forces. These forces are primarily geological. They also contribute to the transport of microorganisms through drainage systems such as rivers and streams.

The biological component, which includes all kinds of living cells, also plays an important role in the dispersal of microorganisms. This occurs because many kinds of microbes can live in a variety of microenvironments in or on the tissues of other living cells, called the host. Thus, the association of microbes with host living cells is often com-

Figure 3.8. Diagrammatic representation of linkages among different habitats in ecosystems. Although the various habitats are different, they are interconnected through meterological, geological, and biological forces, all of which contribute to the passive dispersal of microorganisms.

plex and may be transient or obligate (see Chapter 5). Irrespective of the intimate nature of the association, microorganisms leave host cells by many different portals of exit and enter other habitats. Thus, organisms which serve as host for microorganisms are highly efficient in the dispersal of their microflora. For some microbial species, their survival is dependent on the ability of their host to effectively transmit them to another hospitable environment in a susceptible host.

Key Words

allochthonous   Pertains to an organism or material that is foreign or transient in a particular habitat or region in the biosphere.

aphotic zone   The region in natural bodies of water that lies at depths that receive no significant amounts of light.

autochthonous   Pertains to an organism or material that is native or indigenous to a particular habitat or region in the biosphere.

barophile   An organism that grows optimally in regions of great hydrostatic pressure (e.g., the great depths of oceans).

biodegradeable   Pertains to substances that can be decomposed by physiologically active organisms or their metabolic products.

biomass   Pertains to the weight of living organisms in a designated habitat, region, or even a vessel.

ecosystem   The inclusive region or habitat within the biosphere where interactions between living organisms and nonliving components occur.

eutrophication   Pertains to the nutrient enrichment of natural aquatic habitats directly or indirectly from human-related activities.

facultative   Pertains to an organism that has the ability to grow in the presence or absence of a specific environmental factor.

fastidious   Pertains to the sensitivity of microorganisms, especially with respect to specific nutritional requirements.

halophile   An organism that requires high concentrations of salt for optimal growth.

humus   Residual organic debris that results from the partial decomposition of organic matter.

mesophile   An organism that grows optimally within the temperature range 25 to 40°C.

microaerophile   An organism that requires oxygen for growth but at a tension lower than is present in the atmosphere.

mutation frequency   The speed or rate at which a specific kind of hereditary change (mutation) will appear in a given population of organisms.

osmosis Pertains to the diffusion of water across a semipermeable membrane from a region of low solute concentration to a region of higher solute concentration.

photic zone The region in natural bodies of water that is penetrated by light rays.

psychrophile An organism that grows optimally within the temperature range 0 to 20° C.

saccharophile An organism that requires sugar in high concentrations for optimal growth.

thermophile An organism that grows optimally within the temperature range 45 to 60° C.

xerophile An organism with enzyme systems that enable it to grow optimally in dry places (e.g., deserts).

Selected Readings

1. Adams, A. P., and J. C. Spendlove. 1970. Coliform aerosols emitted by sewage treatment plants. *Science* 169:1218–1220.
2. Ahearn, D. C., and F. J. Roth, Jr. 1966. Physiology and ecology of psychrotrophic carotenogenic yeasts. *Devel. Ind. Microbiol.* 7:301–309.
3. Brock, T. D. 1966. *Principles of Microbial Ecology.* Englewood Cliffs, N.J.: Prentice-Hall, Inc.
4. Brock, T. D. 1967. Life at high temperatures. *Science* 158: 1012–1019.
5. Brown, M. R., Jr., D. A. Larson, and J. C. Bold. 1964. Airborne algae: their abundance and heterogeneity. *Science* 143: 583–585.
6. Edmonds, P. 1965. Selection of test organisms for use in evaluating microbial inhibitors in fuel–water systems. *Appl. Microbiol.* 13:823–824.
7. Horowitz, N. H., R. E. Cameron, and J. S. Hubbard. 1972. Microbiology of the dry valleys of Antarctia. *Science* 176: 242–245.
8. MacLeod, R. A. 1965. The question of the existence of specific marine bacteria. *Bacteriol. Rev.* 29:9–23.
9. Olsen, R. H., and E. S. Metcalf. 1968. Conversion of mesophilic to psychrophilic bacteria. *Science* 162:1288–1289.
10. Sinclair, N. A., and J. L. Stokes. 1965. Obligatory psychrophilic yeasts from the polar regions. *Can. J. Microbiol.* 11: 259–269.
11. ZoBell, C. E., and R. Y. Morita. 1957. Barophilic bacteria in some deep sea sediments. *J. Bacteriol.* 73:563–568.

CHAPTER 4

Microbial interference

- **Competition for survival in nature**
  - *Changes in ecosystems: succession*
  - *Competition for nutrients*
  - *Competition for oxygen*
  - *Competition for space*
- **Role of antimicrobials in nature**
  - *Antibiosis from antibiotics*
  - *Antibiosis from bacteriocins*
  - *Antibiosis from other substances*
  - *Sensitive and resistant species*
- **Key words**
- **Selected readings**

Microbial interference is a concept used to describe those interactions among microorganisms which tend to favor one kind of a population at the expense of another. Interference often results from complex interactions which involve both competitive and antagonistic mechanisms. In all habitats in which microorganisms are metabolically active, populations are regulated to a degree by substances present in the environment that surrounds them. The effects of such substances may be stimulatory or inhibitory. Often these effects are inseparable, because a single substance may have more than one functional role in an ecosystem; for one type of an organism at one concentration, it may be beneficial; but for another type of organism at another concentration, it may be detrimental.

Perhaps the relationship between competitive and antagonistic interactions can be visualized if we assume that some habitats contain only two populations of microorganisms (A and B) that are metabolically active. If population A secretes a substance that is inhibitory to population B, the available resources in that habitat will become more accessible to population A, because the competitive ability of population B will have been reduced. Interactions of this type are referred to as *interference competition*. Our example of a two-population ecosystem represents the simplest type of interaction that may occur in natural habitats. Most often, numerous types of indirect interactions occur simultaneously among all the populations present in a given ecosystem. Therefore, the phenomenon of microbial interference is extremely complex.

123

Furthermore, understanding interference phenomena is made more difficult by the inconsistent use of some basic terms: interference, competition, and antagonism. In this book their use will be consistent with the school of thought that characterizes competitive and antagonistic interactions as subsidiary to the broad phenomenon of interference. Therefore, "microbial interference" will be used to describe all interactions that tend to favor one kind of a population at the expense of another.

Competition for survival in nature

The ability to adapt was described in Chapter 1 as a fundamental characteristic of living things. By *adaptation* we refer to modifications of properties and metabolic processes which enable organisms to function as vital components of ecosystems. Adaptations that result from genetic variations are subject to evolution, so *natural selection* plays a major role in the survival of a species.

In terms of adaptability, microorganisms are the most versatile kind of living thing. They are ubiquitous, and a single population may contain an enormous number of individuals. However, each individual within a population must compete with other individuals for resources that are needed by all. A population of a single kind of microorganism is not a homogenous assemblage of individuals; a marked degree of variation always exists among its members. Consequently, all individuals are not equally likely to survive in any habitat. Those individuals that are less fit, or unable to adapt to environmental conditions, will be eliminated, and those individuals that are better adapted will survive.

Competition will occur when the availability of essential resources in a given habitat is insufficient to satisfy the requirements of the resident microflora. In other words, competition takes place between two or more individuals when their combined needs for a single resource exceed the availability of that resource. Competitive interactions may involve only individuals within a single population, or they may involve individuals from populations of different kinds of organisms. When the struggle occurs among individuals of the same species, the interaction is referred to as *intraspecific competition;* and when the interaction involves individuals from different species it is called *interspecific competition*. Intraspecific competition is more severe than interspecific competition, because organisms with similar metabolic systems are most suited to identical ecological niches. In terms of magnitude, however, interspecific competition is more widespread. While both types of competition play significant roles in determining the population structure of communities, they represent only one aspect of changes that occur in ecosystems. Of equal importance are changes that result from interactions among abiotic components. Thus, the environment of microhabitats may be modified by separate or combined interactions of biotic and abiotic components of ecosystems. In terms of their effect on the resident microflora, modifications may be positive or negative. Positive modifications produce changes that are conducive to organisms residing in a given habitat. In contrast, negative modifications produce changes that are unfavorable to the resident microflora. If negative modifications cause environments to become too hostile for the continual proliferation of the resident microflora, the population will be gradually displaced by organisms that can thrive better in the environment of the modified microhabitat.

Changes in ecosystems: succession

It is impossible to discuss interactions that contribute to microbial interference without considering them as a part of natural changes that occur in all ecosystems. Since interference interactions are universal, their importance will be discussed as an integral part of *succession,* the sequential changes in populations that take place over a period of time in ecosystems. A stable community represents the final stage of succession, and it is composed of different types of organisms which reproduce themselves in a state of dynamic equilibrium called the *climax.*

After being introduced to a new habitat, microbial cells may not respond favorably to the existing environmental conditions. In spite of their remarkable ability to adapt, no single kind of microorganism is versatile enough to proliferate in all accessible habitats of an ecosystem. Therefore, in the new habitat, cells may encounter interactions that tend to function as barriers to colonization. The new habitat may be free of other living organisms (i.e., sterile or virgin), or it may contain an abundance of other living cells. In either case we refer to the initial population of cells as *pioneers* when they are the first to enter a habitat of nonliving things or when they are the first individuals of a particular kind to invade an environment of other living cells.

Pioneers may experience one or more of the following events: (1) when environmental conditions are extremely severe, individuals will be completely destroyed as a result of irreversible damage to their cells; (2) when environmental conditions are less severe, but unfavorable for growth and proliferation, cells may enter a stage of dormancy or exist for a considerable period of time as transients; and (3) when environmental conditions are favorable, most cells will proliferate and become established as part of the resident community.

Perhaps the concept of microbial succession can be visualized if we consider first some general aspects of ecological succession that can be observed with the naked eye. For example, vegetation in climax forest communities in certain regions of the United States is often destroyed by forest fires. Over a period of years after such an event, the vegetation will replace itself in an orderly and sequential pattern: first grasses will appear, next small shrubs, then larger shrubs, and eventually trees will emerge and the ecosystem will become stabilized as the climax forest community. Although the basic patterns of succession are similar, different plant species are involved in different geographical regions. One group of plants modifies the environment, chemically and physically, for the next group, until the climax is reached. Concomitantly, bird species, and species of other organisms, increase in diversity with sequential vegetation changes in a particular region. When vegetation reestablishes itself in areas denuded by forest fires, the sequential changes described above represent *secondary succession,* because the land was previously occupied by vegetation. Obviously, such areas are organically rich. Therefore, from the standpoint of nutrition, many kinds of substrates are available for utilization by heterotrophic microorganisms. Thus, microbial activities are vigorous in denuded areas long before vegetation becomes visible.

Population changes that occur in ecosystems after the initial colonization of a virgin habitat are known as *primary succession.* In view of our earlier statement that microorganisms are potentially everywhere, some readers might be surprised to learn that microbe-free regions do exist in the biosphere. In Chapter 7 we will discuss colonization and succession of microbe-free regions of higher organisms. At this point we shall consider colonization and succession as it would probably occur in nonbiological virgin habitats, regions not previously inhabited by living things. As an example, recently formed igneous rocks have natural surfaces that are free of microorganisms. During the early stages of soil formation such surfaces are nutritionally poor. In this case, pioneer colonizers would be microorganisms that can live in extreme environments. The environment of such habitats selects for chemosynthesizers. The first pioneer colonizers in such microhabitats will most likely be a homogeneous population of chemoautotrophic bacteria, and early successors will probably be some kind of algae. During the period of their development, organic materials and biomass will be synthesized. Subsequently, organisms that utilize only preformed organic matter, such as the fungi, may develop, and at this point lichens often emerge.

Lichens are unique morphological and functional organismic units that form from the union of two separate organisms: an alga and a fungus. The association is relatively constant, and the lichen is distinctly different from either of its members. Lichens are potent colonizers of rocks and can be seen easily with the naked eye. For this reason they are often listed as pioneer colonizers in reports that deal with succession in vegetation. After heterotrophic forms appear in microhabitats, the populations gradually become more diverse, interspecific competiton become a possibility, and numerous other interactions may occur among populations while utilizing nutrients and excreting metabolites. Simultaneously, abiotic forces will be interacting dynamically. The microhabitat will slowly become altered and progressively unfavorable for the continual proliferation of the first arrivals. Thus, individuals among the pioneering population will gradually be displaced by cells that have enzymatic systems which are more compatible with the environmental conditions of the modified habitat. As a result of the entry of more specialized cells, the populations will become more diverse. Then, interspecific competition will occur in the habitat as succession proceeds toward a climax or dynamic equilibrium.

We must remember that in a given habitat the active period of occupation by a pioneering population may not be fleeting. The initial population may proliferate abundantly for a considerable period of time in a luxurious surrounding of resources during the absence of interspecific competitors. However, the initial population will eventually be replaced by more specialized cells as the environment becomes modified. If displacement of the pioneers results from modifications in the habitat by activities of the established populations, the process is called *autogenic succession*. On the other hand, if the pioneering population is displaced by another population as a result of modifications in the habitat by abiotic forces, the process is called *allogenic succession*.

It should be emphasized that changes in ecosystems do not always result in the selection of more specialized cells, with a subsequent increase in species diversity. Occasionally, the pioneers will alter the environment to an extreme extent and in such a manner that only organisms with metabolic systems similar to pioneers can become established in the habitat. In such a case, the initial colonizers remain dominant as the stable community. Such phenomena occur, for example, in the waters of coal-mining regions. Acidophilic bacteria produce sulfuric acid, which increases the acidity of the habitat to pH 2 or below, thereby making the environment more favorable for the continued dominance of the pioneer community and reducing the possibility of competition. When such drainage enters other bodies of water it is toxic to fish and other aquatic organisms. The overall diversity of the aquatic community is thus decreased rather than increased.

Raw milk is an example of a natural ecosystem in which microbial succession has been studied in detail. The term *raw milk* is used to describe the condition of milk as it is removed from the cow's udder. In this state milk from healthy cows contains small numbers of harmless bacteria, although other kinds of microorganisms often enter raw milk from external sources—the air; the surface of the cow's body; the surface of utensils; and from persons in the vicinity. If raw milk is stored in a refrigerator at low temperatures (5 to 10°C), growth of the microorganisms will be suppressed and spoilage will be delayed. On the other hand, if raw milk is allowed to stand at room temperature (25 to 28°C), microorganisms present in it will grow rapidly, and the milk will undergo decomposition or become spoiled. Thus, milk spoilage is actually the result of activities by specific groups of microorganisms that predominate at various stages of decomposition. The sequential population changes result from the fact that microorganisms in milk differ in their ability to compete for various substrates, and also differ in their susceptibilities to the increased acidity.

The first stage of spoilage is initiated primarily by bacteria called *Streptococcus lactis,* which grow rapidly and cause fermentation of the milk sugar (lactose). As the acidity increases (i.e., as the pH is lowered), the environment becomes unfavorable for this group and more conducive for the aciduric group *(Lactobacillus* and *Leuconostoc).* Subsequently, bacteria of these genera proliferate abundantly at the lower pH and continue to ferment the lactose. These groups decline in numbers when the lactose is completely decomposed. Then aciduric yeasts attack the lactic acid and other substrates, thus reducing the acidity (pH increased), and alkaline products accumulate. Finally, as the pH continues to rise, oxidative and putrefactive bacteria predominate and hydrolyze fat and casein. As a result of these degradative processes, end-products that have unpleasant odors are released during milk spoilage.

With the emergence of microbial ecology as a specialized field of science, an increasing number of investigators are applying classical ecological concepts to the study of interactions among microorganisms. Microbial activities in microhabitats of discrete parts of higher plants (roots, stems, and leaves) are highly suitable for such studies, because they can be separated from the parent plant and investigated as isolated ecosystems. The decomposition (rotting) of tree leaves is an interesting example of microbial succession (Figure 4.1). Specific groups of organisms are involved in the sequential degradative process because they compete for different components (substrates) in the leaves. Although several different species may be involved, they are commonly categorized into ecological groups: the sugar fungi, the cellulose-decomposing fungi, and the lignin-decomposing fungi.

The decomposition process is initiated by microorganisms within the first group, the *sugar decomposers.* They excrete enzymes that degrade the simple sugars in freshly fallen leaves. During their period of activity, leaves turn brown in coloration. Members of the second group,

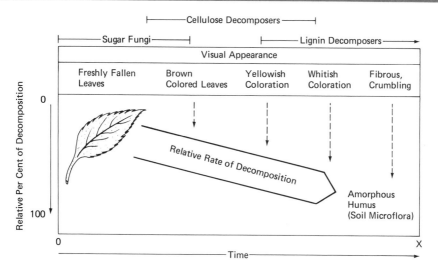

Figure 4.1. Generalized version of fungal succession on leaf litter.

the *cellulose decomposers,* excrete enzymes that degrade cellulose, and the leaves turn yellowish in coloration. The final stage of decomposition involves, the *lignin decomposers,* which excrete enzymes that degrade lignin. The latter group of fungi is highly specialized and somewhat unique, because bacteria and protozoans do not have enzymes that can degrade lignin. During the final stage of decomposition leaves appear whitish in color and can be easily recognized as moldy. After complete decomposition, the fibrous leaves disintegrate into a material called *amorphous humus.* The successional events described above involve not only sequential population changes in microorganisms, but also sequential chemical and physical changes in the structure of the ecosystem. Another interesting point to remember is that microbial decomposition of leaves and other organic materials is a beneficial process, because it releases to the environment important chemical elements for subsequent reuse by other organisms (see Chapter 6).

Ecological succession should not be considered as a simple phenomenon. Competition among populations for essential resources, the inability of populations to adapt to physical and chemical changes, and the susceptibility of populations to antagonistic substances are interference processes that contribute to ecological succession.

Competition for nutrients

When it can be shown that variation in a single environmental component is a determinant of population density, the component is referred to as a *limiting factor.* All environmental factors are potentially limiting, but the concept refers most often to nutrient availability. Since all essential requirements for growth of microorganisms must be available in a usable form, any required nutrient may function to regulate population density if present only in a limited supply or in a form that cannot be used. Basically, microorganisms compete for a substrate in a specific form and under a specific set of environmental conditions.

In some natural ecosystems, competition among microorganisms for nutrients can be observed easily during the early states of succession. For example, the milk sugar, lactose, becomes limited during the natural spoilage of raw milk. During the process mixed populations of microbes compete for the lactose. In a similar, although more complex process, microorganisms compete for nutrients in polluted waters. However, evidence for competition among microorganisms in natural bodies of water is more difficult to obtain. Furthermore, when a utilizable resource is limited in an aquatic ecosystem, the more successful competitors tend to increase in numbers. Therefore, if a required nutrient is limited in a lake, additional quantities of that nutrient will enhance the growth of one or more populations of the indigenous microflora. Conditions of this type prevail in waters that receive pollutants from domestic waste and/or surface runoff from farmland, especially after the application of commercial fertilizer.

We know that quantities of nutrients fluctuate in all natural bodies of water. Aquatic ecosystems that contain high concentrations of certain nutrients also contain large populations of some unicellular green and blue-green algae. For this reason an increase in algal blooms is often considered to be an obvious indicator of water pollution. Consequently, we are concerned about types of materials that may enter natural waters through human activities. When large quantities of organic matter enter water, algae and other microorganisms grow abundantly. Even when the concentration or organic matter is low, inorganic salts such as phosphates and nitrates often contribute to algal growth. Although they may enter natural waters as pollutants, the concentration of phosphorus and nitrogen is often limited, because both are required by aquatic organisms; and at certain periods, they may be utilized almost to the point of exhaustion. Because both are required, the lack of either phosphorus or nitrogen serves to regulate the growth of unicellular algae in aquatic ecosystems. When such a condition occurs, we can assume that indigenous microorganisms are forced to compete for the limited supply of those vital nutrients.

Competition for oxygen

Organisms that have an absolute requirement for oxygen were described in Chapter 2 as aerobes. Such organisms have enzymatic systems that cannot function unless oxygen is present in their environment. Since many kinds of aerobic organisms live in aquatic habitats, their activities are directly related to the availability of oxygen. Thus, when oxygen is limited, competition for this vital gas is one of the most important kinds of interactions that occur among aquatic organisms.

Oxygen, present in the atmosphere at about 21 per cent by volume, dissolves in water at the air–water interface. It is also released into water by aquatic photosynthetic organisms as a metabolic product. However, oxygen may not be equally distributed in the water. It is most abundant near the surface, and is distributed to greater depths by diffusion and circulation. When mixing extends from top to bottom, the oxygen concentration may become uniform. However, the deeper portions of many aquatic systems have oxygen deficits. In such environments, facultative organisms tend to predominate, and at greater depths the environment may be entirely anaerobic.

Aerobic heterotrophic microorganisms grow rapidly and are highly active in waters that receive organic pollutants. Even when such pollutants do not contain toxic chemicals, they are undesirable, because they support a luxurious growth of aerobic microorganisms. These aerobic heterotrophs vigorously attack and decompose the organic pollutants. During the process they utilize oxygen that has been dissolved in the water. This demand for oxygen by aerobic microorganism in aquatic systems is called the *biochemical oxygen demand* (BOD).

As the organic pollutants increase, algae and other aquatic plants grow abundantly, and their dense growth on the water surface inhibits the penetration of light. As a result, photosynthesis is also reduced, and submerged plants die. The dead plants then become an additional supply of organic material to be decomposed by aerobic microorganisms. Thus, the BOD of that body of water becomes greater. Among the many kinds of aerobic organisms that live in water, fish have the highest requirement. Consequently, many fish and aerobic animals die, because they are the weaker competitors for the limited supply of oxygen. When such conditions occur in a body of water, the surviving aquatic organisms are predominantly anaerobic bacteria and some zooplanktons. The decomposition process does not stop at this point. The dead fish and aquatic plants are further decomposed by anaerobic bacteria, with the simultaneous release of foul odors from the metabolic end products of fermentation.

Competition for space

In this section we are concerned with the site in a microhabitat where microbial cells actually localize. Cells that can adapt to the environmental conditions of a locale will subsequently reproduce themselves and become functional members of the microflora in that ecosystem. This process is colonization, previously described as the localization and subsequent reproduction of microbial cells in a specific site. Thus, colonization may give certain microbial cells a competitive advantage over others, especially if the resources in that microhabitat are limited. For example, if the new colonizers can replicate themselves faster than other members of the mixed microflora, they will tend to predominate.

Furthermore, the predominating members of the mixed microflora will be more capable of expressing their physiological attributes. If the secretion of antagonistic substances is among these attributes, the continual growth and/or subsequent colonization of sensitive cells will be prevented. Indirectly, the microbial cells are competing for space, and interactions of this nature comprise interference competition. Therefore, mere colonization can be of considerable advantage to microbial cells in a given microhabitat.

Since microorganisms are widely distributed in the biosphere, they often compete for space when attempting to colonize newly exposed surfaces. This fact was recognized several years ago, and the principle of bacterial interference was utilized in a hospital nursery in an attempt to prevent colonization of newborn infants by pathogenic bacteria. The nasal mucosa and umbilicus of newborn infants were artificially contaminated with nonpathogenic staphylococci (bacteria identified as strain 502A) with the anticipation that colonization by this strain would inhibit subsequent colonization of pathogenic staphylococci. Results from that study showed that the nonpathogenic staphylococci protected the infants from colonization by pathogenic staphylococci. Furthermore, the nonpathogenic organisms spread to other human contacts. As a result of the widespread colonization of other individuals with bacteria from the artifically contaminated infants, the dissemination and colonization of the pathogenic strains of staphylococci in the nurseries were reduced considerably. It is important for us to recognize that the physical presence in a given site is only one aspect of the interaction between nonpathogenic and pathogenic strains of staphylococci. Other factors do contribute to microbial interference phenomena.

Role of antimicrobials in nature

In this section we are concerned with antagonistic phenomena, interactions that are harmful to microorganisms. Since the effects are detrimental to microorganisms, the responsible substances are referred to as *antimicrobial agents*. Types of substances classified as antimicrobials are both numerous and varied. Some of them are non-biological chemicals (i.e., pollutants) which enter ecosystems from external sources, and others are excretory products from metabolic processes. When interactions among microorganisms result in injury, reduction in growth, or inhibition of activity in at least one of the participating members, the process is specifically referred to as *antibiosis*. Note that this term is somewhat restrictive and excludes injury to microbial cells from nonbiological toxic chemicals that enter ecosystems from external sources.

Antibiosis among microorganisms is not a recent discovery. In 1877 Pasteur and his associates observed that the anthrax bacillus, *Bacillus anthracis,* did not grow well in urine contaminated with other kinds of microorganisms. From this observation they speculated that the contaminating microorganisms might be useful as therapeutic agents to inhibit the growth of the anthrax bacillus in infected persons and animals. Subsequently, a few other investigators attempted to prevent the spread of anthrax and tuberculosis in human beings by intentionally infecting them with other kinds of bacteria. Such studies were not very successful at that time. However, in recent years, both *in vitro* and *in vivo* studies of antibiosis have become increasingly popular.

Antibiosis from antibiotics

With the emergence of microbial ecology as a specialized discipline in the broad field of microbiology, reports in scientific journals on *in vitro* studies with mixed cultures are increasing. However, the practice of attaching ecological significance to pure culture studies remains controversial. Although some skepticism may be valid, it is imperative for us to recognize that *in vitro* experiments with both pure and mixed cultures of microorganisms have generated a wealth of knowledge that has ecological implications. The best understood phenomena of antagonism center around *in vitro* studies of antimicrobial action of antibiotics. Thus, our discussion of antibiosis from antibiotics will include a synopsis of the discovery and usage of these important substances.

It is important to recognize that there is no magic in good research and no formula for significant discoveries. Most important scientific discoveries have resulted from two common events: (1) carefully planned and executed experiments by prudent investigators, and (2) a great deal of fortuitousness or simply good luck. The discovery of the antibiotic penicillin was no exception. Alexander Fleming, a research physician at St. Mary's Medical School, London, had been studying the antimicrobial action of a variety of substances, including tears and saliva, for about fifteen years. He had already discovered lysozymes, but continued his search for substances that would kill pathogenic bacteria when injected into the bloodstream of human beings without damaging normal host cells. In other words, he was searching for a substance with selective toxicity, one that could kill bacteria without damaging normal cells. (For a review of this concept, see the discussion of enzyme inhibition by sulfanilamide in Chapter 2.)

Fleming was a meticulous investigator, but his laboratory was usually cluttered with old cultures. Unlike many of his colleagues, he did not clean his work area and discard cultures immediately after completing an experiment. Instead, he would keep the old cultures for a considerable period, for further observation. Eventually, every older culture would be scrutinized for any significant phenomena that were not detected during earlier observations. While carrying out this ritualistic process with some old contaminated bacterial cultures of *Staphylococcus aureus,* he noticed that the bacteria in one plate appeared to be different from the others. The *Staphylococcus* colonies had been dissolved immediately surrounding the mold contaminant. Instead of appearing as typical yellow masses, they resembled drops of dew on the surface of the agar. Fleming quickly suspected that the mold had produced an antibacterial substance. Since it is not unusual for cultures to become contaminated, Fleming's observation of bacterial inhibition by a chance contaminant might have escaped recognition by a person with less curosity. The incident served as a catalyst for subsequent studies that led to the characterization of the first antibiotic. Thus, Fleming's keen intellect and peculiar habit of scrutinizing old cultures proved the veracity of Pasteur's statement that "chance favors the prepared mind."

Fleming knew little about mycology (the scientific study of fungi), but he continued to test the ability of his mold to inhibit different kinds of bacteria. Of particular interest from his studies was the fact that the mold produced substances that inhibited the growth of some pathogenic bacteria. This observation was unusual and significant, because he readily recognized that the substances had potential use as a therapeutic agent. Subsequently, Fleming cultivated his mold in large containers of liquid medium (nutrient broth). It grew primarily on the surface of the liquid and progressed through various shades of color — first gray, then green, and finally black. After several days of growth, the liquid assumed a yellowish color. His next important experiment was to determine if the yellowish liquid would kill bacteria. Indeed, it did, in very dilute concentrations, and he called the yellow substance "mold juice." As the inhibitory power of his mold juice became more interesting, he recognized that the mold should be classified scientifically, and consulted expert mycologists concerning the task of identifying it. Finally, the mold was definitely identified as a fungus with the scientific name *Penicillium notatum.*

From these experiments, Fleming recognized that he was dealing with a phenomenon of antibiosis. This was not a new concept, because Pasteur and coworkers recognized antibiosis in 1877. Therefore, in 1929 the scientific community showed little interest in Fleming's mold juice, because all previous substances that had been demonstrated to kill bacteria *in vitro* also destroyed human cells when injected into the bloodstream. Fleming was not discouraged and continued to conduct experiments with his yellow substance. Although he could not extract the active ingredients from his mold juice, he named it "penicillin" after the fungus *Penicillium notatum* from which it was excreted. About 10 years later, two biochemists from Oxford, England (Howard Florey and E. B. Chain), purified the substance that Fleming had shown to be a powerful antibacterial agent, and confirmed its identity as penicillin. Subsequently, penicillin proved to be highly effective as a therapeutic agent for certain bacterial infections.

In 1943 industry began the mass production of penicillin, and there soon followed a radical change in the therapeutic treatment of many infectious diseases. The importance of Fleming's research was finally recognized, and he received many citations and honors for his role in the discovery of the first antibiotic. Fleming, Florey, and Chain were awarded the prestigious Nobel Prize for their accomplishments. The story of penicillin marks the beginning of the period in microbiology that we now recognize as the "antibiotic era."

Among the eminent investigators of that period was Selman A. Waksman, a soil microbiologist at Rutgers University, who pioneered the systematic screening of organisms that inhabit the soil for their antibiotic-producing ability. His efforts were highly successful, but the first antibiotic that Waksman isolated was too toxic for human use. He later isolated the antibiotic streptomycin from a soil bacterium, *Streptomyces griseus,* and within a short time after the start of clinical trials, it proved to be highly effective against *Mycobacterium tuberculosis,* the bacterium that causes tuberculosis, and quickly gained prominence as an important therapeutic agent. Following the successful discovery of streptomycin, other antibiotics were discovered by Waksman's group. However, streptomycin proved to be the most important, and Waksman was awarded a Nobel Prize in 1952 for the discovery of this potent chemotherapeutic agent.

How, then, can we define an antibiotic? The classical definition, as proposed by Waksman, is as follows: "An antibiotic is a chemical substance, produced by microorganisms, which has the capacity to inhibit the growth and even destroy bacteria and other microorganisms, in dilute solutions." Originally, the term was restricted to natural products of microorganisms. However, with the tremendous advances in organic chemistry, many antibiotics can now be produced by chemical synthesis. Consequently, the term "antibiotic" is widely used to include both the execretory products of microbial metabolism and their synthetic counterparts. Waksman's definition remains adequate, because it is both precise and restrictive. It limits antibiotic substances to products of microbial metabolism, thus excluding antimicrobial substances from higher plants and animals, such as alkaloids, tannins, blood, and tissue fluids. The definition also states that antibiotic substances must exhibit their antimicrobial activity in dilute solution, thus excluding such products of microbial metabolism as organic acids and alcohols.

Does antibiosis from antibiotics play a significant role in natural ecosystems? Unequivocal evidence is not available, because antagonistic phenomena are complex and may involve many kinds of substances. We know that mixed populations of microorganisms interact dynamically in microhabitats such as soil, water, and the gastrointestinal tract of human beings and other animals. Yet, our knowledge about types of associations that occur among microorganisms in such habitats is scanty.

If antibiosis occurs in natural ecosystems from antibiotics, the extent of antibiosis may be influenced by one or more of the following:

1. The chemical nature of the antibiotic substance
2. The number of different kinds of antibiotic-producing organisms that may be excreting antibiotics simultaneously
3. The presence of other substances that may destroy or neutralize the activity of antibiotics
4. The ability of antibiotics to bind to inert materials
5. The ionic strength of the environment
6. The solubility of antibiotic substances
7. The susceptibility of the indigenous microflora to antibiotics that are being excreted

Perhaps numerous other considerations could be added, but these should suffice to emphasize the complexity of the phenomena of antibiosis from antibiotics. Although many avenues for research in this area of microbial ecology remain open, a great deal is known about antibiotic-producing microorganisms. Therefore, some pertinent facts relative to the production of antibiotics are listed in Table 4.1.

We also know that antibiotic substances comprise a widely diverse group of compounds, both in terms of chemical structure and range of activity. The latter is called the *antimicrobial spectrum*. To demonstrate the range of activity for a given antibiotic, it must be isolated,

Table 4.1.   Some General Principles Relative to Antibiotic-Producing Microorganisms

1. The ability to produce antibiotics has been demonstrated in only a small fraction of the enormous numbers of microorganisms that have been characterized.
2. Antibiotic-producing microorganisms excrete only small quantities of antibiotics when cultivated under normal laboratory growth conditions.
3. Antibiotic-producing microorganisms can be manipulated (induced) to form mutations that will produce larger quantities and/or more potent antibiotics in the laboratory.
4. A change in an antibiotic-producing organism's medium (substrate) may cause a change in the nature of antibiotics that it produces.
5. A single strain of antibiotic-producing organism may produce several different antibiotics.
6. A specific kind of antibiotic may be produced by widely different species of microorganisms.
7. Antibiotic-producing microorganisms are usually resistant to antibiotics which they produce.

purified, and assayed under a set of standardized experimental conditions. Preferably, each antibiotic should be tested for its ability to inhibit more than one kind of microorganism. Antibiotic activity is usually expressed quantitatively as the minimum concentration of substances that will inhibit growth of the organisms being tested.

Some antibiotics exhibit only a narrow antimicrobial spectrum, being active only against organisms that are similar in chemical composition or that belong to closely related groups. Other antibiotics exhibit a wide antimicrobial spectrum, being active against microorganisms that have various chemical compositions and belong to widely different microbial groups. For example, penicillin and streptomycin represent antibiotics that have a narrow range of activity. On the other hand, tetracyclines and chloramphenicols represent antibiotics that have a wide range of activity, commonly called *broad-spectrum antibiotics.*

In modern medical practice, physicians routinely request the determination of antibiotic sensitivity patterns on bacteria that are isolated from infected persons. Actually, the physician wants to ascertain both the kind of antibiotic and the concentration of it that will suppress the growth of the bacteria in question. One method for testing bacteria for their sensitivity to antibiotics is shown in Figure 4.2. Note that the bacteria being tested grew confluently on the surface of the agar to which discs impregnated with various concentrations of commercially prepared antibiotics have been added. Zones of growth inhibition (halos) immediately surrounding the antibiotic disc are evidence of antibiotic activity. Although the agar-diffusion method for the determination of antibiotic sensitivity may appear simple, the reliability of the results are related to the composition of the growth medium

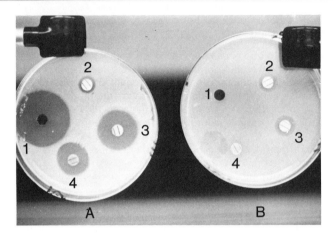

Figure 4.2. Antibiotic sensitivity pattern for two bacterial cultures: **(A)** *Staphylococcus aureus;* **(B)** *Pseudomonas aeruginosa.* **Antibiotics impregnated on paper discs: (1) penicillin, (2) streptomycin, (3) tetracyline, and (4) chloromycetin.**

(substrate), the pH of the growth medium, the solubility of the antibiotic, the ability of antibiotics to bind with agar, the concentration of agar in the medium, the ability of the antibiotic to diffuse in agar, the incubation temperature, and the incubation time.

While antibiotics vary with respect to their activity against different kinds of microorganisms, they also vary with respect to their ability to produce toxicity (side affects) in human beings and other animals. For this reason, of the more than 2,000 antibiotics that have been characterized, only a small fraction have been certified for therapeutic use in human beings.

The biological habitats of the gastrointestinal tract of human beings and other animals represent ecosystems in which antibiosis from antibiotics is known to occur. This phenomenon is evident in human beings during prolonged treatment with antibiotics, especially the broad-spectrum tetracyclines. During extended therapy with such antibiotics, growth of antibiotic-sensitive members of the indigenous microflora is suppressed, resulting in an imbalance of populations within the ecosystem; antibiotic-resistant organisms then predominate. Complications caused by the sudden increase in populations of antibiotic-resistant organisms are referred to as *superinfections.* These may result from indigenous antibiotic-resistant organisms, such as *Candida albicans,* or they may result from exogenous antibiotic-resistant organisms that enter the ecosystem from environmental reservoirs. Complications caused by *Pseudomonas aeruginosa* are an example of the latter.

Superinfections in human beings with *C. albicans* and *P. aeruginosa,* separately or mixed, are currently recognized as serious medical problems. Most often they occur in debilitated persons, those with chronic diseases, those with traumatic injuries such as severe burns, those undergoing prolonged treatment with immunosuppressant agents, and, as previously mentioned, those unndergoing prolonged treatment with broad-spectrum antibiotics.

Evidence in support of the concept that superinfections occur subsequent to antibiosis from antibiotics has been obtained from studies with laboratory animals: chicks, rats and mice. It appears that bacteria in the alimentary tract of healthy animals inhibit the rapid multiplication of *C. albicans*—organisms that are present, but only in relatively small numbers. However, numbers of *C. albicans* increase when animals are administered broad-spectrum antibiotics, and can be recovered from their feces for extended periods. Some explanations for the enhancement of antibiosis by antibiotics include the fact that antibiotics

1. Remove antibiotic-sensitive organisms that compete for nutrients
2. Remove antibiotic-sensitive organisms that secrete antagonistic substances
3. Enhance the growth of *C. albicans* by direct stimulation
4. Depress the animal's (host) immunological system.

The production and widespread use of antibiotics have had a tremendous effect on human life, especially in developed countries. Most of us can appreciate the importance of antibiotics as antimicrobial therapeutic agents in medicine, but we are less familiar with their nonmedical usage. The economic benefits from the antibiotic industry as a whole are significant, and a major portion of these efforts is devoted to the production of antibiotics for use in areas outside clinical medicine (Table 4.2).

Table 4.2. **Some Nonmedical Applications of Antibiotics**

Category of Application	Purpose
Animal nutrition	To stimulate the growth of animals for market
Food preservation	To inhibit the growth of microorganisms that cause spoilage
Plant disease control	To inhibit the growth of plant pathogens
Microbiological research	To make growth media (substrates) selective by inhibiting unwanted microorganisms

Antibiosis from bacteriocins

Unlike the widely known antibiotics, substances called *bacteriocins* are not familiar to most nonmicrobiologists, partly because they have no practical applications that are of value to the general public. By definition, bacteriocins are proteinaceous products of certain bacteria that will inhibit or kill only strains of the same or closely related species. They differ from all other antibiotics primarily in their narrow range of activity (antimicrobial spectra). Some bacteriocins resemble defective bacteriophages, but as intracellular structures they are non-replicating entities. However, they absorb to specific receptor sites on the surface of susceptible cells and kill them through processes that involve lysis.

The ability to produce bacteriocins appears to be widespread, because bacteriocin producers have been found within many widely diverse genera of bacteria. However, bacteriocins are produced only by cells that harbor a specific kind of plasmid called the *c-factor*. The c-factor is an extrachromosomal element in certain cells, which controls the synthesis of bacteriocins. Although their action is specific, some strains that produce bacteriocins have the ability to transfer their c-factor to other non-bacteriocin-producing strains through the process of conjugation or transduction. The recipient strain then produces the same type of bacteriocin as is produced by the donor, while all its other properties resemble the nonbacteriocinogenic parent. Bacteriocin production is a lethal event for the producer strain. Specific cells that synthesize bacteriocins die in the process. However, survivors that possess the c-factor for a particular bacteriocin are immune (protected) from lysis by bacteriocins of that type.

Since bacteriocins represent a special category of antibiotic-like substance, a brief discussion of their nomenclature is in order. The general names for specific groups of bacteriocins have been derived from the species of bacteria that produce them. For example, colicins are bacteriocins produced by *Escherichia coli,* and megacins are bacteriocins produced by *Bacillus megaterium.* Some of the various bacteriocins and their producer strains are listed in Table 4.3.

If antibiosis from bacteriocins occurs in natural habitats, it seems possible that the antagonism would involve intraspecific competition, since microorganisms that are closely related would have similar ecological niches. Although only a small fraction of a given population may produce bacteriocins and die, the remaining members that harbor the c-factor will have a selective advantage to compete for resources, because they are immune to that particular bacteriocin, which may also be produced by other closely related organisms. In addition, the competitive ability of sensitive strains in that habitat will have been

Table 4.3. Nomenclature for Some Groups of
Bacteriocins

Bacterial Producer Strain	Name of Bacteriocin
Escherichia coli	Colicins
*Bacillus cereus	Cerecins
*Bacillus megaterium	Megacins
Enterobacter cloacae	Cloacins
Serratia marcescens	Marcescens
Yersinia pestis (formerly Pasteurella pestis)	Pesticins
*Pseudomonas aeruginosa (formerly Pseudomonas pyocyanea),	Pyocins
*Pseudomonas fluorescens	Fluocins

*Note that different species from within these two genera, produce different kinds of bacteriocins.

reduced by the lethal effect of that bacteriocin to some of its members.

As a result of their narrow range of activity, some bacteriocins have been shown to have merit as diagnostic typing tools for the intraspecific differentiation of bacteria. Such tools are extremely valuable in hospitals when the medical staff encounters outbreaks of infections from bacteria such as *Pseudomonas aeruginosa*. These bacteria are widely distributed in the environment, and they often cause serious complications in debilitated individuals, and in those who have experienced severe burns. *P. aeruginosa* is especially troublesome as the causative agent of hospital-acquired infections. With the aid of pyocin typing, microbiologists have been able to identify sources and routes of transmission for this important "opportunistic pathogen." Several methods for typing other kinds of bacteria on the basis of bacteriocin activity have also been developed, and this area of diagnostic bacteriology is attracting considerable attention.

Antibiosis from other substances

The substances to be considered in this section are excretory products of microorganisms that do not have the characteristics of either antibiotics or bacteriocins. These substances include alcohols, organic acids, and enzymes. Under certain conditions all of them exhibit antagonistic behavior in natural habitats. Consequently, microorganisms that excrete such substances in a given habitat are responsible for antibiosis, because species sensitive to the substance themselves, or to environmental changes caused by such substances, will be suppressed or excluded.

Scientists concerned with the study of fermentations were among the first to observe antibiosis from alcohols and organic acids. Thus, we are indebted to industrial microbiologists for providing us with a better understanding of antibiosis from such substances, because a great deal was learned about the phenomenon during the successful development and subsequent expansion of many fermentation processes, especially those which employ mixed cultures of microorganisms.

The basic concept for using mixed cultures in fermentations is that two successive populations of different kinds of microorganisms can produce a product of greater quality than either of them can acting separately. When laboratory-prepared inocula are used in such processes, both kinds of microorganisms may be introduced simultaneously, or one may be allowed to grow in the medium for a specified time before the second is introduced. Fermentation processes of this nature involve microbial succession: the first microbial population converts the raw material (substrate) into an end product; then the second microbial population converts the end product into a third substance.

The phenomenon of antibiosis is inherent to such processes. This will become evident as we briefly describe the activities of mixed microbial cultures in the commercial production of vinegar or cider. Commercial varieties of grapes or apples are commonly used raw materials (substrates) for the process, which, with only minor variations, applies to the production of either vinegar or cider. The fermentation process is initiated by yeasts *(Saccharomyces cerevisiae)* which grow under anaerobic conditions and convert sugars in the substrate to ethyl alcohol with the concomitant release of carbon dioxide. In the second phase, bacteria in the genus *Acetobacter* grow under aerobic conditions and convert the alcohol to acetic acid (vinegar). The entire fermentation process is regulated by several conditions:

1. The anaerobic phase favors the growth of the yeast.
2. Ethyl alcohol increases and it inhibits the growth of many microorganisms, including the yeast, thus causing the growth of the latter to be self-limiting.
3. Vigorous aeration favors the growth of *Acetobacter*.
4. Acetic acid increases and contributes to antibiosis by lowering the pH (increasing the acidity) of the environment.

Although carefully selected strains of microorganisms are used in many fermentation processes, others depend entirely on the indigenous microflora of the fruit. In the latter, the ecology of microorganisms that mediate the fermentation process is of special interest. Therefore, we will consider reservoirs and modes of distribution (dispersal) for *Acetobacter* species. Such bacteria are not present on healthy fruit until the fruit is harvested, but bacterial spore-forming species, molds, and yeast are present in small numbers during various stages in the development of the fruit. Obtaining high-quality healthy fruit for raw materials (substrates) is a major concern of industry. Consequently, the harvester usually leaves mummified, rotten, and heavily damaged fruit on the ground in the orchard or vineyard. The rejected fruit becomes heavily contaminated with microorganisms, and their juices serve as selective enrichments: yeasts ferment the sugars and produce ethyl alcohol, and the alcohol serves as a substrate for *Acetobacter* species. The bacteria are then transported by vehicles and by vectors (e.g., air, insects) to wildflowers, which become reservoirs. During harvesting, healthy mature fruit becomes contaminated with microorganisms from plant debris and dust. Subsequently, the microflora on the fruit becomes the source of inocula for the industrial process. Extraneous microbial contamination during fermentation is controlled by anaerobic conditions and from inhibition by the alcohol; subsequent contamination is minimized by the increased acidity. The important thing to remember is that antibiosis actually results from the alcohol-producing yeast during the first phase and from acetic acid-producing bacteria in the second phase.

Antibiosis from microorganisms that excrete organic acids is also common in other habitats. Homofermentative bacteria excrete lactic acid during the production of sauerkraut and pickles; and during the natural spoilage (souring) of raw milk. In such habitats, lactic acid causes a reduction in pH, and environments become selective for aciduric organisms. Because of the commercial interest in the products mentioned above, the dominant microbial populations are well known, and their activities during the various stages of the fermentation have been characterized. Thus, many data are available to support the fact that lactic acid secreted by homofermentative bacteria is responsible for their antagonistic role. In the absence of definitive data, it is difficult to differentiate antibiosis (the effects one microorganism has on another) from indirect inhibitions.

Although antibiosis from organic acids is extremely common in processes that involve the production of fermented products, we should not assume that it occurs only in restricted ecosystems. On the contrary, its occurrence appears to be widespread and is believed to be operative in many diverse habitats. For example, antagonism among bacteria commonly found in the intestinal tracts of animals is probably caused by the excretion of fatty acids by some members of the indigenous microflora. Evidence in support of this contention was obtained from *in vitro* experiments with *Klebsiella* (various species) and *Shigella flexneri* grown separately (in pure cultures) and grown together (in mixed cultures). Populations of *Klebsiella* in both pure and mixed cultures were the same in all phases of growth during the entire period of observation. In contrast, the cultures of *Shigella* grew equally well during the early stages; but after approximately 15 hours of incubation, *Shigella's* growth in the mixed culture was suppressed while that of *Klebsiella* continued to increase. The abrupt cessation of *Shigella's* growth in the mixed culture, but not in pure culture, suggested that its inhibition was caused by *Klebsiella*. In an attempt to explain the inhibition, the medium was analyzed. Two metabolites (acetic and formic acids) had accumulated in the medium after the growth of the mixed cultures, and they were also present in the medium in which *Klebsiella* had grown in pure culture. The latter findings demonstrated that the acids were not produced by a *Klebsiella*–*Shigella* synergistic interaction. Furthermore, cell-free filtrates prepared from a 20-hour pure culture of *Klebsiella* were toxic to *Shigella*. In fact, *Shigella's* growth in the filtrate was inhibited in a manner similar to that observed in the mixed culture. Data from those collective experiments provide us with strong evidence that *Klebsiella* produces formic and acetic acids *in vitro,* and that at a given concentration in the medium, those acids become toxic to *Shigella* cells. Thus, a mechanism for antibiosis by *Klebsiella* involves the excretion of volatile fatty acids.

Now, we shall focus on microbial enzymes as contributors to antibiosis. This concept may appear to be inconsistent with our earlier consideration of enzymes as mediators for metabolic processes, but in reality it is only another aspect of enzymatic behavior. We must remember that enzymes, in addition to being vital organic catalysts, are themselves products of metabolic processes. In Chapter 2 we learned that all enzymes are produced within the cytoplasm of cells. Some of them remain in the cytoplasm or attached to membranes, and others are excreted into the cell's environment. Enzymes that function inside cells are called *intracellular enzymes,* and those that function in the environment surrounding cells are called *extracellular enzymes.*

We are familiar with the functions of many common extracellular enzymes, especially those which are utilized by cells to hydrolyze (split or lyse) larger polymers (substrates) such as carbohydrates, fats, and proteins during food-gathering processes. We also know that enzymes exhibit a high degree of specificity for substrates they attack. Such structural components of microorganisms as capsules and cell walls are known to contain polymers that are susceptible to lysis by certain microbial extracellular enzymes. Therefore, in habitats where enzymes with specificities for such substrates are excreted, sensitive cells may be destroyed or converted into osmotically fragile entities.

Antibiosis that involves the action of extracellular lytic enzymes is believed to occur among microorganisms that live in the soil. For example, certain *Streptomyces* species produce extracellular enzymes that cause lysis of cell walls in some species of *Streptococcus*. Also, many *Myxobacter* bacteria produce extracellular enzymes that can degrade cellulose and other complex polymers. *Myxobacter* are abundant

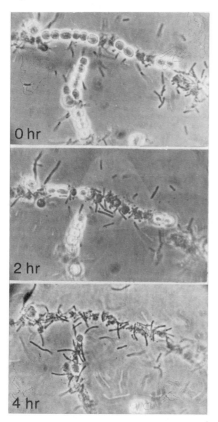

Figure 4.3. Lysis of a blue-green alga *(Nostoc)* by *Myxobacter.* Note the increased disintegration of the alga cells as the contact time between the bacterial cells and the alga cells increase. Almost no intact algal cells can be seen after 4 hours of contact. Observed with phase-contrast microscope (×2,000). [Courtesy of M. Shilo, *J. Bacteriol.* **104:453**, 1970, with permission.]

in the soil, and it is reasonable to assume that their extracellular enzymes contribute to antibiosis in that habitat. Cells composed of structures sensitive to enzymes in their surrounding environment will be weakened, if not completely destroyed. In laboratory experiments, a strain of *Myxobacter* has been shown to cause lysis of *Nostoc,* a blue-green alga (Figure 4.3). In those experiments, only vegetative cells of the alga were destroyed. Heterocysts were not affected by the lytic agent.

Finally, after considering many examples of antibiosis from a wide variety of microbial substances, we might investigate one important question: why individuals within a microbial population respond differently to antimicrobial substances?

Sensitive and resistant species

We know from previous discussions that ecosystems are dynamic, and that populations of microorganisms residing in them are subjected to environmental conditions that change with time. We also know that, as a result of evolution, each currently existing population is endowed with a certain degree of genetic variability—heritable traits that form the basis for adaptations. Therefore, a population that is highly variable is most likely to contain individuals with characteristics that will enable them to survive in adverse conditions that would otherwise lead to their extinction. In other words, the continual existence of a species is dependent upon the adaptability of its individual members. Thus, the predominant organisms in any natural habitat are the ones most suited to that habitat, a common observation that is often overlooked. Such organisms possess favorable variations, and have a greater probability of surviving and reproducing themselves. Individuals without favorable variations will be lost, owing to their inability to reproduce. It is important to remember that evolution, through the process of natural selection, acts only on heritable variations. All heritable variations originate within a cell's DNA and are specifically caused by changes in genes. Such changes are referred to as *mutations,* and they occur naturally in all populations. Consequently, we know that the environment does not produce genetic variations (except those which are artificially induced by specific mutagenic agents such as x-rays) but selects for individuals that already possess them. Genetic variations occur spontaneously in all populations at a low frequency (some occur in bacteria at the rate of one mutation per 10^9 cells), but such variations remain unobservable until the environment becomes favorable to their adaptive expression.

An interesting example of genetic variability among microorganisms surrounds the adaptive expression of resistance to the antibiotic penicillin by *Neisseria gonorrhoeae*. This bacterium, the causative agent of gonorrhea, the most prevalent type of venereal disease, is known for its susceptibility to penicillin. Consequently, this drug is widely used as a therapeutic agent for the treatment of infected persons. The evolution of resistance in these bacteria over the past 25 years has been manifested indirectly by the increasing dosage required to eradicate the infection. The recommended intramuscular dosages have changed as follows: 150,000 units in 1940, 2.4 million units in 1973, and 4.8 million units in 1975.

Apparently, the widespread use of penicillin as a therapeutic drug, coupled with an increasing number of asymptomatic infected persons, have created an environment that selects for the resistant strains. Furthermore, the resistant strains have a greater chance for reproducing themselves through the apparent increase in sexual activity among persons in recent years.

Although the role of antibiosis in natural ecosystems cannot be definitively explained, we assume that this phenomenon contributes to the maintenance of balanced microbial populations. Thus, both sensitive and resistant species of microorganisms must be considered as products of evolutionary processes and represent types of cells that are most suited for their respective ecological niches.

Key Words

acidophilic   An organism that grows optimally in acidic habitats.

aciduric   An organism that has the ability to endure acidic environments.

biochemical oxygen demand (BOD)   The demand for free oxygen in aquatic ecosystems by aerobic organisms.

chemosynthesis   The processes through which certain organisms obtain and utilize energy and carbon from inorganic substances.

immunosuppressant   A compound or substance that retards the functioning of the immune system in animals.

lysozyme   A hydrolytic enzyme that catalyzes the degradation of certain components in bacterial cell walls.

microflora   The sum total of microorganisms in a given habitat.

plasmid   An extrachromosomal piece of DNA that may be present in certain kinds of bacteria but is not essential to growth.

vector   A living organism that has the ability to transmit infectious organisms from one locale to another.

virulence   The degree of pathogenicity or disease-producing capacity of an organism.

Selected Readings

1. Passmore, S. M., and J. G. Carr. 1975. The ecology of the acetic acid bacteria with particular reference to cider manufacture. *J. Appl. Bacteriol.* 38:151–158.
2. Roberts, D. S. 1969. Synergistic mechanisms in certain mixed infections. *J. Inf. Dis.* 120:720.
3. Ryther, J. H., and W. M. Dunstan. 1971. Nitrogen, phosphorus, and eutrophication in the coastal marine environment. *Science* 171:1008–1031.
4. Shinefield, H. R., J. C. Ribble, M. Boris, and H. F. Eichenwald. 1963. Bacterial interference: its effect on nursery-acquired

infections with *Staphylococcus aureus*. I. Preliminary observations on artificial colonization of newborns. *Am. J. Dis. Child.* 105:646–654.

5. Tinnin, R. O. 1972. Interference of competition? *Am. Natur.* 106:672–675.

6. Watson, E. S., D. C. McClurkin, and M. B. Huneycutt. 1974. Fungal succession on loblolly pine and upland hardwood foliage and litter in north Mississippi. *Ecology* 55:1128–1134.

CHAPTER 5

Intermicrobial and extramicrobial relationships

- **Types of symbiotic relationships**
    Mutualism
    Commensalism
    Parasitism
- **Intermicrobial associations in nature**
    Lichens
    Lysogeny
    Miscellaneous
        Protozoan associations
        Bacterial associations: A triad
        Mycoviruses
        Cyanophages
- **Extramicrobial associations in nature**
    Microbial associations with animals and insects
        Microbial-ruminant associations
        microbial-insect associations
    Microbial associations with higher plants
        Mycorrhizae
        Bacterial-legume associations
    Obligate microbial parasites of higher organisms
- **Key words**
- **Selected readings**

In Chapter 4 we dealt with relationships among microorganisms in a restricted manner and considered only types of interactions that tend to promote the survival of one microbial population at the expense of another. In this chapter we shall examine types of interrelationships that may occur among all kinds of living organisms, some of which may actually result from interference phenomena.

Since microorganisms are potentially everywhere, we must take their ubiquitous distribution into consideration when discussing interactions that may occur in nature. Unlike higher forms of life (plants and animals), individual members of the various microbial groups are endowed with characteristics which enable them to grow and reproduce in numerous kinds of widely diverse habitats. Furthermore, microorganisms of some kind are always present in or on the tissues of higher organisms (except those produced and reared in experimentally controlled germ-free environments). For these reasons, infinite possibilities exist in nature for the establishment of intermicrobial and extramicrobial relationships. Microbe–microbe types of interactions are referred to as *intermicrobial,* associations in which all participants are microorganisms. On the other hand, microbe–macrobe types of interactions are called *extramicrobial,* associations in which one partner is some kind of a microorganism and the other is a higher form of life (e.g., a bacterium and a plant).

Types of symbiotic relationships

Types of associations that may occur among living organisms are both varied and complex, because the behavior of each participant is influenced by a host of factors. Thus, the nature of the relationship is determined not only by the manner in which the participating organisms relate to each other, but upon the achievement of an equilibrium balance between them and other components of that ecosystem. In general, symbiotic interrelationships among living organisms can be categorized into three patterns of behavior: mutualism, commensalism, and parasitism. Those patterns of behavior are not separated by clearly defined zones of demarcation, because each type of relationship is dynamic, and the interacting organisms may shift gradually form one type of symbiotic relationship into another.

The term *symbiosis* simply means the living together, with a certain degree of constancy, of two dissimilar organisms. This definition is inclusive, because it takes into consideration all kinds of associations that may occur among two or more dissimilar organisms when living together, and it encompasses those associations which do not require cell-to-cell contact.

Participants in symbiotic relationships are not always represented in a 1:1 ratio. The association may involve only two dissimilar unicellular organisms; several unicellular organisms of one kind and one unicellular organism of another kind; or several unicellular organisms and one multicellular organism. Furthermore, the relationship between partners in symbiotic associations may be specific or nonspecific. The former usually requires some degree of selectivity and occurs most often among intracellular associations—types of relationships in which one partner, the *endosymbiont,* lives inside the other partner, the *host.* In associations of this type, the specificity of the endosymbiont may be for a particular kind of organism or a specific intracellular structure, organ, or kind of cell. In general, less selectivity is found among extracellular associations, types of relationships in which one partner, the *ectosymbiont,* becomes associated with the exterior surface of its host in order to maintain some degree of security. Some ecto-symbionts merely colonize the surface of host tissues, but others commonly utilize some form of attachment to maintain the association.

Mutualism

As the name implies, *mutualistic associations* are beneficial to both partners. In another sense, mutualism may be viewed as an opportunistic association of two parties from which both obtain mutual rewards. The 1975 cooperative endeavor between the United States and Russia which resulted in the successful completion of the Apollo—Soyuz rendezvous in space exemplifies mutualism. During the joint experiment a considerable amount of interaction took place between the participants (Russian and American human organisms), and at specific intervals both partners were interdependent. Furthermore, the rewards from the association were mutually beneficial, in this instance the acquisition of data from outer space.

Mutualistic associations are widespread in nature, and many of them are ecologically important, not only to the participants but to other organisms in that ecosystem. Symbiotic nitrogen fixation that results from the mutualistic association between *Rhizobium* bacteria and certain plants called legumes is an example. Neither the legume nor the bacteria when living separately can utilize atmospheric nitrogen. However, after roots of the legume are invaded by the bacteria, specialized cells acquire the ability to utilize gaseous atmospheric nitrogen, a nutrient that is essential for growth of both the bacteria and the legume. Another interesting example of mutualism is the interaction between termites and certain kinds of microorganisms that live in their intestinal tract, especially protozoans. Termites consume woody material but cannot digest it because they lack the enzyme for degrading cellulose. Protozoan endosymbionts perform this vital task for them. The association is beneficial to both organisms: the termite provides the protozoan with a habitat and certain micronutrients, and the protozoan rewards the termite with food, glucose residues from the degradation of cellulose.

Commensalism

Commensalism is characterized by an imbalanced arrangement between the participants, because one partner benefits while the other one remains unaffected, neither benefited nor harmed by the association. Organisms that obtain benefits from interactions of this type are called *commensals*. They obtain something useful from the association, but they do not contribute to the process or affect their partners in an adverse manner. Furthermore, the host does not benefit from the activities of commensals and remains indifferent to them.

Commensalistic interactions are widespread in nature and are extremely common among microorganisms that live in biological habitats. Types of microorganisms that we refer to as members of the normal or autochthonous microflora of higher forms of life are commensals. Regardless of whether commensals inhabit external or internal regions of higher organisms, they benefit from the association by obtaining provisions from their partner. Some examples of benefits that commensals may derive are a favorable residence, transportation, and a supply of utilizable nutrients. Thus, many commensals live as "hoboes." For example, several different kinds of bacteria normally live in a commensalistic association with human beings: some as inhabitants of the gastrointestinal tract, and others as inhabitants of the skin. Yet, human beings do not receive benefits from them, nor are human beings affected adversely by their presence. Although commensalistic interactions of this type are common, we should not regard microbe—macrobe types of interactions as a requirement for commensalism. Organisms that live in close proximity, without cell-to-cell contact, may also participate in commensalistic associations. In some aerobic ecosystems, anaerobes live with aerobes as commensals. In such microhabitats, aerobes grow abundantly, and as a result of their aerobic metabolism, the oxygen concentration is reduced to a level that anaerobes can tolerate. Thus, the anaerobe benefits from the mere presence of actively growing aerobes. Similarly, commensalism occurs in nature when one kind of microbe excretes a metabolite that can be utilized as an essential nutrient by another organism.

Parasitism

Types of associations that we refer to as *parasitism* differ from mutualism and commensalism in that the relationship is beneficial to one partner and harmful to the other. The symbiont that benefits from the association is called the *parasite,* and the other participating organism is called the *host.* Parasites may be considered as unwelcomed guests, because the host is always damaged to some degree by the interaction.

Usually, we think of parasitism as being an association between organisms of different sizes, with the smaller member being the parasite, and the larger member being the host. Associations of this nature are numerous, but this does not imply that parasitism involves only microbe–macrobe types of interaction. Parasitic types of interrelationships occur within and among all kinds of organisms, and must be considered to be a universal phenomenon.

Parasitism as a way of life is facultative for some organisms and obligatory for others. *Facultative parasites* are characterized by intermittent associations with their host. They may spend a definite period of their development in or on their host, or they may visit their host only periodically for some vital resource. Thus, facultative parasites have flexible habits and may live as independent, self-sufficient entities. On the other hand, *obligate parasites* have deficiencies that prohibit them from functioning in natural ecosystems as free-living entities. Consequently, obligate parasites must utilize other living cells to perform some vital function for them. In terms of the location of their microhabitats, obligate parasites have been divided into two groups: those that live outside cells are referred to as *obligatory extracellular parasites,* and those that live within host cells are called *obligatory intracellular parasites.*

Parasitism is an extremely interesting type of association. The interrelationship is one of intimacy between the parasite and its host, and the degree of dependency ranges from intermittent visits to the host by facultative parasites to complete dependency on the host by obligate parasites. Organisms that live as obligate parasites have evolved with physiological mechanisms that enable them to maintain a delicate equilibrium with their hosts. While the association is detrimental, damage to their hosts is minimal—a precarious relationship that enhances the survival of the parasite. This type of behavior is typical of many organisms that cause chronic and slowly developing diseases in humans and other animals.

In the discussion above, we have presented an overview of the interrelationships that may occur among living organisms in nature. Too often ecological phenomena are considered only from observations of free-living organisms. Evolutionary forces that bring about adaptations in free-living organisms also impinge on symbiotic associations of organisms while living in the various types of interrelationships.

Intermicrobial associations in nature

In this section we shall deal exclusively with microbe–microbe types of interrelationships, symbiotic associations in which both participants are members of the microbial world. Types of associations that have been observed among microorganisms are varied, ranging from mutualism on one extreme to parasitism on the other. Furthermore, each partner in such an association may belong to a different taxonomic group (i.e., one member being an alga and the other a fungus). In this regard, some of the most interesting examples of symbiosis can be found among microorganisms.

In order to discuss symbiotic relationships that may occur among closely related microorganisms in a meaningful way, one must delve into details relative to the physiology, anatomy, and in some instances life cycles of cells that comprise the various microbial groups. Such specialized subject matter is beyond the scope of this book. Therefore, our discussion of interrelationships that may occur among microorganisms that belong to the same taxonomic group will be minimized. However, a few examples of unusual associations that appear to be unique to cells that belong to different microbial groups (e.g., an alga and a fungus) will be presented. We shall begin by examining some common types of interrelationships.

Lichens

As discussed earlier, lichens are formed from a symbiotic association between an alga and a fungus. During the association, both partners undergo morphological and functional modifications and develop into a unique structure, distinctly different from either of the partners.

Lichens are widely distributed in the biosphere, and their plant-like structures (Figure 5.1) can be seen easily with the naked eye on rock surfaces, tree trunks, and in a variety of nutritionally poor habitats. Often their only source of water is fog or mist. They grow in nature as self-reproducing units, but they are not phylogenetically a different kind of organism, merely an association of an alga and a fungus in modified form. Many different species of fungi and of algae may be components of lichens. In most cases the fungal partner is a species within the class called Ascomycetes, and the predominant algal partner is a species from within either of two groups: the eucaryotic green group or the procaryotic blue-green group. For this reason there are many different kinds of lichens. The association between the alga and fungus in a lichen is relatively constant, but not permanent. Lichens can be forced to dissociate into their separate components by cultivating them on organically rich substrates under laboratory conditions. In this manner, both partners in several different kinds of lichens have been isolated. Separated partners have also been cultivated in the laboratory as independent organisms, but their resynthesis into a lichen is not an easy task. However, successful resynthesis of lichens has been accomplished when both partners, growing in close proximity, were subjected to extremely unfavorable environmental conditions.

The lichen association may be viewed as a mutualistic interrelationship. The phototrophic alga supplies the fungus (a saprophyte) with vital growth factors (organic nutrients), and the fungus supplies the alga with inorganic materials and protection from desiccation.

It is also significant that lichens are seldom found in areas near

Figure 5.1. Lichen morphology: (A) crustose lichen;
(B) fruticose lichen. [Courtesy of the Carolina Biology
Supply Company.]

our vast urban centers. Their absence is believed to be related to the
fact that they have tremendous absorptive powers, a feature that is not
restricted to the absorption of minerals; it also enables them to concentrate toxic air pollutants. Thus, if not extinguished, their growth is
suppressed in urban areas where air pollutants are high.

Lysogeny

Lysogeny was characterized in Chapter 1 as one process that may
occur in bacterial cells subsequent to an infection by a temperate bacteriophage. Unlike virulent bacteriophages, temperate bacteriophage
infections seldom lead to the development of virus progeny and the lyt-

ic destruction (death) of host bacterial cells. Instead, the nucleic acid from the temperate bacteriophage becomes integrated into the bacterial chromosome (see Figure 1.18). The bacterial cell is then said to be lysogenized bacteria during normal cell division. Occasionally, a lyso-*prophage.* Thus, prophages are passed to subsequent generations of the lysogenized bacterial during normal cell division. Occasionally, a lysogenized bacterial cell will spontaneously produce a complete virus from its prophages and undergo self-destruction. However, this lethal event takes place only in a small percentage of the lysogenized cells in a bacterial population.

It must be remembered that lysogenized cells have an added piece of genetic material in their chromosome. Consequently, bacterial cells that carry prophages may have a functional capability that is lacking in their nonlysogenized counterparts. Not only is the association between a temperate bacteriophage and a susceptible bacterial cell an interesting phenomenon, lysogeny has far-reaching ecological implications.

As mentioned previously, diphtheria is caused by lysogenic strains of a bacterium *(Corynebacterium diphtheriae).* Diphtheria is known as a toxigenic disease (i.e., pathological lesions are caused by the bacterial toxin), and only strains that carry the prophage have the gene for toxin production. Nonlysogenic strains of *Corynebacterium diphtheriae* are avirulent and harmless, because they are unable to synthesize the potent toxin. It is also significant to note that the ability of lysogenic strains of *Corynebacterium diphtheriae* to produce high concentrations of toxin is related to the amount of iron in the organism's surroundings. Low concentrations of iron seem to enhance toxin production. Thus, in the microenvironment of lysogenic strains, the availability of iron determines the virulence of the bacterium and indirectly the severity of the disease.

The example above will suffice to demonstrate the nature of the interaction that may occur between a temperate bacteriophage and its bacterial host. Lysogeny has been observed in many different kinds of bacteria and is believed to be a phenomenon that is common to all bacteria. Furthermore, it is important to remember that all bacteriophages are obligate intracellular parasites. Even those that develop through the lytic cycle with the ultimate destruction of their host are ecologically important, because they may play an important role in the regulation of population density in susceptible bacterial species.

Thus, the phenomenon of lysogeny may be viewed as a symbiotic relationship that encompasses both commensalistic and parasitic features. Commensalism is exhibited by the lysogenized bacterial cell because it has gained a function (not present in nonlysogenic cells) as a result of genetic information encoded within the prophage. Parasitism is exemplified when the prophage spontaneously shifts to the lytic cycle and undergoes vegetative development that leads to the ultimate destruction of the host bacterial cell.

Miscellaneous associations

Some less commonly discussed types of intermicrobial associations will be examined in this section. They occur frequently in natural ecosystems and involve representatives from among all groups of microorganisms.

Protozoan associations   The protozoa are generally recognized as a highly diverse and widely distributed group of unicellular organisms that exhibit a considerable degree of complexity in structure and behavior. They are equally diverse in respect to the types of intermicrobial associations they can establish. Especially, they are known for their ability to harbor a wide variety of microbial cells as their endosymbionts.

Our discussion will be limited to a few examples of intermicrobial associations that involve protozoan cells that belong to the genus *Paramecium,* an extremely common representative of the ciliates. Ciliates are defined as protozoans that possess hairlike structures (cilia) on their bodies during some period of their life. In addition, they are characterized by having both a micronucleus and a macronucleus. The former plays an important functional role in sexual reproduction, and the latter is concerned primarily with other regulatory functions. As predaceous entities, paramecia feed on a variety of microorganisms. Food material is ingested through an oral groove (mouth) and passes through a tube (esophagus) into a food vacuole, where digestion occurs. Waste from food vacuoles is released from the cell through the anal pore.

Microorganisms that live as endosymbionts of paramecia are present in food vacuoles and also within other regions of the cytoplasm, a fact that supports the symbiotic nature of the relationship. In the absence of a symbiotic relationship, microorganisms ingested as food are present only in food vacuoles. Although food vacuoles change positions or circulate within the cytoplasm of paramecia as a result of cytoplasmic streaming, nonsymbiotic microorganisms remain inside them and are either digested or excreted from the host along with other undigestible materials through the anal pore. Of particular interest are interrelationships that have been observed in *Paramecium bursaria* and *Paramecium aurelia.*

Paramecium bursaria often harbor large numbers of phototrophic green algae *(Chlorella)* in their cytoplasm, a fact that accounts for their intermittent greenish appearance. The symbiotic association is beneficial for the protozoan, because the photosynthetic algae supplies it with carbon materials during periods when extracellular food material is limited. It is also interesting to note that during *Paramecium–Chlorella* associations, the number of algae present in a protozoan cell remains relatively constant. It has been suggested that the regulatory process may involve a mechanism for inhibiting algal cell division; or the equilibrium is maintained by frequent excretion of surplus algae through the anal pore, or by digestion of surplus algae in food vacuoles. The mechanisms by which certain endosymbionts avoid being digested is not known. The interaction between *Paramecium bursaria* and *Chlorella* is a facultative symbiotic relationship, because both organisms can function in ecosystems as independent free-living entities.

Now, let us examine some interesting aspects of the interrelationships that may occur between *Paramecium aurelia* and its endosymbionts. While harboring certain kinds of endosymbionts, *P. aurelia* exhibits "killer" properties. During the symbiotic association, the host protozoan produces a toxin that will kill sensitive strains. Cells with this ability are called killer paramecia. Toxins are released to the host's extracellular environment, from which they are believed to be ingested through food vacuoles by sensitive strains of paramecia. However, other means of toxin uptake by sensitive strains may also be operative. Toxins are generally harmless to their hosts. While harboring endosymbionts, host cells are rendered resistant to the toxins they produce. When freed of their endosymbionts, host cells also lose the ability to produce toxins.

Endosymbionts of *P. aurelia* can easily be seen within the cytoplasm of their host in areas that are clearly distinguishable from food vacuoles (Figure 5.2). Structural details of endosymbionts, and sites of localization within host cells, have been ascertained from electron microscopy studies. These endosymbionts have been given Greek names, and as a result of recent research, they are now recognized as bacteria. Since the phenomenon of killer paramecia is somewhat unusual, we shall make a few brief comments about the classification of these newly described bacteria. With the exception of alpha, all other endosymbionts of *P. aurelia* have been placed in newly described genera. Alpha closely resembles bacteria that have been previously described and is characterized as a new species of the genus *Cytophaga*. The remaining endosymbionts have been given new generic designations (Table 5.1).

The Greek names clearly distinguish the bacterial endosymbionts of *P. aurelia* from other kinds of bacteria that are widely distributed in the biosphere.

Most of these newly described bacteria appear to be obligate endo-

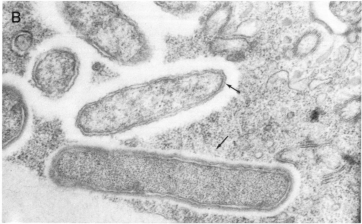

Figure 5.2. *Paramecium aurelia* **and its endosymbionts. (A) Light micrograph of a whole protozoan, showing lambda endosymbionts (dark-stained rods) scattered throughout the cytoplasm (×500). (B) Electron micrograph of a thin section of a** *Paramecium,* **showing large rod-shaped bacteria embedded within the cytoplasm (arrows) (×30,000). (Courtesy of J. R. Preer,** *Bacteriol. Rev.* **38:113, 1974, with permission.)**

symbionts of *P. aurelia* that is, multiplication occurs only within the host's cytoplasm. A few of them are able to maintain a very stable relationship with their host, but under laboratory conditions most endosymbionts become separated from the protozoan cell. One explanation for the separation is the rapid multiplication rate of *P. aurelia* as compared to the slower growth of its endosymbionts. As a result, endosymbionts become diluted during the normal growth of

Table 5.1. Generic Classification of
Bacterial Endosymbionts of
Paramecium aurelia

Genus	Greek Name
*Cytophaga	Alpha
Caedobacter	Kappa, mu, gamma, nu
Lyticum	Lambda, sigma
Tectobacter	Delta

*Denotes a genus previously described.

paramecia and are eventually eliminated from their host. The symbiotic association is not readily reestablished under laboratory conditions.

Obviously, the association is beneficial for the endosymbionts, because their host provides them with a source of nutrients, a conducive habitat, and protection from the external environment. Apparently, the association is also beneficial for the host, because killers have a competitive advantage over their nonkiller counterparts. This competitive advantage results from the fact that host paramecia, while harboring endosymbionts, are resistant to the toxins they produce.

Bacterial associations: A triad   In this section we shall examine a unique three-membered parasitic system. The triad consists of bacteria called *Bdellovibrio,* which are able to function simultaneously as a host for a bacterial virus (bdellophage) and as a parasite of another bacterial cell.

Bacteria that we now recognize as *Bdellovibrio* were first characterized in 1962. They are unique because they can live as parasites of other bacterial cells. In this parasitic state, they obtain all their nutrients and growth factors from the host bacterial cell. When first discovered, they were thought to be obligate parasites, but in recent years a few of them have been grown successfully in the laboratory as host-independent entities. Thus, the parasitic way of life is facultative for most strains of *Bdellovibrio,* but a few can live as host-independent saprophytes. The *Bdellovibrio* are now referred to as two types: parasitic and nonparasitic. Parasitic *Bdellovibrio* cannot reproduce in the absence of host cells or their extracts.

A schematic view of a generalized developmental cycle is shown in Figure 5.3. First, a single *Bdellovibrio* cell becomes attached to the outer surface of its host. Subsequently, it penetrates the periplasmic space of its host and undergoes a series of morphological changes. During the sequential events, the cell becomes elongated; then it becomes fragmented into several daughter cells; each daughter cell produces a single flagellum; and finally daughters are released with concomitent lysis of the host cell. The number of daughter cells produced, and the

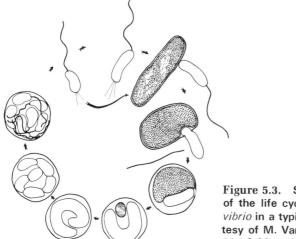

Figure 5.3. Schematic representation of the life cycle of a parasitic *Bdellovibrio* in a typical bacterial cell. [Courtesy of M. Varon, *Critical Rev. Microbiol.* **3:221, 1974,** with permission.]

time for completion of the growth cycle, vary with the species of *Bdellovibrio* and type of bacterial cell being parasitized.

The *Bdellovibrio* bacteria are widely distributed in soil and water habitats, and they can parasitize several different kinds of bacterial hosts. However, host susceptibility is determined by a number of factors. For example, *Sphaerotilus natans* is a bacterium that grows characteristically as filaments of colonial cells enclosed within a sheath (R-type or rough form), but when cultivated on certain kinds of laboratory media, it will dissociate into single cells (S-type or smooth form). Smooth forms of *Sphaerotilus natans* are susceptible to parasitization by *Bdellovibrio bacteriovorus* (Figure 5.4), but rough forms are resistant to the parasite. When in the rough form, the sheath that surrounds the filament of colonial cells protects the cells from being parasitized by *B. bacteriovorus*.

Now, with that overview of *Bdellovibrio,* we shall examine the nature of interactions that occur in the three-membered parasitic system. The first bdellophage (a virulent bacteriophage that is specific for *Bdellovibrio*) was isolated in 1970 from a nonparasitic strain of *Bdellovibrio*. Since that time, several additional bdellophages have been isolated from both nonparasitic and parasitic strains. Furthermore, some bdellophages will infect and lyse both types of *Bdellovibrio*. However, bdellophages do exhibit a high degree of specificity. Their host range appears to be restricted to bacterial parasites that have been characterized as *Bdellovibrio*. Even though they exhibit group specificity, a single bdellophage may lyse several different *Bdellovibrio* bacterial strains. It is important to note that bdellophages interact with cells of nonparasitic *Bdellovibrio* in a manner that is typical for the

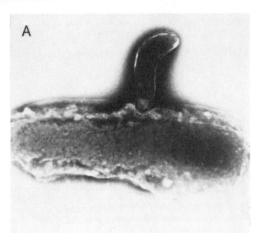
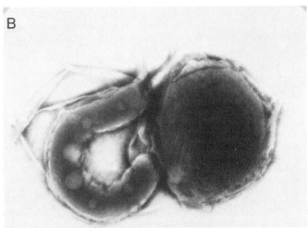

Figure 5.4. *Bdellovibrio* parasite of *Sphaerotilus natans* swarm cells. (A) Host bacterium *(Sphaerotilus natans)* with *Bdellovibrio* attached. (B) Early elongation phase of H-D *Bdellovibrio* within the periplasmic space of *S. natans*. [Courtesy of A. D. Venosa, *Appl. Microbiol.* **29:702, 1975,** with permission.]

development of other virulent bacteriophages in the lytic cycle (see Figure 1.12), but bdellophages do not undergo development in parasitic *Bdellovibrio* in the absence of a susceptible bacterial host or its extracts.

A schematic view of the three-membered parasitic system is shown in Figure 5.5. In order for bdellophages to develop in a parasitic *Bdellovibrio*, the bdellophages must become absorbed to the *Bdellovibrio* cell within a very critical period (immediately prior to the bacterial parasite's attachment or before it has completed its penetration into the host's periplasm).

After absorption, the bdellophage injects its nucleic acid into the *Bdellovibrio*. Then the parasite completes its penetration of the host cell. The empty bdellophage heads remain on the outside of the host cell. Subsequently, the bdellophage-infected *Bdellovibrio* retains its original shape and size, in contrast to the uninfected parasite, which would have become elongated (see Figure 5.3). The bdellophages then undergo development within the *Bdellovibrio,* which itself is a parasite within the periplasm of the host. Finally, the *Bdellovibrio* undergoes lysis within the host, and the fully developed bdellophages are released from the host by mechanisms which are yet unclear, perhaps by enzymatic lysis of host cell walls or through the penetration pore, which is the site of entry by the parasite.

This unique parasitic system was demonstrated under laboratory conditions; but with some variations, interactions of this type may occur in natural ecosystems. However, it seems unlikely that the three

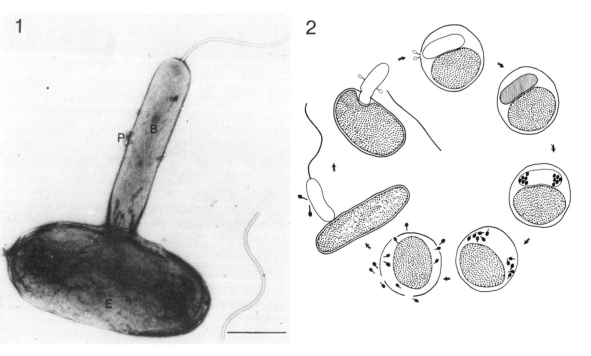

Figure 5.5. Bdellophage development in the three-membered system. (1) Schematic representation of Parasitic *Bdellovibrio* (strain 109J) infected by bdellophage (P), and the *Bdellovibrio* (B) itself attached to the bacterium *Escherichia coli* (E). (2) Schematic representation of bdellophage VL-1 development in a three-membered system. [Courtesy of M. Varon, *Critical Rev. Microbiol.* 3:221, 1974, with permission.]

components would be present at exactly the same time in a natural microhabitat. Other explanations for the phenomenon are also plausible. The natural host for bdellophages might be nonparasitic cells of *Bdellovibrio,* or bdellophages may exist in nature in the lysogenic state. In the former, bdellophages could reproduce in a free-living state, and in the latter, their development could be induced by the establishment of the parasitic system.

Mycoviruses Viruses that live and multiply within fungal cells are called *mycoviruses.* The first definitive evidence in support of a mycovirus was released in 1962 after an extensive investigation of a viral disease in commercially cultivated mushrooms *(Agaricus bisporus).* The presence of virus particles in diseased tissues was confirmed by observations with an electron microscope. Within a few years of that initial report, other types of mycoviruses were found in that species of mushrooms and in other kinds of fungi, namely several species of *Penicillium* molds. Thus, it should not be surprising to learn that the most extensively studied group of mycoviruses is comprised of those that have been isolated from *P. crysogenus,* a mold used in the commercial production of penicillin.

In terms of ecological consequences, interrelationships that may occur between fungi and their mycoviruses are both numerous and varied. It is significant to note that a particular fungus can simultaneously be the host for a mycovirus and a parasite for a higher plant. In this situation, the mycovirus may antagonize the fungus and contribute indirectly to the well-being of the plant; or the mycovirus may interact with the fungus in a manner to increase its virulence and contribute to the destruction of the plant. It should be mentioned that mycoviruses do not have a host range wide enough to multiply both within a fungus and within a higher plant. However, if a virus is non-infectious for a fungus, but a parasite of a higher plant, the fungus can serve as its vector and contribute indirectly to the destruction of the plant.

In terms of host infectivity, most mycoviruses appear to be relatively latent and not lytic, because high concentrations of viruses have been observed in healthy mycelium. Furthermore, old hyphae and fungal spores often contain virus particles. Release of virus particles is believed to occur frequently by autolysis, but occasionally lytic plaques have been observed in some fungal species. With regard to the latter, future research may reveal that some mycoviruses are worthy of consideration for use as agents in the biological control of fungal parasites of higher plants.

Cyanophages   Viruses that infect and replicate in blue-green algae are collectively known as *cyanophages.* In terms of intracellular replication and lysis, they behave in a manner analogous to virulent bacteriophages. In other aspects the alga–cyanophage association is somewhat unique, because the alga partner is a photoautotroph. Furthermore, these procaryotic organisms carry out photosynthesis in a manner that is analogous to photosynthesis in eucaryotic plants. Thus, the alga–cyanophage system is a unique biological model that can be studied from many aspects.

Blue-green algae are ubiquitous in freshwater systems. Many species exhibit a characteristic cyclical growth during different seasons of the year, recognizable by algae blooms, foul odors, and large fish kills. Waters that contain blue-green algae also contain cyanophages. They have been isolated from fresh waters and sewage oxidation ponds in many geographical locations throughout the world. Of significance is the fact that some cyanophages have been found consistently in a given aquatic habitat throughout the year. Even in aquatic habitats where the water flows rapidly, high numbers of cyanophages that attack three kinds of blue-green algae *(Plectonema, Phormidium,* and *Lyngbya)* remained relatively constant. This observation is indicative of continuous virus replication. It is also important to note that the three genera of algae mentioned are found worldwide in aquatic habitats but are not known to produce blooms. This may be due to the fact that these groups of blue-green algae cells are being continuously lysed by cyanophages. Thus, while proliferating, their populations are simultaneously being reduced. These findings strongly suggest that cyanophages play an important ecological role as regulators of blue-green algal growth in nature. Thus, the intentional use of cyanophages to control algal blooms in fresh waters may become practical.

Extramicrobial associations in nature

The term *extramicrobial* is used to describe microbe–macrobe types of interactions. In associations of this nature, one or more of the participants will always be some kind of a microorganism, and the other participant will be a higher form of life (animals, plants, and insects). Microbe–macrobe types of associations are widespread in nature, because in natural habitats microorganisms are intimately associated with other forms of life. In this regard we shall present an overview of some classical types of symbiotic associations. Other types of interactions that may occur between microorganisms and higher forms of life will be discussed in Chapters 7, 8, and 9.

Microbial associations with animals and insects

Although numerous kinds of microorganisms can be found in and on the tissues of higher forms of life, those types that participate in symbiotic associations often exhibit some type of selectivity for their partners. Furthermore, the selectivity is often governed by some kind of a deficiency (in one or in both partners) that can be satisfied by the relationship.

Microbial–ruminant associations Animals classified as *ruminants* have several interesting characteristics. All are mammals with hooved even-toed feet, and all have a stomach with four compartments. Most familiar to us are domesticated ruminants (cattle and sheep), but goats, deer, antelopes, and giraffes are also included in this category.

Ruminants live primarily on grasses and foliage of other plants, all of which have a high cellulose content. Yet, ruminants are unable to digest such materials and utilize them as a direct source of nutrients, because they do not have enzymes for degrading cellulose. A mixed population of microorganisms which live as endosymbionts in their compartmentalized stomach (Figure 5.6) perform this vital task of cellulose digestion for them. For this reason ruminant symbiosis has been well studied by animal physiologists and microbiologists. Furthermore, because of their anatomical features, the largest compartment in the stomach of ruminants (the rumen) has become a convenient chamber for studying microbial interactions in a natural ecosystem (Figure 5.7). Experimentally, a surgically prepared opening is made through the animal's side and into the rumen (a fistula) from which fluids can be removed as desired. Animals utilized for such experiments do not experience severe complications. Actually, the rumen may be viewed as a large continuous-culture fermentation chamber somewhat analogous to a chemostat, an artificially operated *in vitro* culture system manipulated in a manner to maintain a constant population of microorganisms.

Ruminants are noted for their peculiar eating habits. Large quantities of bulky plant materials are consumed rapidly and swallowed with only a minimum of chewing. The ingested material passes through the esophagus and directly into the *rumen,* where it remains for several hours. During this holding period, the materials undergo microbial fermentation become distributed between the first two compartments (rumen and *reticulum*). Subsequently, the animal relaxes and leisurely "chews its cud," a process that consists of regurgitating small portions of the mixture from the rumen into the mouth where it

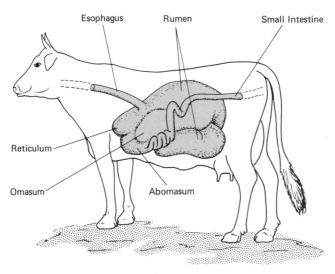

Figure 5.6. Schematic diagram of the compartmentalized stomach, and other portions of the digestive system, of the cow.

Figure 5.7. Fistulated sheep. (Courtesy of R. W. Dougherty, the National Animal Disease Laboratory, Ames, Iowa.)

is chewed again, this time thoroughly. The well-chewed material (mixed with salivary juices and microorganisms) is swallowed again. After the second swallowing, the material passes down a different route in the digestive system. It bypasses the rumen and enters the third compartment (the *omasum*), where it is further mixed, and finally it enters the fourth compartment (the *abomasum*), an organ similar to the stomach of nonruminants. True digestion begins in the abomasum and continues in the small and large intestines.

Microorganisms present in the rumen play a significant role in ruminant nutrition. The resident microflora of the rumen is a mixed population of bacteria and protozoans. As in any other ecosystem, microbial activities are influenced by the environmental conditions of that habitat. In terms of size, the rumen is the largest compartment in the stomach of ruminants; it has a relatively constant temperature (approximately 39°C), a pH range of approximately 5.0 to 7.0 and with respect to gases, its environment is anaerobic. As a result of those conditions, the rumen environment selects for cellulolytic anaerobic microorganisms (Figures 5.8 and 5.9). Obviously, the interactions that may occur in the rumen are complex. We shall not attempt to describe them but will consider the manner in which some end products from the fermentation of cellulose are used by the indigenous microflora and/or by the host animal.

The first metabolic change that occurs in the plant material is the degradation of cellulose. Cellulose-digesting microbes excrete enzymes that degrade cellulose into its constituent subunits: the diasaccharide cellobiose and the monosaccharide glucose (see Figure 2.4). Subse-

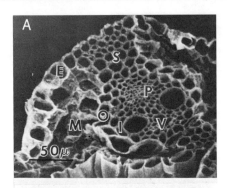

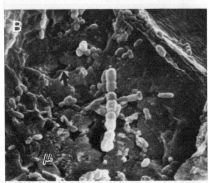

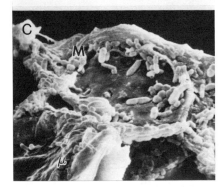

Figure 5.8. *In vitro* degradation of plant tissue by a mixed population of bacteria obtained from rumen fluid. (A) Control leaf section, observed by scanning electron microscopy after 72 hours of incubation in buffer without microorganisms. Vascular tissue (V), phloem (P), inner (I), and outer (O) bundle sheaths, mesophyll (M), epidermis (E), and sclerenchyma (S) are all intact (×240). (B) Scanning electron micrograph of mesophyll 6 hours after incubation with rumen bacteria. Nonuniform areas in tissue (arrows) show zones of degradation by bacterial extracellular enzymes, because they are free of bacteria (×2,080). Note the large number of cells in the vicinity. (C) Scanning electron micrograph of epidermal tissue 6 hours after incubation with rumen bacteria (×2,080). Note the large number of bacteria that surrounds depressed zones of degraded tissue. M, shows bacterial adhesion between cells. [Courtesy of D. E. Akin, *Appl. Microbiol.* **29**:692, 1975, with permission.]

quently, the sugars are fermented, and a mixed group of organic acids is produced: primarily acetic, butyric, and propionic. Simultaneously, the gases carbon dioxide (CO_2) and methane (CH_4) are released. Some of the newly formed organic acids pass through the rumen wall and into the blood stream of the animal. Subsequently, the newly absorbed short-chained fatty acids are utilized by the animal for energy and for the biosynthesis of long-chained fatty acids and glycogen.

While degrading cellulose, microbes in the rumen utilize some of

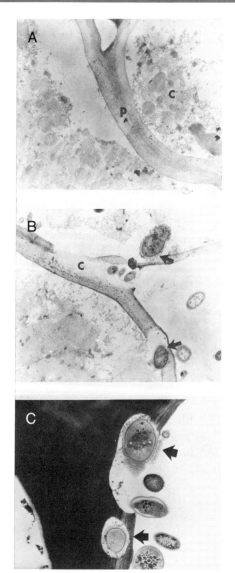

Figure 5.9. *In vitro* degradation of plant tissue by mixed populations of bacteria obtained from rumen fluid as observed by the transmission electron microscope (TEM). (A) Control leaf section after 72 hours of incubation in buffer without microorganisms: phloem tissue (P) is intact, and in area "C" chloroplasts are visible (×8,000). (B) TEM observations of mesophyll tissue after 4 hours of incubation with bacteria. Note the attachment of bacteria at two sites on the tissue (arrows) by extracellular substances (×8,000). (C) TEM observation of epidermal tissue after 6 hours of incubation with rumen bacteria (×9,250). Very sharp zones of degradation surround the bacterial cells which appear to be attached to the plant tissue by extracellular substances (arrows). [Courtesy of D. E. Akin, *Appl. Microbiol.* 29:692, 1975, with permission.]

the substrates for their own biosynthetic processes. As a result, microbial populations in the rumen are extremely large. This huge population of microorganisms competes with the animal for some substrates, but in the end, the microbial cells (biomass) serve as proteins for the animal. The microbial biomass, plus the partially digested plant material, are further digested as they pass through the animal's digestive tract. During the digestion of microbial cells, the nitrogenous compounds that are released from the degradation of protein constituents are absorbed by the ruminant. For this reason, ruminants do not have an external dietary requirement for amino acids and can survive optimally on a diet that is composed primarily of cellulose.

Newly born ruminants, like newly born human infants, do not have a normal microflora and encounter microorganisms for the first time during or immediately after birth (see Chapter 7). Consequently, the rumen microflora must be established. It is believed that young ruminants become inoculated with microbes present in the rumen of adults through exchanges of saliva during mouth-to-mouth contact between older and newly born ruminants.

Now, let us review the major aspects of rumen symbiosis:

1. Ruminants benefit from the association by having a nondigestible material (cellulose) converted into a utilizable substrate by their endosymbionts.
2. Microbes compete interspecifically with the animal for end products of cellulose digestion.
3. Ruminants obtain their dietary proteins from the digestion of microbial cells (biomass).
4. Microorganisms benefit by having a conducive place to live and access to an array of utilizable substrates.

Microbial–insect associations Types of interrelationships that have been observed between microorganisms and insects are numerous, and the microbial partners may behave as ectosymbionts or endosymbionts. Furthermore, some of the mutualistic arrangements are rather unique. For example, endosymbionts of wood-eating termites (a mixed population of protozoans), behave in a manner analogous to the cellulose-digesting microorganisms that live in the stomach of ruminants. In other associations, certain kinds of insects cultivate fungi for their food. Our discussion in this section will be limited to interactions that occur between certain ants and the fungi they cultivate, and to the interactions that occur between termites and their endosymbionts.

Certain kinds of ants and the fungi they cultivate live in a very interesting type of mutualistic relationship. This relationship is peculiar because the ants eat only the fungus they cultivate, and the fungus grows only in the narrow ecological niche of the ant's nest — the fungal garden. Why? Apparently, this type of association is highly specific for both partners, and it involves a number of interrelated factors.

The association between the ants and the fungus they grow is obligatory and mutually beneficial to both of the very different kinds of organisms. These fungal-growing ants, commonly known as attine ants, are widely distributed in the biosphere. The attine nest or fungal garden may be a large above-ground hill, an array of underground craters, or a small hill indistinguishable from a pile of trash near the base of a tree. Fungal gardens may be prepared from a variety of substrates, such as insect carcasses or leaves and/or flowers from live plants. Ants forage for such materials, and there appears to be some specificity among the various kinds of fungus-growing ants for a particular kind of substrate material. After collecting the materials, ants very meticulously prepare the substrate (a soil) for planting by chewing or gnawing the fungus into small fragments. Then the ants deposit fecal droplets on all substrate fragments. Subsequently, the ants distribute small tufts of fungal mycelium over the substrate in a characteristic manner. Apparently, the anal secretions contain fungal-growth-promoting substances, micronutrients and protein-digesting enzymes (proteases). In spite of the organic-rich substrate, only one kind of a fungus will grow in a normal garden while ants are present. However, when ants are removed, gardens quickly become overgrown with other kinds of fungi and bacteria.

Laboratory studies indicate that the viability of the fungus is directly related to the presence of the ants. Such fungi can be isolated from fungal gardens easily with commercially prepared media such as Sabouraud's dextrose agar. In sterile media the fungi grows very slowly and only for a short time in the absence of the ants. On the other hand, when ants are present the fungus can be maintained indefinitely on commercially prepared media, even when deliberately exposed to external contamination.

During early growth it appears that the competitive ability of the fungus is enhanced by nutrients present in anal secretions. Thus early growth of the fungus is accelerated. Furthermore, proteases in anal exudates facilitate the utilization of proteinaceous substrates by the fungi. Of interest is the fact that the fungus which the ant grows is a cellulose decomposer. Therefore, as an ectosymbiont, the fungus degrades cellulose for the ants. This specific, and somewhat peculiar, mutualistic relationship between Attine ants and the fungus they grow can be described as a metabolic alliance in which the ants contribute enzymes for degrading proteins and the fungi contribute enzymes for degrading cellulose. The relationship also has other ecological implications, because it enhances the decomposition of organic matter, a vital process that releases organically bound elements, thus increasing their availability in ecosystems.

Wood-eating termites and their endosymbionts also live by a mutualistic arrangment. Termites (sometimes referred to as white ants) feed primarily on woody substrates, but they cannot digest cellulose. Protozoans that live in their hind gut perform this vital task for them.

Cellulose digestion proceeds through a fermentation process within the protozoan that is analogous to the fermentation of cellulose by rumen microorganisms.

The protozoans can be freed from termites by starvation, because the protozoans and their host have differential death rates. During starvation in the laboratory, the protozoans die first and defaunated termites can survive only for a few weeks, even if fed a diet of wood. Experiments of this nature led to an understanding of termites' obligatory dependency on the association with cellulose-digesting protozoans. It was also learned that these cellulose-digesting protozoans seldom form cysts. Consequently, their external survival in fecal pellets is relatively short. It is believed that transmission (dispersal) of protozoan endosymbionts from termite to termite is accomplished by cannibalism and through proctodaeal feeding, feeding directly from the anal opening. Although the suggested methods of transmission are without proof, direct transmission is essential, because defaunated termites can live on a diet of wood for only a few weeks, and many of the protozoans cannot live in the external environment for long periods. Therefore, the symbiotic association appears to be mutually beneficial to both the termite and their protozoan endosymbionts.

Microbial associations with higher plants

Interrelationships that have been observed between microorganisms and higher plants are both numerous and varied, and they are widespread in nature. In some instances, a particular association is essential for the survival of one or both partners; but in other associations, the partners appear to be engaged in only a casual relationship. In addition to being beneficial to the participants, some interactions between microorganisms and higher plants yield indirect benefits that are both ecologically and economically important. We shall review some well-known examples of such interrelationships in this section.

Mycorrhizae   The symbiotic association between certain kinds of fungi and the roots of higher plants is referred to as *mycorrhiza*. This type of fungal–plant symbiosis is widespread in terrestrial habitats, and the manner in which the associated organisms interact is an extremely interesting phenomenon. In general, two types of mycorrhizae are recognized: ectomycorrhizae and endomycorrhizae. In *ectomycorrhizal associations,* the fungal partner grows on the outside surface of plant roots as a sheath. Extensive invasion of the plant is uncommon, but rudimentary hyphae from the fungus can penetrate the outer region (epidermis layer) of the roots. In *endomycorrhizal associations,* the fungal partner lives entirely within the tissues of plant roots.

Ectomycorrhizae are most common in mineral-deficient soils. Plants with mycorrhizal roots absorb minerals more efficiently than their counterpart species with nonmycorrhizal roots. Mycorrhizal roots tend to absorb mineral ions rapidly from the soil and selectively regulate their passage into plant tissues. This process is extremely beneficial to plants, because it enables them to have a balanced supply of mineral ions during periods of the year when ions in the soil are in short supply. Fungal partners in ectomycorrhizae also benefit from the association, because they absorb soluble sugars and micronutrients from plant tissues. Since some of the fungal partners are unable to metabolize complex carbohydrates (e.g., cellulose and lignin), their dependency on plants for soluble sugars is believed to be obligatory. Although the relationship appears to be obligatory for some fungi, they exhibit little specificity for the kinds of plants they infect.

Endomycorrhizae exhibit different characteristics. During these associations the fungal partner lives within the root tissues of plants that they infect. A well-known example of endomycorrhizae is the symbiotic association that occurs between certain fungi and orchid plants. Orchids are widely distributed flowering plants of unusual beauty. The most striking feature of orchids is the variety of colors that they display. As a result of their beauty, orchids are economically important and are grown by florists throughout the world.

Orchid seeds are extremely small and somewhat deficient in endosperm, a nutritive tissue formed within the embryo sac. Consequently, the developing seed must be supplied with nutrients from external sources. Under laboratory conditions, seeds will develop and grow if supplemented with micronutrients; but in natural ecosystems, orchids depend upon certain fungi to satisfy this need. In natural habitats, only fungal-infected seeds develop, a process that may take several years. During the process, certain fungi invade the seeds and form coils of hyphae within the tissue. While undergoing this slow developmental process, infected seedlings grow as saprophytes and obtain carbon and micronutrients from the fungi. Orchids develop normally when the endomycorrhizal relationship exists in a state of equilibrium. The equilibrium balance is maintained by orchid antifungal substances produced in response to fungal invasion. In an imbalanced relationship, the fungus may behave as a destructive parasite and kill the orchid; or the orchid may destroy the fungus by a digestive process — indirectly, a suicidal event for the orchid.

Fungal partners in orchid endomycorrhizae can utilize complex carbohydrates, cellulose and lignins. Consequently, they are nutritionally more self-sufficient than fungal partners in ectomycorrhizae. Furthermore, many of the fungal partners in orchid endomycorrhizae can grow as free-living saprophytes. In this regard it is evident that the association is obligatory for orchid development but facultative for the fungal partner.

Bacterial—legume associations The bacterial—plant *Rhizobium*—legume mutualistic association is a widely known classical type of symbiosis. The bacterial partners, one of several species of *Rhizobium,* are common types of bacteria that live in the soil, and the other partner is one of several species of plants called *legumes*—soybeans, alfalfa, and clovers are common examples. When roots of a susceptible legume become infected with an appropriate strain of *Rhizobium* bacteria, the roots undergo a modification and form structures called *nodules* (Figure 5.10). Subsequently, specialized cells within the bacterial-infected nodules convert gaseous nitrogen (a form in the atmosphere that cannot be used by either the bacteria or the plant) into a combined form of nitrogen that can be used by both types of organisms. The process is called symbiotic nitrogen fixation. Both partners in the association (the legume and the bacteria) can grow separately in the soil as independent free-living organisms, but in the nonsymbiotic state, neither organism can fix atmospheric nitrogen.

Although the association is facultative, some strains of *Rhizobium* exhibit a high degree of specificity for the roots of certain kinds of leguminous plants. In other words, a specific strain of *Rhizobium* can only infect susceptible species of legumes. For example, strains of *Rhizobium* that cause nodule formation in soybeans are unable to stimulate nodule formation in other legumes, such as clover, alfalfa, and garden beans. Some researchers believe that lectins present on the surface of soybeans are responsible for the selectivity.

Since soybeans and other legumes have emerged as important agricultural field crops, we shall mention some practical aspects of the *Rhizobium*—legume symbiosis. When farmers plant seeds of leguminous plants in soil that have large populations of *Rhizobium,* the seeds will probably become infected by the indigenous bacteria. On the other hand, if legume seeds are planted in soil where legumes have not been grown for several years, *Rhizobium* bacteria may be absent or present in very small numbers. Thus, natural infection and subsequent nodule formation may not occur. To avoid this possibility, it is now a common practice for farmers to purchase from commercial suppliers legume seeds preinoculated with the appropriate strain of *Rhizobium*. The preinoculation process is somewhat complicated. In some instances the *Rhizobium* bacteria are vacuum-impregnated into the seeds, and in other procedures they are applied to the exterior surface of the seeds and sealed within a protective coating. The survival rate of bacteria on seeds prepared by either process is relatively high. Roots of soybean plants that developed from uninoculated and from inoculated seeds are

Figure 5.10. Well-nodulated soybean roots
after being incoulated with *Rhizobium*
bacteria. [Courtesy of the Nitragin Company,
Milwaukee, Wisconsin.]

shown in Figure 5.11. Plants that developed from preinoculated seeds
have well-nodulated roots, but uninoculated seeds developed into
plants with practically no nodules on their roots. Furthermore, plants
with well-nodulated roots also have larger structures and more foliage,
and as a result produce more beans (Figure 5.12). Thus, the increased
yield is the economical reward for the farmer. Indirectly, there are
other benefits:

1. Preinoculated soybeans will have a greater chance of producing good yields in nitrogen-poor soils.
2. Decomposition of plants from preinoculated seeds will add
   nitrogenous compounds to the soil.

3. The need for applying commercial fertilizers rich in nitrogen will be negated or diminished.

The end result will be a reduction of high-nitrogen agricultural run-offs. Such runoffs enter aquatic habitats and contribute to eutrophication (see Chapter 11).

Recently, another bacterium–plant association was found to fix nitrogen symbiotically. The bacterium *Spirillum lipoferum* can live symbiotically in the roots of several nonlegume tropical grasses and was reported to fix nitrogen at a rate comparable to the *Rhizobium*–legume system. However, the anatomical relationship appears to be somewhat different. Plants infected with *Spirillum lipoferum* do not form nodules on their roots. The bacteria in the infected root live within the cells of the root cortex, the layer of cells immediately under the epidermis.

The association between *Spirillum lipoferum* and nonlegume grasses appear to be facultative, because the bacterium can fix atmospheric nitrogen nonsymbiotically when grown in the laboratory on an

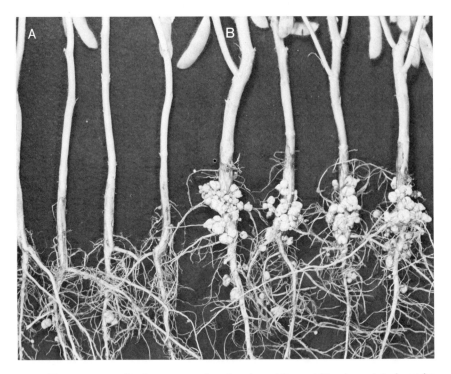

Figure 5.11. Soybean roots that developed from: (A) uninoculated seeds; (B) *Rhizobium*-inoculated seeds. (Courtesy of the Nitragin Company, Milwaukee, Wisconsin.)

Figure 5.12. Soybean plants developed from uninoculated seeds and from *Rhizobium*-inoculated seeds: plants in the center developed from un-inoculated seeds, and those to the left and right developed from inoculated seeds. [Courtesy of the Nitragin Company, Milwaukee, Wisconsin.]

appropriate medium. The association between *Spirillum lipoferum* and nonlegume grasses which results in the symbiotic fixation of atmospheric nitrogen is extremely important because of its potential impact on the agricultural production of cereal crops.

Obligate microbial parasites of higher organisms

We characterize obligate parasites as those microorganisms with deficiences that prohibit them from proliferating as independent free-living organisms when separated from their hosts. In this category we must include all viruses (see Chapter 1). Other kinds of parasites are widely distributed within and among the different groups of micro-organisms. Generalizations are difficult to state, but all obligate parasites are restricted to the microhabitats of their hosts, and their behavior is governed by environmental forces that exist in those micro-habitats. For this reason, obligate parasites are dependent on their hosts not only for some essential factor, but for their transmission to another susceptible host. Without such linkage to a susceptible host, obligate parasites are destined to a state of dormancy or to destruction when separated from their host.

We generally divide obligate parasites into two broad groups: destructive parasites and balanced (or prudent) parasites. *Destructive parasites* are seldom successful because their activities lead to the destruction of their host, a process that will limit their dispersal and ultimately lead to self-annihilation due to lack of a place to live. On the other hand, *prudent parasites* are able to live in a compatible relationship with their host. Although the relationship is detrimental, damage to their host is minimized, thus survival of the parasite is enhanced. Prudent or balanced parasites are etiological agents of chronic and slowly developing diseases.

Parasitism is a universal phenomenon and has many ecological implications. In terms of the obligate parasite, the host–parasite association is an evolutionary mechanism that provides a means of survival of the parasitic organism. During the association, the host provides the parasite with some essential resource, a conducive place to live, and the means for dispersal. Simultaneously, the association tends to contribute to the regulation of the host populations. Pressures exerted on the host during the association often result in the elimination of the weaker members of the host population, and others may become modified in a manner to reduce their reproductive potential and ability to compete.

Other aspects of parasitism will be considered in Chapters 7, 8, and 9. For details of specific kinds of parasites and their behavior, the reader should consult one of the advanced references at the end of this chapter.

Key Words

bdellophages The group of bacterial viruses that have a host-range specificity limited to bacteria within the genus *Bdellovibrio*.

ciliates The group of protozoans characterized by having hairlike (ciliary) appendages on their surface.

commensalism The type of symbiotic relationship from which one partner in the association receives benefits, while the other partner is neither benefited nor harmed by the association.

cyanophages The group of algal viruses that have a host-range specificity limited to the bluegreen algae.

cyst   A dormant or resting stage of certain kinds of organisms, often formed in response to unfavorable environmental conditions.

defaunated   The removal or clearing of animal life from a region or habitat (i.e., the removal of protozoan endosymbionts from the gut of termites).

endosymbiont   Refers to the microbial partner that lives symbiotically within the cells or tissues of its host.

lichen   A unique and distinct morphological organismic form that results from the mutualistic association of an alga and a fungus.

mutualism   The type of symbiotic relationship from which both partners in the association receive benefits.

mycoviruses   The group of viruses that have a host-range specificity limited to the fungi.

periplasmic space   The area between the cell wall and the plasma (unit) membrane of a bacterial cell.

symbiosis   The living together of two or more dissimilar organisms with some degree of constancy.

Selected Readings

1. Margulis, L. 1972. Symbiosis and evolution. *Scientific American* 224:48–57.

2. Moulder, J. W. 1974. Intracellular parasitism: life in an extreme environment. *J. Inf. Dis.* 130:300–306.

3. Padan, E., and M. Shilo. 1973. Cyanophages—viruses attacking blue-green algae. *Bacteriol. Rev.* 37:343–370.

4. Preer, J. R., L. P. Preer, and A. J. Jurand. 1974. Kappa and other endosymbionts of *Paramecium aurelia*. *Bacteriol. Rev.* 38: 113–163.

5. Varon, M. 1974. The Bdellophage three-membered parasitic system. *Crit. Rev. Microbiol.* 3:221–241.

6. Weber, N. A. 1966. Fungus-growing ants. *Science* 153: 587–604.

7. Wistreich, G. A., and M. D. Lechtman. 1976. *Microbiology and Human Disease* 2nd ed. Beverly Hills, Calif.: Glencoe Press.

CHAPTER 6

Transformations in geochemical cycles

- **Dynamic aspects of microbial populations**
    Determinants of population size
        Unicellular growth
        Population growth patterns
        The significance of measuring microbial activity
    Energy conversion and biosynthesis
- **Role of microorganisms in geochemical cycles**
    Carbon cycle
    Nitrogen cycle
    Phosphorus cycle
    Sulfur cycle
- **Key Words**
- **Selected readings**

We shall consider broad-scale phenomena in this chapter and examine those interactions which provide for the continuous flow of energy and materials in the biosphere. The biosphere is composed of an infinite number of subunits called *ecosystems*. All of them are dynamic, because living (biotic) and nonliving (abiotic) components interact continuously. Within all ecosystems, microorganisms are an integral part of the biota, and the role they play in the biosphere is far more important than is sometimes thought.

The biosphere encompasses all life-supporting regions of the universe, although the location and/or scope of such regions is a matter of conjecture. In view of current speculations that extraterrestrial microbial life may exist on other planets, we should consider the biosphere as a dimensionless concept rather than a restrictive zone that surrounds our planet. The unprecedented experiments conducted in 1976 by Viking I in its search for microbial life in Martian soil are evidence that such speculations are within the realm of scientific thinking.

Earth is a major life-supporting region of the universe and, when compared to other known planets, has several unique characteristics. Among them are an abundance of water in the liquid state, continuous movements (energy and matter are being exchanged continuously between its interior and its other components—oceans, outer surface or crust, and atmosphere), and a position in the solar system which allows it to receive a considerable amount of radiant energy from the sun.

In a functional sense, interactions in ecosystems represent an exchange of energy and materials between organisms and the environment that surrounds them. Energy flows unidirectionally through ecosystems, and materials (chemical elements) are cycled within them.

Dynamic aspects of microbial populations

Directly or indirectly, all aspects of microbial behavior are manifestations of growth. The ability to colonize in a given habitat and the ability to excrete high concentrations of metabolites are related to population size. Furthermore, a change in population size will, in general, reflect a change in an organism's functional abilities. This phenomenon is of interest to us both from an economical and an aesthetic point of view. When the functional abilities of organisms are desirable (e.g., the invasion of soybean roots by *Rhizobium*), attempts are made to enhance bacterial growth and increase population size. On the other hand, when the functional abilities of organisms are undesirable (e.g., the production of algal blooms in a freshwater lake), attempts are made to inhibit algal growth and reduce population size. Thus, naturally occurring and artificially induced conditions that tend to regulate the size of microbial populations have far-reaching ecological implications.

Determinants of population size

Resident populations of all habitats are subjected to, and influenced by, operative environmental factors, forces that tend to impinge directly on the organism. Such interaction contributes to natural selection processes, because the indigenous microflora of a given habitat tend to be those organisms that are best adapted to that habitat. Even when environmental conditions are extreme, the environment selects for organisms that are the most fit. High temperatures in hot springs favor colonization by thermophiles; low pH in acid-mine waters favors colonization by acidophiles, and high salt concentrations in Great Salt Lake favors colonization by halophiles. Environmental factors that are relatively constant contribute to natural selection processes; whereas environmental variables tend to function as determinants of population size. Such variables include, but are not limited to, the following: (1) nutrient concentration in aquatic systems, (2) moisture and temperature fluctuations in terrestrial habitats, and (3) the concentration of antimicrobial agents in biological habitats. The latter consideration is important in humans when undergoing prolonged therapy with broad-spectrum antibiotics.

The role that microorganisms play in infectious diseases is well known. Since the early acceptance of Koch's postulates as standards for defining pathogenicity, other criteria have emerged for differentiating significant pathogens from casual contaminants that may be present in or on tissues of infected individuals. As a result of such criteria, the agents that cause most infectious diseases in human beings have been described. In addition, virulence mechanisms—the specific manner by which organisms cause damage to people—for many of them have also been elucidated. We are saying that pathogenicity and virulence are functional attributes of microorganisms that can be measured. Without the ability to measure those attributes, our knowledge of infectious diseases would have remained in its infancy. Progress in delineating the roles that microorganisms play in nonbiological natural habitats has been slow, because we are lacking in the ability to measure their functional attributes.

Our meager knowledge of the roles that microorganisms play in nature can be attributed to our inability to simulate natural habitats. Measurable responses obtained for a given kind of microorganism under laboratory conditions do not imply that similar responses will occur in nature. We can justly assume that nature abounds with many kinds of microorganisms that have never been characterized, because most taxonomic criteria are based on characteristics that are observable under laboratory conditions. Thus, gaps in our understanding of microbial behavior in nature exist because of our dependency on isolating and studying them under artificial conditions in the laboratory. In an attempt to circumvent these problems, a great deal of attention is being directed toward techniques for measuring microbial activity under field conditions. Although aimed at detecting metabolically active cells, each technique or assay is based on the premise that such cells represent populations that are actually growing. For this reason it is imperative that we understand thoroughly the concept of microbial growth. After discussing this concept, we shall examine briefly some of the assays that are being used to measure microbial activity.

Unicellular growth Previously, growth was defined as an orderly increase in the chemical constituents within a cell. Implied within that definition is the biosynthesis of macromolecules and the subsequent increase in cell numbers (biomass). In multicellular organisms, growth is characterized by the following: (1) an increase in cell numbers, (2) a differentiation process in which cells acquire specialized functions, and (3) an overall increase in the size of the individual. The growth of unicellular organisms is characterized by the biosynthesis of macromolecules, a process that leads to an increase in cell size to a fixed point, at which each cell undergoes a fission process. In preparation for the fission process, each cell becomes elongated. Then fission is initiated by an inward growth of the cytoplasmic membrane, a process that involves the participation of structures called *mesosomes* (Figure 6.1). This is followed by an inward growth of the cell wall and the formation of a transverse septum. Afterward, the two newly formed cells separate. The overall process is referred to as *binary fission*. Each newly formed cell may then repeat the process. Thus, we refer to the growth of unicellular organisms in terms of population size rather than individual size. As a result of binary fission, bacterial cells grow by a logarithmic or exponential increase in cell numbers. For example, a single cell will grow as follows: $1 \rightarrow 2 \rightarrow 4 \rightarrow 8 \rightarrow 16 \rightarrow 32$. After five generations, 32 new cells will have been produced. The actual time

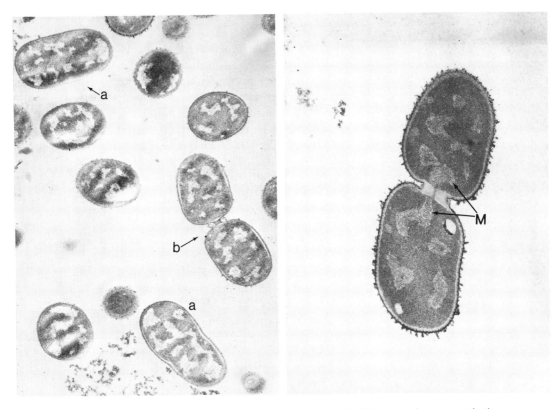

Figure 6.1. Electron micrograph of bacterial cells (*Achromobacter* species) undergoing division. *Left:* (a) elongated cells in preparation for division, (b) A late stage in division with two newly formed cells nearly separated (about ×24,000). *Right:* Final stage of cell division, in which complex mesosomes (M) can be seen on each side of the partition that barely connects the two newly formed cells (about ×27,000). [Courtesy of G. B. Chapman, *J. Bacteriol.* 95:1862, 1968, with permission.]

that is required for a single bacterial cell to divide is referred to as its *generation time,* the time required for a population of bacterial cells to double in size. The generation time is relatively constant for a species, but it varies among types of bacteria and is influenced by environmental conditions. The generation time for *Escherichia coli* or *Pseudomonas aeruginosa* may be in the range of 10 to 20 minutes when conditions for growth are optimum, but a slow-growing bacterium such as *Mycobacterium tuberculosis* may have a generation time of 18 to 24 hours. As a result of their ability to increase exponentially and to have relatively short generation times, bacterial populations can become extremely large within a few hours if environmental conditions are favorable for growth.

   Population growth patterns   Since unicellular organisms multiply exponentially, their populations develop through phases that are now recognized as being a characteristic feature for unicellular organisms as a group (Figure 6.2). When bacterial cells are inoculated into a liquid medium and placed in environmental conditions that are favorable for growth, they do not divide immediately but undergo a period of adaptation. This period is called the *lag phase.* There is no detectable cell division, but the cells are actually preparing themselves biosynthetically for the fission process. The *logarithmic phase,* which begins when cells undergo their first division, is characterized by exponential growth of most cells within the population and continues until

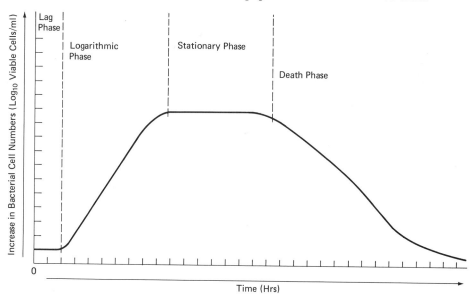

Figure 6.2.   Generalized bacterial growth curve.

environmental conditions begin to deteriorate (e.g., exhaustion of a nutrient or buildup of a toxic substance). At this point, exponential growth occurs only in some of the cells, and the population enters the *stationary phase,* a period in which new cells are being produced at a rate equal to the number of cells that are dying, resulting in no overall increase in population size. This period of equilibrium may continue for a considerable time, but if environmental conditions continue to become more intolerable, cells enter the *death phase,* a period in which cells die or become dormant. During the phase of gradual death, cell numbers appear to decrease logarithmically, and the population is referred to as an old culture, in contrast to the logarithmic growth of a young population.

The discussion so far has pertained to growing populations of unicellular organisms in liquid media. Many kinds of unicellular organisms form colonies when growth occurs on solid substrates. When a single cell or clumps of cells undergo repeated divisions at one site, the newly formed progeny accumulate into a pile or solid mass that can be seen by the naked eye. The mass of cells is called a *colony.* Colonies of two different kinds of microorganisms are shown in Figure 6.3.

We should emphasize the fact that colonial growth of microorganisms is not a phenomenon that occurs only on a solid substrate. Certain kinds of microorganisms are recognized for their ability to produce colonial growth in liquid laboratory media, and also in natural water. Bacteria within the genus *Zoogloea* are somewhat unique in this respect. These bacteria are common inhabitants of polluted waters. During growth, their cells aggregate into free-floating colonies called *flocs.* In the biological phase of municipal sewage treatment plants, *Zoogloea* cells play an important ecological role (see Chapter 11). *Zoogloea* cells have slimy surfaces, a property that enables the floc to serve as a matrix for the entrapment of other microorganisms. The mass, containing a mixed group of microbial cells, will eventually settle to the bottom and become a component of sludge.

Numerous other kinds of microorganisms produce colonial growth in municipal water supplies. Unlike *Zoogloea* species, colonial growth of filamentous forms are often troublesome, because their colonies become attached to the sides of pipes and to tank walls, and clog filters. Such conditions result from the growth of many kinds of molds; several kinds of bacteria, especially *Sphaerotilus* species; and several species of procaryotic and eucaryotic algae. Examples of filamentous procaryotic algae that are commonly found in polluted waters are *Lyngbya, Phormidium,* and *Tolypothrix,* and representative filamentous eucaryotic algae in such environments include *Stigeoclonium* and *Ulothrix.*

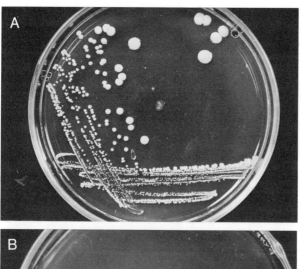

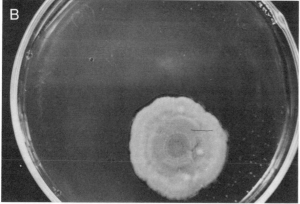

Figure 6.3. Representative colonies of microorganisms: (A) isolated colonies of *Staphylococcus aureus;* (B) mold colony (*Penicillium* species).

The significance of measuring microbial activity. Previously, we alluded to the fact that attempts to simulate natural microhabitats in the laboratory have not proved highly successful. Thus, many of their roles in nature remain elusive. To gain an understanding of their behavior in nature, emphasis has recently shifted toward measurements of metabolic activity. Actively growing populations will exhibit high metabolic activity. Therefore, such measurements will contribute to our understanding of the roles that microorganisms play in problems such as biofouling.

Among techniques and assays that are being used widely to measure the metabolic activity of cells in natural habitats are: (1) radioactive materials as tracers, and (2) the *Limulus* assay for lipopolysaccharides (constituents of bacterial cell walls). In techniques that employ radioactive materials, substrates that microorganisms use are labeled with radioactive isotopes such as carbon (^{14}C), phosphorus (^{32}P), or sulfur (^{35}S). Such techniques are highly specific in that a particular metabolic process can be measured. For example, glucose (a six-carbon compound) can be labeled with (^{14}C) on the first carbon, or labeled uniformly. Actively metabolizing heterotrophic organisms will incorporate each of the labeled substrates equally, but end products from glucose degradation in each case will contain the radioactive ^{14}C in different fragments. By using ^{32}P, the rate at which organisms utilize phosphorus in an aquatic system can be measured. In a similar manner, ^{35}S can be used to measure the rate at which sulfate (^{35}SO$_4$) or another sulfur compound in nature is converted to inorganic sulfur (^{35}S). From these examples we can see that techniques for utilizing radioactive tracers are highly specific. In all of them, an appropriate radiation-detection device must be used to detect the radioactivity in the various kinds of end products.

The *Limulus* assay is based on the use of an aqueous extract from the blood of the horseshoe crab, *Limulus polyphemus*. The extract contains an enzyme that is activated by lipopolysaccharides. Since lipopolysacchrides represent about 10 per cent by weight of constituents in many kinds of bacteria, this assay can be used to measure microbial biomass. Furthermore, it is extremely sensitive; as little as 10^{-11} gram of lipopolysaccharide per milliliter of solution can be detected.

The scanning electron microscope, especially when used in combination with tests that measure microbial activity, has enhanced ecologists' ability to study microbial behavior in nature. This approach was used recently in a study of the ability of microorganisms in Woods Hole harbor to become attached to various substrates submerged in marine waters. Several morphological forms of bacteria were observed attached to the surface of nickel and polystyrene substrates (Figure 6.4). The ability of marine organisms to become attached to various surfaces has far-reaching economical and environmental implications. All seagoing vessels, water pipes, and other structures that are submerged in the ocean are subjected to biofouling by microorganisms. For this reason, a great deal of research is directed toward reducing the susceptibility of various materials to microbial degradation. Biocontrol approaches involve the incorporation of antimicrobial agents into the structures themselves, and the fabrication of structures that are less susceptible to microbial attachment. Evaluation of such tests is greatly enhanced by observation with the scanning electron microscope. Such observations will also contribute to our understanding of morphological types that inhabit marine waters, especially those that do not grow under laboratory conditions.

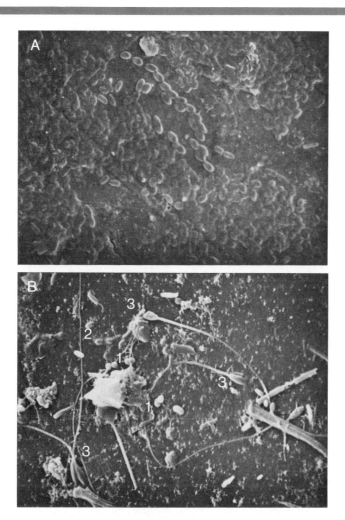

Figure 6.4. Observations of microorganisms in a natural habitat of Woods Hole harbor. (A) Rod-shaped bacteria attached to a nickel substrate after being submerged for 14 days. (B) Several morphological types of bacteria attached to a polystyrene substrate: (1) cocci, (2) rod-shaped, and (3) stalked forms. [Courtesy of S. C. Dexter, *Appl. Microbiol.* **30:298**, 1975, with permission.]

Energy conversion and biosynthesis

In order to function as an integral part of ecosystems, microorganisms must acquire and utilize energy in some form. Different forms of energy and types of organisms that utilize them were described in

Chapter 2. Regardless of the diversity among metabolic systems used in the procuring of energy, organisms with different physiological capabilities tend to interact coordinately in ecosystems with respect to the manner in which energy is exchanged.

Solar radiation is the ultimate source of all energy in the biosphere (see Figure 2.15). Certain types of electromagnetic rays are used by the mixed group of organisms called photosynthesizers. Photosynthetic organisms have the ability to capture and utilize light energy for the synthesis of carbohydrates. Because of this capacity, they are referred to as primary producers of energy in ecosystems. Actually, photosynthetic organisms do not produce energy; light energy is transformed within their cells to chemical energy (carbohydrates).

Theoretically, the amount of energy produced in a given ecosystem can be calculated, but a discussion of the quantitative aspects of primary production is beyond the scope of this book. However, a generalized understanding of the manner in which energy flows and materials are cycled within ecosystems can be obtained by examining the linkages between photosynthetic organisms and types of organisms that are lacking in that capacity. In reality, there is a codependency between photosynthetic and nonphotosynthetic organsims, because enzymes that mediate the reactions of photosynthesis are dependent on the availability of certain trace elements. The trace elements are released from organic matter and from inorganic reservoirs through mineralization, a process that is carried out by nonphotosynthetic organisms. Perhaps such codependencies can be visualized if we examine the continuous and sequential nature of events that occur in ecosystems, beginning with photosynthetic organisms and ending with decomposers, organisms that release organically bound elements for reuse by producers. Collectively, these interactions provide mechanisms for sustaining the biosphere as a life-supporting region in the universe. We refer to those interactions as *food chains* (Figure 6.5).

In food chains, the mixed group of photosynthetic organisms (higher green plants, eucaryotic algae, blue-green algae, and photosynthetic bacteria) may be considered as power generators for ecosystems, because they, collectively, transform radiant energy into chemical energy. Simultaneously, another group of organisms, the *chemolithotrophic bacteria,* contribute to the initial process. Thus, we recognize two groups of organisms as primary producers (photoautotrophs and chemolithotrophs). Through the activities of those groups, energy-rich organic compounds in the form of cellular constituents (biomass) become available for use by heterotrophs. As previously defined, heterotrophs derive their energy from preformed organic matter.

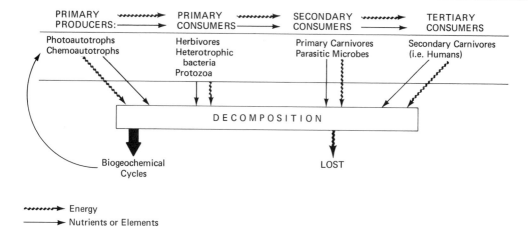

Figure 6.5. Generalized version of food chains. Energy flows unidirectionally, and nutrients or elements cycle within the system.

Their survival is completely dependent on the preexistence of autotrophs. They are an extremely large group of organisms, and we generally refer to them in broad categories as herbivores and carnivores. *Herbivores* obtain their energy solely from plants and are referred to as *primary consumers,* and carnivores obtain their energy from herbivores and are referred to as *secondary consumers.* Most secondary consumers are carnivores, animals that eat animals. From this point many variations may occur in food chains. Some of them contain tertiary consumers (secondary carnivores), and *omnivores,* animals that eat both plants and animals.

Our discussion of food chains will be generalized. In food chains, energy and nutrients are moved from primary producers to subsequent consumers sequentially, but at each link (trophic level) a considerable amount of energy is lost through respiration and decay. Furthermore, the energy loss continues at each successive transfer along food chains. Thus, we refer to energy flow as being *unidirectional.* Simultaneously, organisms at each trophic level are dying. Their remains (dead organic biomass) would constitute a dead-end reservoir of organic matter were it not for the existence of another group of heterotrophs, which we call *decomposers.* Thus, nutrients in animal remains represent only a temporary loss from food chains, because decomposers, predominantly bacteria and fungi, obtain their energy from dead or decaying organic materials. While acquiring nutrients from the remains of other cells, they degrade macromolecules into their inherent components. Simultaneously, decomposers replicate and make more of themselves (biomass). The processes that decomposers mediate release organically bound elements to the biosphere for reuse by other living cells. As a result of the unique array of interactions, ecosystems are self-perpetuating units through which energy flows unidirectionally, and chemical elements are recycled within them. An important fact to remember from this is that microorganisms of some kind contribute to the exchanges that occur at each link in a food chain.

Perhaps the complexity of food chains can be visualized if we consider some aspects of food chains that involve human beings. Most of us pay little attention to an animal eating grass (herbivory), but some people react emotionally when they see one animal devouring another. Both processes are means through which energy and materials are transferred in ecosystems. Thus, feeding relationships among organism within ecosystems involve the gathering or capturing of food and its consumption, directly or indirectly, by a succession of organisms at different feeding (trophic) levels. We shall now examine some interesting aspects of a modern food chain that leads to human beings:

$$\text{autotrophic plants} \longrightarrow \text{domestic cattle} \longrightarrow \text{human beings}$$

In this example it is important to remember that the production of food in modern societies involves many intermediate activities, all of which require inputs of energy. For this reason modern food production is directly related to the availability of fossil fuels to support our technology. The burning of fossil fuels also contributes to the cycling of elements in the biosphere, an aspect to be discussed later.

Nevertheless, food chains that lead to human beings involve some interesting and diverse linkages. We can reason from our example above that autotrophic plants are the primary producers for the cattle (herbivores). From this point, some interesting questions can be asked: (1) Technically, are cattle primary consumers or secondary consumers? Although cattle are considered to be herbivores, they are lacking in the ability to acquire energy directly from the autotrophic plants they consume. Microorganisms within their rumen degrade the plant material (cellulose) into a mixed group of fermentation products (organic acids) which the cattle use for energy. (2) Does biomass (beef) and its products represent energy and nutrients from a primary or a secondary carnivore? In addition to obtaining energy from microbial fermentation products, ruminants digest the microbial cells (biomass) as their major source of proteins. (3) What is the trophic level of human beings? Since energy and nutrients in the form of beef biomass have passed through an array of linkages, the present example does not fit the simple food-chain model.

Role of microorganisms in geochemical cycles

Cycles through which chemical elements move within the biosphere are rather complex. All chemicals from which cells are constructed move through them. Key elements, such as carbon, nitrogen, phosphorus, and sulfur, while being essential for life, also exist within the biosphere in nonliving reservoirs. The major nonliving reservoir of carbon and nitrogen is the atmosphere. Pools of these elements exist in combined form as atmospheric gases, carbon dioxide and molecular nitrogen. The other two elements (phosphorus and sulfur) are stored within the earth's crust as components of sedimentary materials. Thus, the recycling of these elements involves biological–geological types of interactions, often called *biogeochemical cycles.*

Carbon cycle

The *carbon cycle* is an integral part of the energy cycle. The reactions of photosynthesis are essential to both, because through these processes, atmospheric carbon dioxide is incorporated into components of living cells. Subsequently, a portion of organic carbon is transported within ecosystems through the interaction of food chains, but the cycle as a whole involves some exchanges between nonbiological components. Our purpose here is to present an overview of these movements (Figure 6.6).

The cyclical process begins with the incorporation of atmospheric carbon dioxide into a specific carbohydrate molecule (ribulose diphosphate) by photosynthetic organisms. Then, through normal metabolic processes that occur in all organisms, carbon dioxide is released from cells and reenters the atmospheric reservoir. Since the element carbon is a constituent of all organic compounds, a portion of it is organically bound in all living organisms (biomass). When organisms die, organically bound carbon is then released from biomass through decomposition. However, all of it does not reenter the atmosphere. Some remains organically bound and stored as fossil fuel, such as coal and petroleum. Although organic carbon in fossil fuels represents a loss from ecosystems, carbon within them is eventually released through volcanic activities and their burning by people. It is important to remember that fossil fuels represent organic carbon that has been out of circulation for eons of time. In recent years the general population has become more aware of fossil fuels, because of the controversial discussions among world leaders and major industries about its availability and the cost of recovery from major stockpiles.

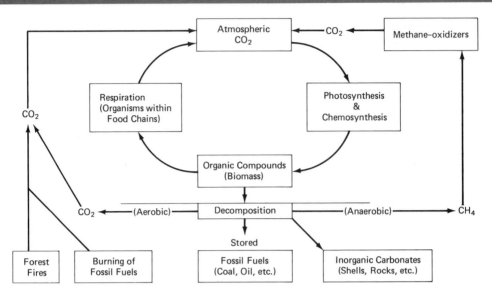

Figure 6.6. Generalized version of the carbon cycle.

Another important aspect of the carbon cycle is its nonbiological movement. Exchanges among dissolved carbon dioxide (CO_2), carbonates (CO_3^{2-}), and bicarbonates (HCO_3^-) are common in aquatic habitats. Under certain conditions, inorganic carbonates precipitate and contribute to the formation of limestone deposits.

It is also important to note that photosynthetic processes in aquatic habitats are mediated by eucaryotic algae, blue-green algae, and photosynthetic bacteria. These groups assimilate dissolved carbon dioxide and release oxygen to their surroundings. The released oxygen favors the rapid growth of aerobic organisms, and many of the small aquatic animals incorporate carbonates into their shells. Thus, their shells eventually contribute to limestone deposits. Remember that photosynthetic processes in aquatic habitats are most vigorous in the upper portion of the photic zone. In deeper portions of aquatic habitats, both photosynthesis and aerobic respiration decrease. The rate of decomposition also decreases in the anaerobic aphotic zones of aquatic habitats, and partially degraded organic matter tends to accumulate as humus. Carbon is released from humus and rocks through mineralization processes.

In the discussion above, we have presented a simplified overview of the carbon cycle. Although the carbon dioxide concentration in the atmosphere is only 0.03 per cent by volume, photosynthesis is considered to be the largest naturally occurring chemical reaction in the biosphere. In terms of importance, all forms of life, except the chemolithotrophic bacteria, are dependent upon it. Thus, from the standpoint of energy and carbon, it is vital to the functioning of ecosystems.

On land, agricultural crops and forests are the major photosynthesizers. It is important to note that forest ecosystems are the largest reservoirs of organically bound carbon. Periodically, as a result of forest fires, some of the carbon dioxide is released to the atmosphere. However, a considerable amount of carbon in forest communities remains tied up in various kinds of woody materials.

We can see that the carbon cycle is continuous and essential for life. What effect, if any, will human disturbances have on it? Will broad-scale consumption of fossil fuels affect the distribution of carbon dioxide among its major reservoirs: atmosphere, biological organisms, and inorganic and organic deposits?

Nitrogen cycle

The chemical element nitrogen is essential for life. All living cells require it for biosynthetic purposes. Within living cells, nitrogen is a major constituent of all amino acids, nucleic acids (DNA and RNA), and many other important molecules. Molecular nitrogen (N_2) is also abundant in the atmosphere. In terms of volume, gaseous nitrogen represents approximately 79 percent of the atmospheric gases. Yet, in the gaseous form (N_2) it cannot be utilized by most organisms. Like atmospheric carbon dioxide, it must be "fixed." Only certain kinds of microorganisms have the ability to carry out the process of nitrogen fixation. Some of them can perform this task as free-living entities, but others fix atmospheric nitrogen only while living symbiotically with other organisms. The latter process is called symbiotic nitrogen fixation. Thereafter, nitrogen becomes a part of living cells and enters food chains in ecosystems. Nitrogen availability in ecosystems is then determined by the manner in which it is cycled among biological components and its nonbiological reservoir (Figure 6.7). The nitrogen cycle involves a series of rather complex processes, and most of them are mediated by microorganisms.

The nitrogen cycle begins with the fixation process, during which molecular nitrogen is incorporated into constituents of living cells. In the past, organisms with this ability have been divided into two distinct groups: symbiotic nitrogen fixers and nonsymbiotic nitrogen fixers. In view of some recent research, it may become necessary to add a third category, the intermediate or facultative symbiotic nitrogen fixer. Briefly, each group can be described as follows.

The *symbiotic nitrogen fixers* are microorganisms with the ability to live as endosymbionts in the roots of legumes or other plants. The

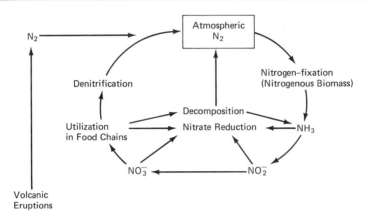

Figure 6.7 Generalized version of the nitrogen cycle.

fixation process is mutualistic and occurs only while the two organisms are associated. Neither partner can mediate the process when living separately. Well-known bacterial representatives within this group are species of *Rhizobium*.

The *nonsymbiotic nitrogen fixers* are those types of organisms that can incorporate molecular nitrogen into their tissues while living independently of other organisms. Species of *Clostridium* and *Azotobacter* are well-known bacteria with this capability. Nitrogen can also be fixed nonsymbiotically by blue-green algae within the genera *Nostoc* and *Anabaena*.

The *intermediate* or *faculative symbiotic nitrogen fixer* is *Spirillum lipoferum*. A Brazilian scientist, Johanna Dobereiner, reported recently that *Spirillum lipoferum* was capable of fixing atmospheric nitrogen symbiotically when living in the roots of tropical grasses (nonlegumes) and was also capable of fixing atmospheric nitrogen nonsymbiotically when grown in the laboratory on an appropriate substrate.

Subsequently, a variety of organisms contribute to the cycling of nitrogen within the biosphere. Organically bound nitrogen is released from living organisms in many forms (e.g., excreta and organic metabolites), and from dead organisms through decomposition, called *ammonification* when carried out by aerobic microbes and *putrefaction* when carried out by anaerobes. The latter process accounts for the unpleasant odors that are associated with decaying animals or degradation of their excreta (urine and manure). The odor is the smell of urea, amines, and ammonia, products released from protein deomposition. In the soil, ammonia may become bound to particulates or assimilated into plant tissues, a reentry to the food chain. However, a major portion of the ammonia is oxidized by two groups of chemolithotrophic soil bacteria: *Nitrosomonas* and *Nitrobacter*. Species of *Nitrosomonas* oxidize ammonia (NH_3) to nitrite (NO_2^-), and species of *Nitrobacter* oxidize nitrite to nitrate (NO_3^-). The latter is extremely soluble and is the form of nitrogen that is used by most plants, especially agricultural crops. It is also the form of nitrogen that is leached from soils easily. Remember that these chemolithotrophic bacterial groups, while oxidizing ammonia and nitrate, are also producing more of themselves (biomass).

To this point we have only considered processes that add nitrogen to ecosystems. The last step in the nitrogen cycle is called *denitrification*. It occurs under anaerobic conditions, but many facultative microorganisms contribute to the process, the removal of nitrogen from ecosystems. Although denitrification may seem undesirable, it is essential for maintaining a balanced ecosystem. The specific process is reductive: nitrates are reduced to nitrites, and nitrite to molecular nitrogen (N_2). A summary of the processes involved in the nitrogen cycle is given in Table 6.1.

In addition to the role that microorganisms play in the nitrogen cycle, nonbiological processes add nitrogen to ecosystems. Volcanic activity releases nitrogen from sedimentary deposits, and automobile exhausts release oxides of nitrogen to the air. In the atmosphere, oxides of nitrogen are subjected to many photochemical reactions, and eventually some of the nitrogenous products are returned to ecosystems through precipitation. Finally, commercially produced fertilizers are used widely on agricultural land to supplement biological nitrogen fixation. While the use of commercial fertilizers contributes to increasing yields of agricultural crops, indiscriminate use also accelerates the process of eutrophication in natural waters through land runoff. In this regard efforts should also be made to maintain some degree of balance.

Table 6.1. Key Transformations in the Nitrogen Cycle

Process	Representative Microorganism(s)	Reaction(s)	
		Substrate	Product
Nitrogen fixation			
Nonsymbiotic	*Azotobacter*	Molecular	Nitrogenous
Symbiotic	*Rhizobium*–legume	nitrogen (N_2)	biomass
Ammonification	*Clostridium*	Nitrogenous	Ammonia (NH_3)
	Proteus	biomass	
Nitrification	*Nitrosomonas*	Ammonia (NH_3)	Nitrite ion (NO_2^-)
	Nitrobacter	Nitrite ion (NO_2^-)	Nitrate ion (NO_3^-)
Dentrification	*Pseudomonas*	Nitrate ion	Nitrite ion
	Bacillus	(NO_3^-)	(NO_2^-), N_2

Phosphorus cycle

Phosphorus is vital to all living cells. It is present in the structure of all nucleotides, and we know from Chapter 2 that adenosine triphosphate (ATP) is intimately linked to energy-conversion processes in living organisms. Furthermore, phosphorus is stored in large reservoirs within sedimentary materials. From such materials, phosphorus is released through weathering processes — among them is solubilization, caused by nitric acids formed during nitrification.

Because of the role that it plays in eutrophication, phosphorus and phosphate detergents have become popular subjects for discussion among environmentalists. Widespread attention was attracted to phosphorus when domestic waste, containing high concentrations of phosphate detergents, was linked to foaming in lakes. In nonpolluted waters, phosphorus exists in very low concentrations. An explanation for low concentrations of phosphorus in oligotrophic lakes is the fact that phosphorus reacts readily with particles in the soil, especially those that contain calcium, aluminum, and iron to form highly insoluble compounds. Such insoluble materials form precipitates and become a part of the sedimentary deposits.

We can readily see that phosphorus exists in the biosphere in several forms: some of it is present in aquatic systems in soluble form, some of it is bound to macromolecules in living organisms, and a large portion of it exists in inorganic sedimentary deposits. The manner in which phosphorus is exchanged from among the various sources that we mentioned above is shown in Figure 6.8. It should be recognized that phosphorus enters ecosystems from inorganic reservoirs through the assimilation of soluble phosphates by all physiological groups of microorganisms and by higher plants. Subsequently, the element is moved through food chains in the same manner as was described for other elements. Then through decomposition processes, the element is again released from its organically bound state.

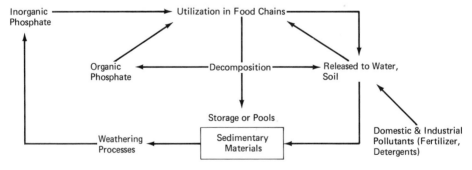

Figure 6.8. Generalized version of the phosphorus cycle.

Sulfur cycle

Sulfur is also an element that living organisms require for biosynthetic purposes. Within cells it is found as constituents of certain amino acids (cystine, cysteine, methionine) and in structures that contain a sulfhydryl group (-SH). Sulfur also exists in the biosphere in large reservoirs, primarily fossil fuels and sedimentary materials. The element sulfur moves through more complex cycles than those described for phosphorus (Figure 6.9). It enters ecosystems through microbial biomass (some bacterial cells concentrate it) and through the assimilation of soluble sulfates by plants. Subsequently, it moves through food chains as other elements.

Sulfur compounds in the biosphere undergo many complex oxidation and reduction reactions, and most of them are mediated by a mixed group of microorganisms that live in aquatic habitats. Among the microorganisms that contribute to the transformations of sulfur are a variety of heterotrophs. This group of organisms is primarily the decomposers. Through their activities, sulfur is released from organic materials in the form of hydrogen sulfide (H_2S) gas; this accounts for the offensive odor that most people recognize when proteinaceous materials are undergoing decomposition.

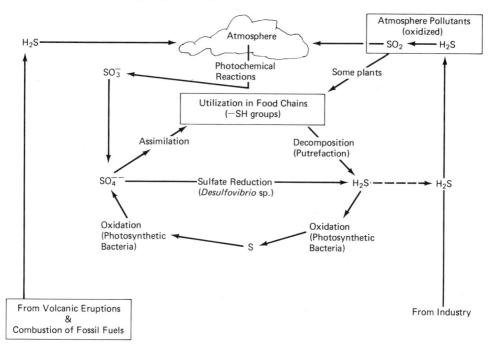

Figure 6.9. Generalized version of the sulfur cycle.

Two groups of photosynthetic bacteria are involved with the transfer of sulfur compounds: the purple sulfur bacteria *(Chromatiaceae)* and the green sulfur bacteria *(Chlorobiaceae)*. Organisms from within the different groups have been observed to carry out identical reactions. Some of them oxidize hydrogen sulfide (H_2S) to elemental sulfur (S), and others oxidize elemental sulfur to sulfuric acid (H_2SO_4). H_2S and S serve as electron donors for CO_2 reduction in the groups above, whereas water serves this purpose in photosynthetic higher plants (see Chapter 2).

Chemolithotrophs also contribute to the transformation of sulfur in ecosystems. Reactions that are mediated by some species of *Beggiatoa* and *Thiobacillus* are of interest. The former are common in domestic sewage-treatment plants in which hydrogen sulfide is present. Such organisms grow in long filaments of undifferentiated cells and move by gliding motion. They oxidize hydrogen sulfide and store elemental sulfur particles in their cells. *Thiobacillus thiooxidans* and *T. ferrooxidans* also mediate some interesting processes. The former oxidizes elemental sulfur to sulfuric acid, and the latter can oxidize several inorganic materials (hydrogen sulfide, elemental sulfur, and iron) to produce sulfuric acid. Both of these types of bacteria are extremely aciduric. They live in habitats where pH values may be 2 or less. In such environments, acid drainage plays an important role in the solubilization of sedimentary rocks, thus releasing minerals such as phosphorus. In nonpolluted waters, phosphorus is usually present in very small quantities as compared to nitrogen.

Finally, we should note that sulfates are utilized by both microorganisms and higher plants. Under anaerobic conditions, bacteria in the genus *Desulfovibrio* reduce sulfates to hydrogen sulfide. When this occurs in waterlogged soils, it is undesirable, because hydrogen sulfide is toxic to many kinds of agricultural plants.

The atmospheric phase of the sulfur cycle consists primarily of hydrogen sulfide and sulfur dioxide. The former is released from decomposition of organic biomass, and the latter enters the atmosphere from the burning of fossil fuels, a serious problem in many urban areas because it contributes to air pollution. Volcanic activities also release sulfur dioxide. Subsequently, those gases may undergo photochemical reactions and produce products that return to the earth in precipitation. Such precipitates have been found to be harmful to many kinds of vegetation and a variety of structural materials in buildings.

From the above we can see that microbial interactions in the sulfur cycle are complex. Although some of the reactions appear to be undesirable to us, they contribute to the transfer of sulfur materials within ecosystems.

Key Words

biofouling   Pertains to the deterioration or pollution of a
   structure by living organisms and/or their metabolic products.
biota   The sum total of living organisms within a region or
   habitat.
colony   A population of cells growing on the surface of a solid
   medium; can be seen with the naked eye.
generation time   The amount of time required for a population
   of any kind of organism to double in number.
mesosome   The membranous structure in the cytoplasm of
   bacteria, associated with the formation of cross-walls during
   the process of cell division.

Selected Readings

1. Clapham, W. B., Jr. 1973. *Natural Ecosystems* New York:
   Macmillan Publishing Co., Inc.
2. Dexter, S. C., J. D. Sullivan, Jr., J. Williams III, and S. W. Watson. 1975. Influence of substrate wettability on the attachment
   of marine bacteria to various surfaces. *Appl. Microbiol.* 30:
   298–308.
3. Hutchinson, G. E. 1970. The biosphere. *Scientific American*
   223:45–53.
4. Peroni, C., and O. Lavarello. 1975. Microbial activities as a
   function of water depth in the Ligurian Sea: an autoradiogra-
   phic study. *Marine Biol.* 30:37–50.
5. Soffen, G. A., and C. W. Snyder. 1976. The first Viking mission
   to Mars. *Science* 193:759–766.
6. Woodell, G. M. 1970. The energy cycle of the biosphere. *Scientific American* 223:64–74.

CHAPTER 7

The human body: A natural ecosystem

- **Inhabitable anatomical regions**
    The skin
    The gastrointestinal tract
    Miscellaneous areas
- **The indigenous microflora**
    Normal ecological niches
    Sterile (forbidden) zones
- **Defense mechanisms**
    Mechanical barriers
    Immune barriers
        Phagocytosis
        Antibodies
    Types of immunity
        Natural
        Acquired
- **Germ-free animals**
- **Key words**
- **Selected readings**

Human beings *(Homo sapiens)* exhibit behaviorial and technological capabilities that are unmatched by other kinds of living organisms. Yet, we cannot escape the fact that the human organism is an integral part of nature. In spite of our abilities to modify the environment, human beings cannot survive independently of other living things. Those previously described interactions through which energy flows and materials are cycled sustain the biosphere as a region that is conducive to human life.

In this chapter we shall not concern ourselves with the broad aspects of human ecology—interactions among human populations and ecosystems in which they are a part. Instead, we shall concentrate on the human individual with specific emphasis on the manner in which the human body interacts with microorganisms. From birth until death, all human beings harbor an enormous number of microorganisms in or on tissues of their bodies. Furthermore, microorganisms are present in the air we breathe, the water we drink, and the food we eat.

The human body can be viewed as an open ecosystem that functions in nature under steady-state conditions. Like all living systems, the human body is referred to as "open" because energy and materials are exchanged between its internal and external environments (i.e., nutrient materials enter from the external environment and waste materials are released to it). Living systems also function in steady-state conditions, because their internal environments remain relatively constant. In human beings the internal environment is the fluid that surrounds the cells that constitute the various tissues and organs of the body. In spite of the fact that food enters and waste leaves intermittently, the internal environment remains in a dynamic state of equilibrium. Exchanges of energy and materials between the internal environment and the outside world are mediated through processes which involve the skin, the respiratory tract, the gastrointestinal tract, and the genitourinary tract. Those structures also have surfaces that are highly suitable for microbial colonization.

Inhabitable anatomical regions

Normally, the human body encounters microorganisms for the first time during the process of birth; infants born naturally acquire them from their mother's birth canal (vaginal orifice), and infants born by cesarean section (surgical removal from the uterus) acquire them from contact with environmental sources (the air, clothing, personnel, etc.). In either instance, the skin of infants becomes colonized by microorganisms during or immediately following birth. Progressively, as infants develop, the mucosal surfaces of the respiratory, gastrointestinal, and genitourinary tracts become colonized with microorganisms. However, the structures above are not colonized uniformly with a homogeneous microflora, because environmental conditions (pH, temperature, availability of gases, nutrient concentration, etc.) vary from one part of the body to another. For this reason, regions of the body that are accessible to microorganisms can be considered as separate ecosystems (i.e., the skin and mucosal surfaces of the respiratory, gastrointestinal, and genitourinary tracts). In the absence of infections, entry of microorganisms to other regions of the body is obstructed by anatomical barriers and physiological mechanisms.

The skin

The outermost part of the human body is the *skin*. It expands over the entire surface of the body and extends into the openings of the body's orifices. Unless damaged, it is an effective barrier that protects the interior of the body from direct exposure to toxic substances and microorganisms. The skin should not be regarded as an inert veneer or shield for the body. It is the body's largest organ, and it carries out many vital functions. Incorporated within the skin are mechanisms that enable the body to communicate with the outside world — thermal regulators and sensory nerve endings are among them. Furthermore, the skin is not a homogeneous structure. Its characteristics and functions vary from one part of the body to another, and the skin's microflora reflect these variations. For this reason we should have some understanding of the skin's structure prior to our consideration of it as an ecosystem.

The skin is composed of two distinct layers (Figure 7.1). The outer layer is the *epidermis* and the inner layer is the *dermis*. Beneath the dermis is subcutaneous fat. The epidermis is composed of a thin layer of epithelial cells, and they are held together tightly by an intracellular cement. There are no blood vessels in the epidermis. Therefore, the outermost layer of cells receive little nourishment and die rapidly. However, they live long enough to produce a protein called *keratin,* which accumulates and makes them tough. These dead cells are collec-

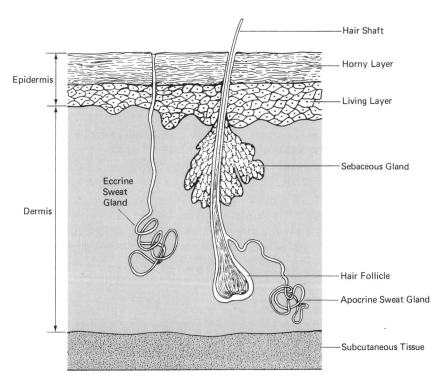

Figure 7.1.   Schematic view of a section through human skin.

tively called the *horny layer* (stratum corneum) and form the body's major protective covering. The dead cells that make up the horny layer are continually being removed from the body. Simultaneously, the epithelial cells are being renewed by rapidly dividing cells near the dermis. This process continually pushes cells in the horny layer outward.

The dermis or lower layer of the skin is composed of connective tissue and is the active portion of the skin. Embedded within the dermis are blood vessels, nerves, muscles, hair follicles, and glands (sweat, sebaceous, and mammary). Of particular interest is the fact that the glands vary considerably with respect to their distribution on the body and the composition of secretions.

Sweat glands are of two types: eccrine and apocrine. *Eccrine glands* are widely distributed over the body and their secretions pass through ducts which open on the outer surface of the epidermis. Secretory products from eccrine glands are primarily salty water, but small quantities of organic materials may be present. Those present on the palms of the hand and soles of the feet are under psychic control and are active at times of emotional stress. Those present on other parts of the body are under thermal control and are extremely active when temperatures increase. *Apocrine glands* are present only in specific regions of the body: areas of armpits, nipples, navel, anus, external genitalia, and external ear and nasal passages. Ducts from apocrine glands do not lead to the outer surface of the epidermis, but open into canals from which hair grows. Apocrine glands are not heat-sensitive; they are entirely controlled by psychic factors. Secretions from these glands are organically rich. Proteins, sugars, and lipids are present in them. As a result of such products, secretions (sweat) from apocrine glands have a distinct odor of volatile fatty acids. In the absence of frequent bathing, odors result from putrefaction (decay), microbial degradation of secretory products. In modern societies such odors are reduced considerably by the use of deodorants.

Sebaceous glands are usually associated with hairs. Their activity seems to be entirely controlled by hormones. Oily secretions from them are rich in waxes, fatty acids, cholesterol, and cellular debris, collectively called *sebum*. Ear wax is an example. When secretory products from sebaceous glands accumulate, especially during puberty, we recognize the effect as acne.

Mammary glands secrete milk through ducts which open on the outer epidermis and they are controlled entirely by hormones. Mammary-gland development in females corresponds to hormonal changes at the onset of puberty, and secretory functions correspond to hormonal changes that occur during pregnancy.

Hair, an outgrowth of skin, forms in the dermis. The shaft that extends from the epidermis is dead, but new hair is constantly being produced from underlying cells in a pit at the base of the shaft (follicle) which push the older keratinized shaft outward. In contrast to other mammals, the skin of modern human beings is sparsely covered with hair. However, hair that grows in certain regions of the body plays important protective roles. For example, orifices of the nostrils and ears contain hairs that trap foreign materials and obstruct their entry into our bodies through those passageways. Hairs that surround the anogenital orifices function in a like manner. Also, eyebrows and eyelashes help to keep foreign materials from entering our eyes.

In view of these anatomical and physiological considerations, human skin can be visualized as a unique structure, composed of a number of widely diverse microhabitats in which microorganisms can live luxuriously. The environmental conditions that prevail in those microhabitats tend to be highly selective for chemoorganotrophs, microorganisms that obtain their energy and carbon from preformed organic materials. However, a great deal of diversity can be found among organisms within this physiological group. Thus, the skin microflora is extremely varied, both qualitatively and quantitatively.

In general, the most prevalent members of the skin microflora are aerobic bacteria, and the staphylococci are probably the best known representatives. One member of this group *(Staphylococcus aureus)* is often associated with a number of skin infections (boils and carbuncles). Numerous other aerobes also live on the skin, but they occur less frequently. The skin microflora of the face often reflects types of aerobes that normally inhabit the mouth, nose, and throat. Organisms from the gastrointestinal tract often exist in large numbers on the skin in the anogenital region of the body. Similarly, a very distinct microflora exists in areas of the body that are hairy and/or moist (armpits, groin, and between the toes). Yeasts and molds are common in such areas. For example, the custom of wearing shoes with heavy socks keeps the skin between the toes moist, a factor that contributes to the frequency of "athlete's foot" (a skin infection caused by fungi).

Although the skin is continually exposed to air, its microenvironments are not entirely aerobic. The anaerobic *Corynebacterium acnes* (often called the acne bacillus) lives luxuriously in the anaerobic zone of hair follicles. In that region secretions from sebaceous glands are abundant, and these organisms may contribute to the development of acne lesions, common during puberty, when the sebaceous glands are extremely active. In most instances, the skin microflora will reflect the habits and activities of the individual and the overall general state of health. Although never sterile, the skin microflora is in a constant state of flux. Surface microbes are removed normally with the flaking of dead cells from the horny layer, by washing, and by other abrasive activities.

The gastrointestinal tract

The human gastrointestinal tract, like the skin, is an interface through which the body interacts with the outside world. It is composed of the mouth, esophagus, stomach, small intestine, large intestine, and the anus. Collectively, those structures form a continuous tube that passes through the body, extending from the mouth on one end to the anus on the other (Figure 7.2). However, the gastrointes-

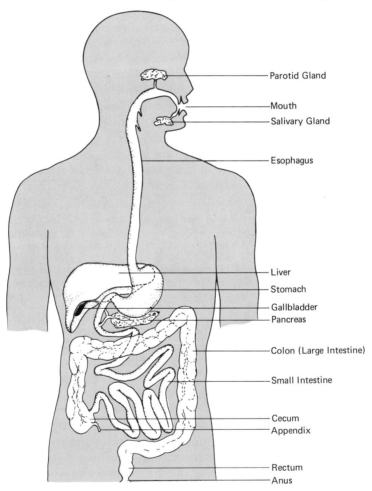

Parotid Gland
Mouth
Salivary Gland
Esophagus
Liver
Stomach
Gallbladder
Pancreas
Colon (Large Intestine)
Small Intestine
Cecum
Appendix
Rectum
Anus

Figure 7.2. Schematic view of the human gastrointestinal tract.

tinal tract is only a portion of the digestive system. Other associated components are the salivary glands, the gallbladder, the liver, and the pancreas, all of which pour digestive juices into the tube.

Some familarity with the overall functioning of the digestive system is essential to a discussion of the gastrointestinal tract as a microbial ecosystem. The major digestive processes involve hydrolytic reactions; and as a result of them, macromolecules (carbohydrates, fats, proteins) are enzymatically broken down into their constituent subunits (see Chapter 2). Then the subunits are absorbed by cells within the mucosal lining of the gastrointestinal tract.

Food materials and a host of microorganisms enter the gastrointestinal tract through the mouth. While food is being chewed, enzymes released from the salivary glands initiate the digestive processes. After being swallowed, food passes through the esophagus and into the stomach. Proteolytic enzymes are extremely active in the stomach and, when coupled with a high acidity, food materials are rapidly degraded. Subsequently, through muscular contractions or *peristalsis,* the partially digested food is forced into the small intestine, where digestive processes are completed. As a result of many coils and folds, the small intestine has a considerable amount of surface area in contact with nutrients. Furthermore, its mucosal lining is covered with extremely small fingerlike projections called *villi.* Cells within the villi actually absorb the nutrients. From those cells, nutrients enter the body's circulatory system. Then the residual materials pass into the distal portion of the large intestine (called the rectum) until removed from the body as feces.

While food is undergoing the various degradative processes as it traverses the body, microorganisms are also being dispersed in the gastrointestinal tract. Intermittently, large numbers of them are released from the body in feces (called fecal microflora), but others tend to become colonized in specific regions. At birth, the human gastrointestinal tract and the initial colonizers (pioneers) reflect the early feeding habits. In breast-fed infants, the pioneers are organisms from the mother's skin microflora. In bottle-fed infants, the pioneering populations are more heterogeneous. In either case, the gastrointestinal tract becomes contaminated with microorganisms within a short time following birth. Although the populations are mixed, many factors tend to influence colonization. For this reason, the intestinal microflora varies with age, diet, eating habits, and the overall state of an individual's health. Furthermore, microorganisms tend to colonize the various regions of the gastrointestinal tract selectively.

Although large numbers of microorganisms enter the gastrointestinal tract through the mouth, this region has a unique microflora. While functioning as the major portal of entry to the gastrointestinal tract, the mouth is also connected to the respiratory system. Yet, within the mouth and nasopharynx, microhabitats vary considerably, and their microenvironments tend to be highly selective for certain kinds of microorganisms. In predentulous (before the development of teeth) healthy infants, members of the streptococci, staphylococci, and yeast are commonly present. After the eruption of teeth, and with increasing age, kinds of microorganisms that reside in the mouth tend to become more complex. In addition to the groups above, the adult mouth and nasopharynx harbor several kinds of anaerobes (members of the lactobacilli), *Streptococcus pneumoniae,* and *Corynebacterium diphtheriae.* The latter two organisms are of special interest, because serious diseases are often associated with them. *Streptococcus pneumoniae* often causes bacterial pneumonia, and lysogenic strains of *C. diphtheriae* (previously discussed in Chapter 5) are responsible for producing diphtheria.

The esophagus, stomach, and small intestine harbor few microorganisms when food material is not present in them, but a large number of microorganisms are constantly present in the large intestine. In that region, populations are extremely heterogeneous and include both aerobic and anaerobic types. *Escherichia coli* is one of the best-known members of this group. The group of bacteria that live in the large intestine are often referred to collectively as *enterics.* Within the group two genera *(Salmonella* and *Shigella)* are extremely important, because they contain types that are often associated with food-poisoning episodes and outbreaks of bacillary dysentery. The large intestine also harbors a diverse group of anaerobes. Some of the most prevalent types of anaerobes are species within two genera: *Clostridium* and *Bacteroides.*

Actually, microorganisms that represent any of the microbial groups may be found in the large intestine at one time or another, some as transients and others as members of the established microflora. Interactions that occur among the latter and their significance will be discussed in a subsequent section.

Miscellaneous areas

Areas of the human body where microorganisms live in large numbers may be considered as reservoirs from which other areas of the body can become contaminated. Thus, in the absence of infections, the external orifices of the ear reflect the skin microflora of the head and face. In a like manner, the microflora of the external genitalia usually reflect types of microorganisms that escape from the gastrointestinal tract and types that are commonly present on the skin of the anogenital region. However, members of the *Mycoplasma,* and yeasts in the genus

Candida, are somewhat more prevalent in the anogenital region. In females, the flagellated protozoan *Trichomonas vaginalis* is a common inhabitant of the external genitalia, and it often causes vaginitis. Areas of the body not mentioned above have environments that are usually hostile to microorganisms.

The indigenous microflora

The indigenous or *autochthonous* microflora are those types of microorganisms that live on or in the tissues of the body in a permanent status, and is composed of organisms that proliferate (grow) optimally in the microenvironments of the healthy human body. On the other hand, microorganisms that reside on or in tissues of the body in a temporary status are called *transients* or the *allochthonous* microflora. Although our association with the indigenous microflora is continuous, it is also dynamic. Some members of the indigenous microflora are beneficial to us, some are indifferent to us, and under certain conditions, some are harmful to us. Thus, our close and continuous association with the indigenous microflora may be mutualistic, commensalistic, or parasitic.

In spite of the fact that the indigenous microflora interact dynamically with the human body, those microorganisms also interact constantly among themselves to maintain a balanced ecosystem.

Normal ecological niches

Previously, we characterized ecological niches as the functional aspects of microorganisms in nature, as opposed to habitats that are places of residence. In other words, when we speak of "normal" ecological niches, we are concerned with the activities of the indigenous microflora.

The indigenous microflora play an important role in the prevention of certain infections because of the phenomenon of microbial interference. Mere colonization of a given region confers a specific advantage to the pioneering population. After becoming established, pioneers are better able to interact with subsequent invaders. Of significance is the fact that the indigenous microflora tend to function as regulators, thereby maintaining an ecological balance between and among the various populations of microorganisms that live on and in tissues of our bodies. Many observations support this contention. In Chapter 4 we described conditions under which harmless strains of staphylococci protected the surface of skin on newborn infants from subsequent colonization by disease-producing strains. In a similar manner, the indigenous microflora of the gastrointestinal tract tend to keep organisms that have pathogenic potential from overpopulating the region. When the body's indigenous microflora are altered, serious infections often develop from organisms that usually live as commensals. For example, prolonged therapy with the broad-spectrum antibiotics often causes a marked reduction in the indigenous microflora of the gastrointestinal tract. Under such conditions, antibiotic-sensitive types are suppressed, and antibiotic-resistant types increase and produce infections. Complications that often result from the yeast *(Candida albicans)* and the bacterium *(Pseudomonas aeruginosa)* are typical examples. Such organisms are often referred to as "opportunists," because they are usually harmless unless the ecological balance is altered in some way.

The human indigenous microflora is also beneficial in other ways. Certain species that live in the large intestine synthesize vitamin K, a blood-clotting factor that cannot be synthesized by mammals. Other types of microorganisms synthesize micronutrients (biotin, riboflavin, etc.) that the human body uses in normal metabolic processes.

In recent years, a great deal has been learned about the activities of the gastrointestinal microflora of experimental animals. The indigenous microflora is believed to play a role in maintaining the normal structure of the intestines. Several investigators have observed changes in the size of the cecum in mice when the indigenous flora is altered by the administration of various drugs. Also the composition of microbial communities and their location within the gastrointestinal tract of experimental animals have been ascertained from studies with the scanning electron microscope. Specific microbe–epithelial attachment sites have been revealed (Figure 7.3). The populations are heterogeneous and seem to be associated with materials on the surface of each other and on the surface of the host epithelium.

Members of the indigenous microflora may also enhance processes that do not contribute to the welfare of the host. This is especially true with oral streptococci. Different species are known for their ability to preferentially colonize specific microhabitats within the human mouth. Environmental factors that influence colonization are age, diet, and the presence of teeth. *Streptococcus mutans* favors the tooth surface, and *Streptococcus salivarius* favors the epithelial surface of the tongue. The latter is among the first bacteria to colonize the mouth of infants, an observation that correlates with its affinity for sites on epithelial surfaces. Colonization of teeth by certain kinds of streptococci is a prerequisite to the development of dental caries, tooth decay. A point of interest is the fact that different strains of streptococci vary in their ability to adhere to surfaces of structures within the mouth. Those types that do not become attached to surfaces (teeth and/or epithelial

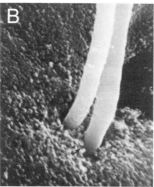

Figure 7.3. Scanning electron micrographs of microorganisms in the intestinal tract of the rat. (A) Rod-shaped bacteria can be seen attached to the keratin layer by their ends and sides. A few cocci are also visible 2,700×. (B) Shows the attachment of two filamentous microbes to the rat ileal epithelial cells by their ends 9,000×. [**Courtesy of C. P. Davis,** *Appl. Environ. Microbiol.* **31: 304, 1976, with permission.**]

cells) are simply washed out of the mouth by fluids and swallowed. The adhesion between bacteria and teeth or between bacteria and epithelial cells involves interactions among components present on the bacterial cell and components of the surfaces. Dietary sucrose is known to influence the ability of certain streptococci to colonize teeth surfaces. In the absence of dietary sucrose, *Streptococcus mutans* is generally unable to attach to teeth. Sucrose in the mouth is the substrate from which *S. mutans* synthesizes extracellular polysaccharides, a substance that facilitates bacterial attachment to teeth surfaces. Alterations of either the bacterial cell or attachment sites of oral surfaces would most likely alter colonization.

Many of the indigenous streptococci have trypsin-sensitive outer layers which facilitate their attachment to surfaces (Figure 7.4). After removal of the trypsin-sensitive coat, the cell's ability to become attached to epithelial surfaces is lessened. Recent observations have shown that numbers of *Streptococcus miteor* present in the mouth seem to be related to their ability to attach to epithelial surfaces. This particular bacterium forms aggregates with salivary glycoproteins, a component that adheres both to tooth surfaces and to bacteria. Aggregates of this type are believed to influence the formation of *dental plaques,* complex matrices consisting of bacterial cells, salivary glycoproteins, and epithelial cells that adhere to tooth surfaces.

Dental caries, a related condition, results from activities of the indigenous microflora, specifically members of the lactobacilli and *S. mutans.* Simply stated, dental caries is a localized and progressive

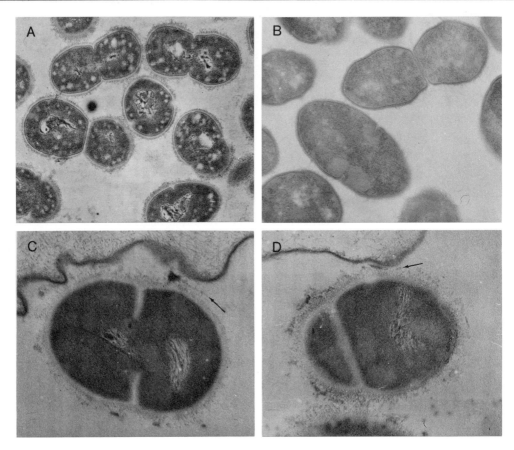

Figure 7.4. *Electron* micrograph of *Streptococcus miteor (mitis)* and its adherence to epithelial tissue. (A) Untreated bacterial cells with their highly visible trypsin-sensitive "fuzzy" surface coat. (B) *S. miteor* after trypsin treatment, which removed the fuzzy coat. (C) and (D) show the attachment of *S. miteor* cells to the epithelial membrane of germ-free rat cheek cells. It appears that the fuzzy coat mediates the attachment (arrows). [Courtesy of W. F. Liljemark, *Inf. Immunity* 6:852, 1972, with permission.]

deterioration of teeth as a consequence of bacterial action. If not corrected, it progressively destroys most of the tooth under attack, and produces an infection of tissues surrounding the tooth. During the process, bacteria ferment dietary carbohydrates, especially sucrose, and produce lactic acid, a substance that demineralizes the enamel surface of teeth. The condition is an undesirable ecological problem that involves several interconnected processes: specific members of the oral microflora, the presence of dietary sucrose, and adherence of bacteria to surfaces. As mentioned above, adherence is a key factor, without which bacteria are simply washed away and swallowed. However, the accumulation of products from bacterial utilization of sucrose facilitates colonization on surfaces. Indirectly, sucrose is the culprit. Thus, dental caries can be prevented by elimination of dietary sucrose, by brushing teeth, and by using dental floss.

In addition to the benefits previously mentioned, some members of the indigenous flora appear to enhance the body's defense against bacterial infections by providing a continuous antigenic stimulus when they accidentally enter the bloodstream (aspects of which will be discussed later).

Sterile (forbidden) zones

The indigenous microflora live in and on surfaces that are directly or indirectly exposed to the outside world and are restricted to such regions by the body's anatomical and physiological mechanisms. Thus, certain regions or zones of the healthy individual are considered to be sterile. They are the blood, spinal fluid, the deeper portions of the respiratory tract, the internal portions of the genitourinary tract, the interior portions of the reproductive system, the peritoneum cavity, the sinuses, and the interior region of other body tissues. These areas are not actually sterile, however. Occasionally, microorganisms do enter them through accidental openings, but the body's defense mechanisms interact with them in an adverse manner. In other words, these environments are hostile toward microbes, and we refer to them as *forbidden zones*. When microbes successfully colonize them, the body is adversely affected and is recognized to be in a diseased state. Under certain conditions any microorganism may interact with the body in a parasitic manner, but those that have the potential to produce disease in healthy individuals we call pathogens.

Defense mechanisms

The human body is well equipped with a defense system to protect it from hostile forces in the environment, among which are toxic substances and microorganisms. The system is composed of both mechanical and physiological mechanisms. In healthy humans, the various components of the defense system function in a very effective manner, although their effectiveness is related to a number of factors: age, nutrition, previous exposure to specific agents, and the overall state of an individual's health. When the system is impaired or overpowered in some way, the body becomes compromised, and must rely on externally induced mechanisms for protection.

Mechanical barriers

Healthy intact skin and its associated structures function as *mechanical barriers* to prevent foreign materials from entering the body. As mentioned previously, the entire surface of the body is covered by the skin, and orifices to the major passageways of the body are protected by skin appendages (hairs). In addition, secretions from sweat and sebaceous glands contain saturated and unsaturated fatty acids that will inhibit the growth of certain kinds of microorganisms. Other surface secretions (tears and saliva) contain lysozyme, an enzyme that disintegrates cell-wall components of some bacteria. Collectively, the intact skin and its associated components form the body's first line of defense against foreign materials.

Immune barriers

Immune barriers consist of several processes which become operative when certain kinds of foreign materials penetrate the body. These barriers are the body's reserve forces or second line of defense, and involve the action of cellular and soluble components of the blood. The cellular components are *white blood cells* or leukocytes, and the soluble components are *antibodies*. White blood cells function as scavengers to remove foreign materials from the body through a process called *phagocytosis*. Antibodies are formed in the soluble portion of the blood in response to the entry of a specific material, and for the purpose of neutralizing the harmful effects of that material.

Phagocytosis   The process of phagocytosis is generally nonspecific, because white blood cells engulf foreign materials in a nonselective manner. Inert particles and microorganisms are engulfed with equal efficiency by polymorphonuclear neutrophils, which are white blood cells with multilobed nuclei and a granulated cytoplasm. Such cells are referred to as *phagocytes* (Figure 7.5). They exhibit ameboid movements and engulf foreign materials by the action of cytoplasmic projections, the pseudopodia, which completely encircle the particles.

The phagocytic process is initiated by the entry of foreign materials. Immediately after the penetration of the initial barriers by foreign materials, phagocytes become mobilized and migrate to the site of entry, in which state they are called *wandering phagocytes*. Processes through which they destroy microorganisms involve several complex mechanisms: contact, ingestion, killing, and digestion. Of major importance is the attachment of phagocytes to bacterial cells. Rough surfaces facilitate the process. This accounts for the fact that most nonencapsulated microorganisms are readily destroyed by phagocytes. There-

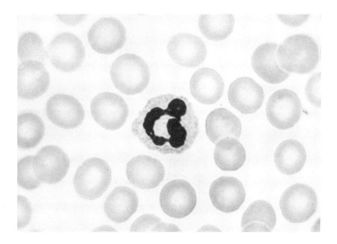

Figure 7.5. Normal white blood cell or leukocyte. Note the multilobed nucleus. [Courtesy of the Carolina Biological Supply Company.]

fore, low virulence is usually associated with them, because they are destroyed and digested by phagocytic enzymes. For example, nonencapsulated types of *Streptococcus pneumoniae* are avirulent, whereas encapsulated strains of the same species produce pneumonia, because they have smooth surfaces and are more difficult to engulf. It is important to note that some organisms can avoid being destroyed by phagocytes even after they have been engulfed. They have mechanisms which enable them to live within the cytoplasm of the phagocyte and for this reason are called *intracellular parasites*. Bacteria with this capability are often responsible for chronic diseases. Tuberculosis is a typical example, because the bacterium *(Mycobacterium tuberculosis)* lives within the cytoplasm of polymorphonuclear neutrophils.

The body is also equipped with *nonwandering phagocytes*. They are located (fixed) within the tissues of the liver, bone marrow, spleen, lymph nodes, and the linings of other vessels. Although fixed, they engulf and destroy microbes in the same manner as wandering phagocytes. In these various tissues, fixed phagocytes act as scavengers, clearing the blood of debris and foreign materials as it circulates through the body.

Antibodies   One of the remarkable features of the body is its ability to discriminate between self and nonself. Within the nonself category are all microorganisms and numerous kinds of other materials. When materials (microorganisms or other substances) from the nonself category enter body tissues, they are immediately recognized as being foreign, and the body responds by synthesizing substances that will neutralize their harmful effects. Materials that the body recognizes as being foreign are called *antigens,* and substances produced by the body to neutralize them are called *antibodies.* Antigens vary widely in terms of chemical composition and structural configuration. Thus, various constituents of a bacterial cell are antigens (e.g., flagella, cell-wall polysaccharide, capsular materials). Such materials are actually composed of a variable number of antigenic sites, which gives them their specificity. Antibodies are highly specific in their activity, and will interact with only one type of antigen or with several similar antigens.

Antibodies are synthesized through complex mechanisms within lymphoid tissues and are secreted by plasma cells. Following an antigenic stimulus, plasma cells multiply rapidly and actively secrete antibodies, soluble proteinaceous substances that circulate in the blood. Antibodies belong to a special class of proteins called *immunoglobulins* (Ig). Five classes of immunoglobulins have been recognized, and each class is believed to carry out specific functions (Table 7.1). IgM antibodies are secreted first, but activity from this class of antibodies is of relatively short duration. Next, IgG antibodies appear. These are referred to as the classical antibody, comprising about 80 to 85 per cent of the total immunoglobulins formed in response to an antigenic stimulus. IgG antibodies also play an active role in protecting newborn infants, because they can pass through the placental wall from the mother to the developing infant prior to birth. IgA antibodies are present both in serum and in mucous secretions (saliva and mucosal linings) in the gastrointestinal and genitourinary tracts. As a result of its presence in secretions, this class of antibodies contributes to the body's defense against microbes that enter through the body's major passageways. IgE antibodies exhibit behavior distinctly different from the other classes of antibodies. They are responsible for the numerous types of human allergies: to hair, dust, foods, ragweed and other pollens, and on and on. Those individuals who experience allergies of one type or another have higher concentrations of IgE antibodies than nonallergic individuals. The functional role of IgD antibodies has not been clearly defined.

It is important to remember that humans do not have circulating antibodies (except those that were acquired from the mother through

Table 7.1.   Classes of Immunoglogulins

Class	Representative Function
IgM	First to appear after stimulus (bactericidal)
IgG	Protects newborn infants
IgA	Protects membranes (present in mucous secretions)
IgE	Responsible for allergies
IgD	Ill-defined

the placental barrier) for a particular antigen until after exposure to the particular antigenic substance. Furthermore, an individual's response to an antigenic stimulus is influenced by a host of factors: age, nutrition, general state of health, presence of drugs in the body, the nature of the antigenic material, concentration of the antigenic material, entry route of the antigenic material, and the number and frequency of exposure to the antigenic material. As a result of such influences, antibody production in response to a specific antigenic stimulus varies considerably from one individual to another. A generalized version of the primary and secondary response to an antigen is shown in Figure 7.6. The primary response is characterized by the initial synthesis of antibodies after the first stimulus from a single kind of antigenic material. The period of antibody synthesis varies and is influenced by the factors mentioned, but maximum synthesis may occur within a few hours or it may take several days. After this, antibody degradation occurs (a natural process), and the level of antibodies in the blood declines. On subsequent exposure to the antigen that provoked the *primary response,* the body responds by synthesizing antibodies at a faster rate. The maximum concentration of antibodies is higher, and they remain circulating in the blood for a longer period. This is called the *secondary response.*

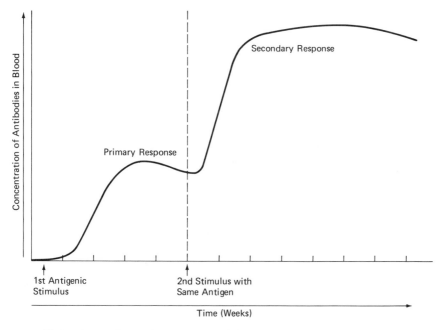

Figure 7.6.   Generalized version of the primary and secondary immune response to a single antigen.

Types of immunity

Immunity is a state or condition in which human beings and other animals are protected in varying degrees from the harmful effects of certain microorganisms and/or their toxic products. In other words, immunity is a relative state of resistance to infectious agents, ranging from a highly immune state on one extreme to a nonimmune state on the other. Furthermore, immunity may occur as a result of natural phenomena (e.g., possession of antibodies without prior stimulation); it may occur from the body's synthesis of antibodies in response to stimulation, or it may occur as a result of the artificial administration of preformed antibodies.

Natural immunity The state of *natural immunity,* which could also be referred to as *nonsusceptibility,* is conferred upon human beings and other animals by physiological and anatomical features peculiar to the species. In other words, natural immunity is an inheritable state of resistance. Human beings are naturally immune to many infectious diseases of lower animals. For example, the infectious agents of hog cholera and canine distemper do not cause similar diseases in people. Similarly, animals do not develop typhoid fever or gonorrhea. In another sense, natural immunity is species resistance. Probably during the course of evolution, various phylogenic groups emerged with physiological differences that are responsible for natural immunity.

Acquired immunity *Acquired immunity* is either active or passive. *Active immunity* results from an individual's own production of antibodies in response to an antigenic stimulus. In many instances such stimuli may occur during a natural infection, and after recovery the blood contains antibodies to the particular infectious agent. Thus, the individual in such an instance would be actively immune to that infectious agent, although the infection occurred naturally. Two examples of diseases from which long-lasting immunity can be actively acquired are German measles and typhoid fever. The former is caused by a virus; the latter, by a bacterium.

In modern medicine, the artificial introduction of antigenic materials into the body, called *vaccination,* is a common practice. A *vaccine* may consist of live but weakened (attenuated) infectious agents, dead microbial cells, or inactivated microbial toxins (toxoids). Regardless of vaccine type, the vaccinated individual's body will respond by producing antibodies to the material. Subsequently, when exposed to the infectious agent, the individual will be immune or protected to some degree. In other words a vaccination simulates a natural infection and often results in long-lasting immunity.

In addition to these methods of acquiring immunity, human beings may become passively immunized through two processes. A natural process through which an individual can become immunized is the passage of antibodies (IgG) from the mother through the placental wall to the developing infant. Thus, the infant acquires protecting antibodies *passively* from the mother.

Human beings may also be immunized *passively* with preformed antibodies. In such instances the antibodies were originally obtained from another human being or animal, prepared into a vaccine, and administered to the first person. Thus, the vaccinated individual obtains antibodies that give protection to the original antigen that stimulated their production. The widely used tetanus and rabies vaccines consist of preformed antibodies. One major advantage of artificial passive immunization is to effect immediate protection to the individual by the administration of potent preformed antibodies.

Germ-free animals

What are the consequences of being born into an environment free of microorganisms? On first thought it might appear to be an ideal situation for one to live in such environment; but apparently such conditions would be far from ideal. Current research findings suggest that germ-free animals have many features (anatomical and physiological) that differ from other members of their species, called "conventional." As we have learned, healthy conventional animals contact microorganisms for the first time while being born and live throughout life in an intimate relationship with these microorganisms.

Techniques and procedures for obtaining, rearing, and studying germ-free animals are well established. It is not uncommon to find college students conducting research with germ-free small animals, because both the animals and the isolators can be purchased commercially. A typical germ-free isolator is shown in Figure 7.7. Regardless of whether it is used for small or large animals, the isolator and all items that enter it must be sterile. Air, water, food, and all other materials introduced into the chamber must be sterile. Furthermore, all personnel must use sterile procedures when working with germ-free animals.

A colony can be started by purchasing animals from an established germ-free colony (generally available from an animal supply house) or by obtaining an infant, just prior to birth, through a cesarean operation. Subsequently, with proper management, germ-free colonies can be maintained for long periods. Examples of animals that have been successfully reared under germ-free conditions are mice, rats, guinea pigs, rabbits, pigs, monkeys, and chickens.

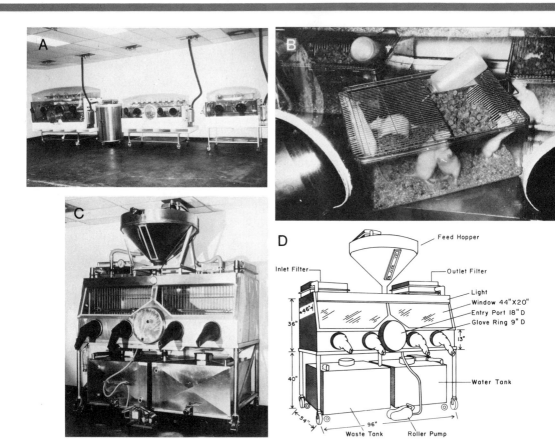

Figure 7.7. (A) Three small-animal germ-free flexible film isolators; (B) Twelve-week-old germ-free Sprague-Dawley rats; (C) a large-animal stainless steel isolator; (D) schematic view of a large-animal stainless steel isolator. [Courtesy of E. Balish and C. E. Yale.]

In general, germ-free animals have characteristics that are distinctly attributable to their lack of contact with microorganisms: (1) their antibody-forming system is undeveloped (2) the cecum (that portion of the intestinal tract that harbors large numbers of microorganisms) is often distorted, and (3) they exhibit a dietary requirement for vitamin K. Conventional animals have vitamin K supplied to them by commensalistic bacteria that live in their gastrointestinal tract.

We should mention the fact that a great deal of new knowledge about the immune response to microorganisms was derived from the study of germ-free animals. Such studies involve contaminating germ-free animals as desired with microorganisms to which they have had no prior contact. These studies are called *gnotobiotic*, because the investigator knows the type or types of microorganisms that the animal is being exposed to.

As a result of many studies with germ-free animals, we have un-equivocal evidence that microorganisms in our environment con-tribute significantly to the well-being of human beings and other animals.

Key Words

antibody A soluble proteinaceous substance produced in the body tissues of animals in response to the introduction of an antigen; will combine specifically with the antigen that stimulated its formation.

antigen Any substance (usually foreign) that when introduced into the body of an animal will stimulate the formation of specific antibodies.

bone marrow The central tissues within bones that give rise to blood cells.

endotoxin Any toxins contained within the interior of cells that produce them; or that are an integral constituent of cellular structures and are not released until the cell disintegrates.

exotoxin Any toxin produced within cells but excreted from intact cells into the surrounding environment.

gamma globulin The protein fraction of blood serum that contains antibodies.

hydrolytic reaction Pertains to the addition or insertion of a water molecule at the site where enzymes catalytically break chemical bonds.

immunoglobulin (Ig) The specific class of proteins that contains antibodies.

pathogen Any organism or agent that has the ability to produce or cause a disease.

phagocyte A specific kind of white blood cell capable of ingesting foreign material.

phagocytosis The process through which phagocytes ingest and/or destroy microorganisms and foreign material.

proteolytic enzyme An enzyme capable of catalyzing reactions that degrade proteins.

toxin A substance produced by living organisms that produces injury to tissues and/or alters the functions of another organism.

trypsin A specific kind of proteolytic enzyme.

Selected Readings

1. Cooper, M. D., and A. R. Lawton III. 1974. The development of the immune system. *Scientific American* 231 (No. 5):59–72.
2. Finland, M. 1970. Changing ecology of bacterial infections as related to antibacterial therapy. *J. Inf. Dis.* 122 (No. 5): 419–431.
3. Marples, M. F. 1969. Life on the human skin. *Scientific American* 220 (No. 1): 108–115.
4. Montagna, W. 1965. The skin. *Scientific American* 212 (No. 2): 56–66.
5. Poole, D. F. G., and H. N. Newman. 1971. Dental plaque and oral health. *Nature* 224:329–331.
6. Prier, J. E. and H. Friedman. 1974. *Opportunistic Pathogens.* Baltimore, Md.: University Park Press.
7. Rosebury, T. 1970. *Life on Man,* New York: Berkley Publishing Corporation.
8. Savage, D. C., and R. V. H. Blumershine. 1974. Surface associations in microbial communities populating epithelial habitats in the murine gastrointestinal ecosystem: scanning electron microscopy. *Inf. Immunity* 10:240–250.

CHAPTER 8

Epidemiology of human microbial diseases

- **Sources of environmental pathogens**
    Living reservoirs
    Inanimate reservoirs
- **Modes of transmission for pathogens**
    Direct transmission
    Indirect transmission
        Vehicle-borne
        Vector-borne
        Airborne
- **Epidemiological investigations**
    Retrospective studies
    Prospective studies
- **Human infectious diseases**
    Selected bacterial diseases
        Boils and carbuncles
        "Strep" sore throat
        Pneumonia
        Tuberculosis
        Typhoid fever
        Cholera
    Selected viral diseases
        Smallpox
        Polio
        Measles
        Hepatitis
        Influenza
    Selected fungal (mycotic) diseases
        Dermatomycoses
        Systemic mycoses
    Selected protozoan diseases
        Amebiasis
        Giardiasis
    Selected venereal diseases
        Gonorrhea
        Syphilis
- **Prevention and control of infectious diseases**
    Chemotherapy
    Prophylactic immunization
    Environmental sanitation
- **Key words**
- **Selected readings**

In communities of wild animals, infectious agents are natural forces that contribute to maintaining the balance of nature by eliminating the weak or less-fit individuals from within populations. Consequently, only the stronger members survive and continue to reproduce. Historical evidence suggests that agents responsible for human diseases tend to play similar roles. Records of the great epidemics (bubonic plague in the Middle Ages; cholera in 1850–1860; and influenza in 1918–1919) attest to the fact that infectious agents have the ability to behave as lethal weapons, especially when widely dispersed among susceptible individuals. To a great extent, modern societies have learned ways to minimize the devastating effects of such episodes, but we have not succeeded in preventing them.

Our present understanding of infectious diseases and methods for controlling them have emerged from the study of epidemiology. *Epidemiology* is concerned with the determinants of diseases and with factors that influence their frequency and distribution within populations. The severity and distribution of any infectious disease are influenced by the manner in which the agent, the host, and factors within the environment interact (Figure 8.1). In order to understand such interrelationships, each component in the triangle of causation must be considered separately. In this regard we must consider the virulence of the agent, host susceptibility, and environmental factors that influence them. Interactions that were discussed in previous chapters are basic to our discussion of epidemiological relationships. In this chapter we shall emphasize those kinds of microbial diseases that are common among human beings.

Epidemiology originated from the study of disease outbreaks or

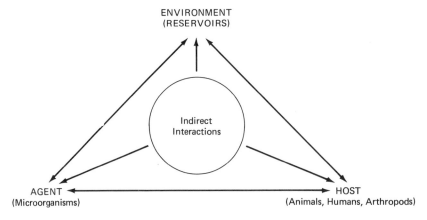

Figure 8.1. Generalized version of the triangle of causation for any kind of communicable disease.

epidemics among human populations. The term *epidemic* is used to refer to the sudden occurrence of a specific disease in large numbers of individuals within a population of human beings over a limited time period. It is important to note that numbers of diseased individuals are referred to in that definition in a relative sense, because a sudden occurrence of a specific disease in only a few individuals may constitute an epidemic if the infectious agent is usually absent from that community. For example, smallpox is currently considered to be an obsolete disease within the United States, but when two or three individuals within the same community develop that disease within a limited time period, it constitutes an epidemic. Two other related terms are pertinent to this discussion: endemic and pandemic. *Endemic* conditions exist when a specific pathogen is harbored by a few individuals within a given geographical area at all times. The term *pandemic* is used to describe a specific disease when it occurs in a series of epidemics that transcend geographical boundaries within a given time period. The classical outbreak of influenza in 1918–1919, and subsequent episodes of that disease, were pandemics.

Sources of environmental pathogens

Previously we learned that microorganisms are the most versatile and most widely distributed kinds of living things. As such, they replicate themselves or merely survive in habitats that are extremely diverse (see Chapter 3). Thus, áll places that support the maintenance of viable microorganisms represent reservoirs from which they can be dispersed.

Living reservoirs

All forms of life (except those that are reared in germ-free environments) harbor microorganisms in and on tissues of their bodies. Consequently, they are living reservoirs from which microorganisms can be dispersed.

The most obvious reservoirs of pathogens are individuals with demonstrable symptoms of microbial diseases. From their bodies, virulent organisms are shed into the environment through various kinds of exudates. Subsequently, such virulent organisms can be transmitted directly or indirectly to susceptible hosts.

Although pathogens vary considerably with respect to the manner by which they actually cause injury to host tissues, the course of events that occur during the development of an infectious disease is fairly characteristic for all pathogens. The chain of events involves the following:

1. Escape from the living reservoir by portals of exit (i.e., body orifices, lesions, and secretions)
2. Entry into the host's external environment
3. Transmission (directly or indirectly) to a susceptible host
4. Entry to a host's tissue by invasion or through portals of entry (i.e., body orifices and lesions such as an injury or bite)
5. Establishment of disease in the new host

After a pathogen gains entry to the human body, competitive interactions are initiated between the body and the pathogen. If the body (by action of its defense mechanisms) wins the contest, no ill effects of the pathogen's presence will be experienced. On the other hand, if the pathogen wins the contest, infection or disease will develop. Infectious diseases are usually recognized by the onset of characteristic symptoms, and confirmed by the isolation and/or identification of the pathogen by appropriate laboratory techniques

When symptoms of a particular disease are apparent, control measures can be directed toward the infectious agent. However, some individuals harbor and shed disease-producing organisms without being affected by their presence. Such individuals are called *carriers*. An individual may also shed pathogens during the pre- and postsymptomatic stages of a disease. The former is called an *incubation carrier,* and the latter is known as the *convalescent carrier*. In all cases carriers represent living reservoirs that are difficult to control.

Inanimate reservoirs

The environment is laden with inanimate, nonliving, reservoirs of microorganisms. In addition to the nonbiological habitats described in Chapter 3, numerous other materials that are associated with human activities may serve as inanimate reservoirs. For example, all food items have associated with them an indigenous microflora. For this reason such items must be processed and/or stored in a manner to protect the consumer from pathogens that may be associated with them. In a like manner, human excreta and other kinds of contaminated materials must be disposed of properly and/or decontaminated in a sanitary manner because they represent potential reservoirs from which pathogens can be dispersed (see Chapter 11).

Modes of transmission for pathogens

In a broad sense we are concerned with dispersal mechanisms. This involves escape from microhabitats in reservoirs, transport to a susceptible host, and colonization in the new environment. These aspects were discussed previously, but we shall summarize them here as processes that involve direct and indirect transmissions.

Direct transmission

For some microorganisms, direct contact between a source and a susceptible host is essential for survival. The obligate intracellular parasites are in this group. Examples of direct contact methods through which infectious agents may be transferred are body contact with infected tissue (open sores and/or exudates), biting, licking, and exchange of mucous secretions through short distances (coughing and sneezing).

Indirect transmission

Vehicles, vectors, and air are all processes that mediate contact between a source of infection and a susceptible host. Through such processes microorganisms can actually be transported from one place to another.

Vehicle-borne   Vehicle-borne processes are those responsible for the transmission of pathogens from reservoirs to susceptible hosts by contact with contaminated inanimate objects (fomites). Food, water, and oral medications are three vehicles that can effectively transmit environmental pathogens from reservoir to susceptible host. During such processes, the infectious agent may or may not multiply.

Fresh meat obtained from healthy animals and raw vegetables prepared for human consumption are always contaminated with microorganisms. However, the level of contamination is usually low, and such products are not a threat to human health unless they are handled in an unsanitary manner and/or stored improperly prior to being eaten. Thus, it is imperative that food handlers exercise good personal hygienic practices at all times. Also, food products should be properly stored under refrigeration or preserved by other appropriate techniques.

Contaminated water can also serve as a vehicle for the transmission of pathogens. Consequently, water used in the preparation of various kinds of food products and for the cleaning of utensils should be free of pathogens and of drinking water quality (see Chapter 11).

Vector-borne   Most often the mediator or vector is an arthropod (e.g., tick, mite, flea, mosquito), but it can be any other kind of living organism. The organism may carry the infectious agent directly, on its feet, wings, or body, from reservoir to susceptible host. In such instances where multiplication of the pathogen does not occur, the arthropod is called a *mechanical vector*. The common housefly is a typical example of a mechanical vector. On the other hand, when the infectious agent multiplies within the body of the organism that transports it from source to susceptible host, the arthropod is called a *biological vector*. The female *Anopheles* mosquito, which transmits malaria, is an example of a biological vector.

Airborne   Many viable microorganisms are introduced into the air from environmental reservoirs as a result of activities that are associated with humans and other animals. Once airborne, the infectious agents can be transported for considerable distances as aerosols and droplet nuclei. *Aerosols* are moist suspensions of airborne materials. Normal activities of humans (coughing, sneezing, walking, dressing) generate aerosols. When such microbial-laden materials become dry and remain airborne, they are called *droplet nuclei*. Other sources of pathogenic aerosols are laboratories that handle infected materials, and slaughterhouses and meat-processing plants.

Epidemiological investigations

Epidemiological investigations are undertaken in an attempt to understand the relevant interactions between and among agent, host, and environment that contribute to the occurrence of a specific disease in a given population. Many widely differing kinds of epidemiological investigations may be undertaken to study the causation and prevention of a single kind of disease, but we can generally group them into two broad categories, retrospective studies and prospective studies.

Retrospective studies

In a *retrospective epidemiological study,* a population group that have been diagnosed as having a particular disease is compared with a population group that does not have that disease. The first group constitutes the *cases;* and the second, the *controls.* The purpose is to determine if the two population groups differ in respect to a specific variable. The variable to be studied should be selected with great care. Such a study is called retrospective because it compares cases and controls with regard to the presence of an element in their past experience. Definite criteria must be established in the experimental design to clearly define the population group under study with respect to age, sex, geographical location, and other relevant data. Then, careful consideration should be given to such information when selecting the control group, so that they are similar to the population group in terms of those characteristics. Information is then collected for the variable and analyzed by appropriate tests to determine if the two populations differ in respect to the variable. It is important to note that when a variable of any kind is used for matching the groups, its influence on the outcome of the disease cannot be investigated, because cases and controls will automatically be alike with respect to that characteristic.

Prospective studies

A *prospective epidemiological study* begins with a single population group in which every member is free of a particular disease, but certain individuals differ with regard to a particular variable. The group is then studied over a period of time to determine differences in the rate at which the disease develops in each member in relation to the variable being evaluated.

It is important to note that the essential difference between retrospective and prospective studies is not in the time sequence but in the way the groups are assembled. In retrospective studies, diseased (cases) and nondiseased (controls) members are selected and compared in respect to a variable factor. In prospective studies, we begin with a single population group in which individuals are free of the disease under consideration. They are classified as positive or negative for that variable and observed over a period of time to determine the development of the disease in each individual. It is important to remember that epidemiological data are extremely useful to the understanding of infectious diseases, but the information obtained from any study is related to the experimental design and the method of analyzing the results. When carefully planned and collected, epidemiological data in conjunction with clinical data enhance the researcher's ability to determine the nature of interrelationships that exist among the pathogen, the host, and environmental factors.

Human infectious diseases

Microorganisms that produce infectious diseases in humans are categorized as pathogens, because they have the ability to injure body tissues and/or alter body functions. In general, pathogenic microorganisms express their disease-producing properties through two kinds of mechanisms: (1) the invasion of tissues, and (2) the production of toxins. Such mechanisms enable pathogens to interact with susceptible host tissues in an adverse manner. In this regard all pathogens are parasites. Organisms that are *invasive* have the ability to penetrate and spread themselves within host tissues, and those that are *toxigenic* produce poisonous substances, toxins, that cause injury to host tissues. These pathogenic mechanisms are possessed by certain strains within all major groups of microorganisms. However, pathogenic microorganisms within the same genus or species vary in terms of their ability to produce infectious diseases. We use the term *virulence* to describe the degree of pathogenicity. Virulence is a quantitative term: some pathogens are slightly virulent, some are moderately virulent, and some are highly virulent.

Epidemics are usually caused by highly virulent pathogens, because the frequency of their transmission from one person to another maintains them in a state of high virulence. This phenomenon is well known and can be easily demonstrated. The virulence of many kinds of pathogens can be enhanced by the serial passage of the infectious agent through the body of susceptible laboratory animals. We see that high virulence and communicability are common features of infectious agents that cause epidemics, but we should remember that all infectious diseases are not communicable. *Communicable infectious diseases* are caused by pathogens that are transmissible (directly or indirectly) under natural conditions from an infected host to another host. *Noncommunicable infections* are not transmissible from an infected host to another; tetanus is such a disease. Tetanus is caused by an anaerobic spore-forming bacterium *(Clostridium tetani)* that normally inhabits the soil. The bacterium gains entrance to the body through breaks in the skin, usually during puncture wounds with objects contaminated with endospores of *Clostridium tetani*. Within the deep tissues of the body, where the oxygen tension is low, the spores germinate and multiply. Subsequently, the bacterium excretes a potent neurotoxin that produces the most characteristic symptom, violent spasmatic contractions of the voluntary muscles. Although this infectious disease may be severe, it is not transmissible from one infected person to another.

Pathogens do not unconditionally express their disease-producing potentials. Regardless of whether virulence is due to invasiveness or to toxigenicity, the pathogen must have in its surrounding environment essential factors for the organism's proliferation, and other essentials for the expression of virulence; the two requirements may be different. For example, virulent strains of *Corynebacterium diphtheriae* must be associated with a prophage (i.e., be in the lysogenic state), and its surrounding environment must contain a very low concentration of inorganic iron. If either of these requirements is negated, the organism cannot synthesize the toxin necessary to produce diphtheria.

Invasive pathogens have the ability to produce and excrete one or more kinds of extracellular enzymes (Table 8.1). As a result of this property, they cause injury to host tissues. In general, organisms that excrete invasive enzymes are better able to spread themselves within host tissues, and, after colonization, they may also produce toxins.

Pathogens that cause disease by producing toxins are many and varied. However, bacterial toxins that have been characterized are of two kinds: exotoxins and endotoxins. *Exotoxins* are produced within certain kinds of bacteria and excreted into their surrounding environment. They are proteins in chemical composition, and are relatively specific in terms of damage to the host. In terms of potency, some exotoxins are among the most powerful poisons known. *Endotoxins,* on the other hand, are complex polysaccharide cell-wall components of certain kinds of bacteria. They are not released until the cell disintegrates naturally or by artificial means. Although relatively heat-stable, endotoxins are less specific in their actions and less potent than exotoxins. Some selected toxigenic pathogens are listed in Table 8.2.

Infectious diseases caused by microorganisms are many and varied, and the discussion to follow does not pretend to be inclusive. Inclusion of a particular disease does not imply importance with respect to severity, but simply that the disease is common and representative of pathogens in the microbial world.

Infectious diseases can be discussed from many perspectives, among them: primary sites of infection (the skin, gastrointestinal tract, respiratory tract, genitourinary tract, etc.); primary modes of transmission (contact diseases, airborne diseases, foodborne diseases, waterborne diseases, etc.); or type of causative agent. Our discussion will emphasize the latter and will include an overview of the epide-

Table 8.1. Selected Invasive Pathogens

Bacterium	Representative Extracellular Enzyme	Damage to Host
Streptococcus pyogenes	β-Hemolysin	Causes lysis of red blood cells
Staphylococcus aureus	Coagulase	Causes blood plasma to clot
Clostridium perfringens	Collagenase	Disintegrates collagen in tissues
Streptococcus pneumoniae	Leukocidins	Destroys white blood cells (leukocytes)

Table 8.2.   Selected Toxigenic Pathogens

Bacterium	Toxin and/or Damage to Host
Exotoxin producers	
Clostridium botulinum	Neurotoxin (paralytic)
Shigella dysenteriae	Neurotoxin (paralytic)
Corynebacterium diphtheriae	Modifies enzymes
Endotoxin producers	
Salmonella typhi	Causes intestinal hemorrhage
Neisseria gonorrhoeae	Acts on mucosal membrane of genitourinary tract
Vibrio cholerae	Acts on intestinal mucosa

miological aspects of selected diseases caused by pathogens within the bacteria, viruses, fungi and protozoa.

Selected bacterial diseases

Types of bacterial diseases that will be considered in this section range from superficial skin infections to extremely severe systemic diseases.

Boils and carbuncles   Boils and carbuncles are infections caused by *Staphylococcus aureus*, a species of bacteria found on the human body (skin, mouth, nose, nasopharynx) as a member of the normal (autochthonous) microflora. Pathogenic *S. aureus* produces several kinds of extracellular enzymes, and all of them contribute to its virulence. The production of coagulase (an enzyme that causes blood plasma to clot) is generally considered as the most reliable laboratory evidence that a given strain of staphylococci is pathogenic. The coagulase test is performed by inoculating a tube of citrated plasma (human or rabbit) with the staphylococcal bacterium. After a period of incubation under the appropriate conditions, formation of enzyme coagulase by the bacteria will cause the plasma to clot. Coagulase production correlates with the ability to produce hyaluronidase an enzyme that enables bacteria to penetrate the skin and spread within body tissues.

Boils (skin abscesses) are the most common type of staphylococcal infection and reflect the prolific pus-forming ability of this bacterium. Boils are nothing more than a collection of pus within a limited encircled area of skin.

Carbuncles are formed from the collective drainage of boils through separate openings in subcutaneous tissues, resulting in a clinical condition that may be extremely severe. Sometimes carbuncles become a progressive disease with ulcerations in subcutaneous tissues accompanied by the invasion of the bloodstream by the staphylococci and the establishment of a septicemia. Thus, untreated carbuncles can develop into a fatal disease. Among other kinds of severe systemic diseases that may develop from staphylococcal infections are endocarditis and osteomyelitis.

Staphylococcal infections are most common among individuals in a debilitated condition: those that are suffering from chronic diseases such as cancer and diabetes, those that are being treated therapeutically with immunosuppresant drugs or broad-spectrum antibiotics, and those that are suffering from trauma (severe burns or other skin lesions). Thus, in the hospital environment, the prevention and control of staphylococcal infections is a formidable task.

Staphylococcal pathogens can be transmitted effectively by a variety of means. Of utmost importance is person-to-person contact, both from those with overt disease and from asymtomatic carriers. The later plays an extremely important role in the spread of staphylococcal infections by personnel in hospitals and nursing homes.

Since *S. aureus* is among the microorganisms normally found on the human body, it is impossible to eliminate the reservoir or potential source of the pathogen. Consequently, prevention and control of these infections depends to a great extent on maintaining good personal hygienic practices and the implementation of good sanitation practices by staff in hospitals and nursing homes. Serious infections can often be managed therapeutically with antimicrobial drugs, but with the emergence of multiple drug-resistance strains, the problem of therapeutic control has become increasingly more complicated, as we shall see later.

"Strep" sore throat   When tissues in the throat region become inflammed as a result of a streptococcal infection, the clinical condition is called *strep throat*. This infection is caused by beta hemolytic *Streptococcus pyogenes*. The high virulence of this pathogen is associated with its ability to produce several kinds of toxic extracellular products, such as hemolysins, hyaluronidase, leukocidin, and the erythrogenic toxin. Among such products are two kinds of beta hemolysins, streptolysin-*S* and streptolysin-*O*, both of which destroy red blood cells by lytic reactions. Often, beta hemolysis of red blood cells is used as the only criterion for differentiating the highly virulent *S. pyogenes* from other streptococci, because this property can easily be demonstrated under laboratory conditions. When *S. pyogenes* is cultivated on blood agar plates under appropriate conditions, the production of beta hemolysins is characterized by the appearance of clear hemolytic zones around bacterial colonies, called *beta hemolysis* (Figure 8.2). Streptolysin-*S* is oxygen-stable and is responsible for the lysis of red blood cells that sur-

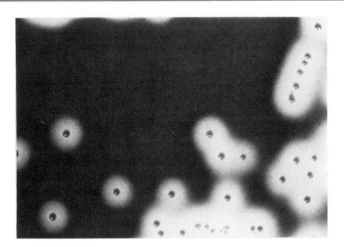

Figure 8.2. *Streptococcus pyogenes* colonies on a blood agar plate, showing clear zones of beta hemolysis around each colony. (Courtesy of T. J. Kloeckl.)

rounds surface bacterial colonies. Streptolysin-*O* is oxygen-sensitive. Therefore, lytic action from this substance can only be seen around colonies that are growing in the deep layers of the blood agar medium.

The beta hemolysins are antigenic. Infections with streptococci that produce the extracellular streptolysin-*O* usually provoke the body to make antistreptolysin-*O* antibodies. Thus, the laboratory detection of such antibodies in a person's blood serum has emerged as an important diagnostic test, called the antistreptolysin-*O* titer. A positive test demonstrates the presence of antistreptolysin-*O* antibodies in an individual's blood serum and is indicative of a past or current infection with beta hemolytic streptococci. It is important to remember that such antibodies neutralize the hemolytic action but do not protect an individual from subsequent attacks of beta streptococcal infections. Immunity to scarlet fever is an exception, because these antibodies (antitoxins) are provoked by each of the specific types of erythrogenic toxins.

Low-virulent streptococci also produce hemolysins and are characterized by their action or lack of action on red blood cells. The green discoloration around colonies of streptococci on blood agar plates results from the incomplete lysis of red blood cells and is called *alpha hemolysis*. When no hemolysis or color change can be seen surrounding the colonies of streptococci on blood agar plates, the phenomenon is called *gamma hemolysis*.

Although similar in cellular morphology, the streptococci comprise an extremely variable group of bacteria with respect to natural habitats and pathogenicity. The more virulent strains and other nonpathogenic species are natural inhabitants of the human body (mouth, nose, throat, and upper respiratory tract), but a number of species are more prevalent on the bodies of lower animals, especially in their gastrointestinal tracts. Thus, human infections are most likely to be transmitted by person-to-person contacts and by fomites.

Strep throat should be regarded as a serious infection, because it can progress into a fatal disease if not properly treated. Because of the variety of extracellular toxic products that *S. pyogenes* can produce, this single species may be responsible for a number of diseases. Persistent or recurrent throat infections may be followed by scarlet fever or rheumatic fever. The beta hemolytic streptococci are generally recognized as the causative agents for human streptococcal diseases, but other, low-virulent species are frequently the cause of endocarditis, urinary infections, wound infections, and a variety of complications in debilitated persons.

Prevention and control of streptococcal infections can be accomplished by a combination of measures. First, overt infections can usually be managed therapeutically with antimicrobial drugs. These pathogens are highly susceptible to chemotherapeutic agents. Infections caused by them can be suppressed by sulfa drugs and a variety of other bacteriostatic agents, and the pathogens can be eliminated by penicillin antibiotics, which are bacteriocidal for the streptococci. Other control measures include the practice of good personal hygiene, and good sanitation in institutions.

Pneumonia   Classically, *pneumonia* is an infectious disease of the air sacs (alveoli) within the lungs and associated bronchial structures. The acute disease is characterized by the accumulation of exudates, composed of fibrin, red blood cells, and polymorphonuclear (PMN) leukocytes in one or more lobes of the lungs, resulting in consolidation, hence the name *lobar penumonia*. Individuals experiencing lobar pneumonia exhibit the characteristic symptoms of chills; fever (temperatures of 102°F and above): rapid, shallow breathing; and rust-colored sputum, caused by the presence of fibrin and red blood cells. Often, lobar pneumonia can be diagnosed on the basis of clinical symptoms.

Acute cases of lobar pneumonia are generally caused by the pneumococci *(Diplococcus pneumoniae),* recently classified as *Streptococcus pneumoniae* (Figure 8.3). The virulence of *S. pneumoniae* is associated with encapsulated strains of these bacteria. On the basis of structural components present in the polysaccharide capsules, between 75 and 100 immunological types of *S. pneumoniae* have been reported. However, certain types are more virulent than others. Nonencapsulated cells are avirulent.

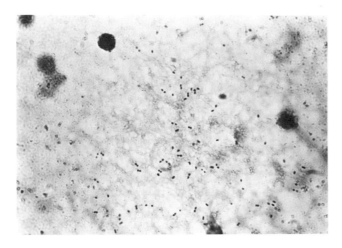

Figure 8.3. Light microscope photo of *Diplococcus (Streptococcus) pneumoniae* cells. Note the characteristic arrangement of the cocci or ovoid cells in pairs. (Courtesy of T. J. Kloeckl.)

Bronchial pneumonia, unlike lobar pneumonia, is characterized by infections that are scattered widely within the bronchial structures. This type of pneumonia may be caused by many different kinds of bacteria and viruses. In some cases the disease is caused by a single kind of microbe, and in others the disease results from the interaction of a combination of pathogens. Frequently, bronchial pneumonia develops as a secondary infection rather than as a primary disease. Serious complications from bronchial pneumonia often develop subsequent to influenza and other conditions that lower the body's resistance. Consequently, bronchial pneumonia is often the terminal disease responsible for death of persons suffering from chronic illnesses.

The pneumococci are generally spread by direct contact (coughing and sneezing) and by fomites. Therefore, pneumonia is especially common in institutions (hospitals, barracks, dormitories, prisons, etc.) where susceptible persons reside in close contact with each other. Convalescent carriers are also responsible for spreading these pathogens. Thus, a combination of measures must be used to prevent and/or control this disease: (1) because of the frequent occurrence of resistant mutants, chemotherapeutic drugs must be used prudently in the management of infections; (2) active cases must be isolated, if possible; and (3) strict personal hygienic practices and rigorous environmental sanitation measures must be exercised when caring for active cases.

After recovery from lobar pneumonia, the individual has a short-lasting immunity to the specific type of causative organism, and no immunity to virulent strains of other types. In general, pneumonia can be prevented by implementing the many measures that contribute to the maintenance of good health, because human beings have a rather high natural resistance to the pneumococci.

Tuberculosis Since ancient times, *tuberculosis* has been recognized as a serious human disease, and it is currently a leading cause of death in countries where sanitary conditions are poor and malnutrition is common. The socioeconomic status of a population is directly related to the prevalence rate of the disease within a community. In terms of magnitude or spread, tuberculosis occurs on a worldwide scale. Within the United States, the disease is more prevalent among those that live in overpopulated and unsanitary areas within urban communities than those that live in suburban or rural environments.

During the period when microbiology was being developed as a science, Robert Koch demonstrated that tuberculosis was caused by a rod-shaped bacterium, *Mycobacterium tuberculosis*. In 1882 he established the disease experimentally in guinea pigs, and by adhering to criteria that he had developed earlier, called Koch's postulates, the causative nature of the disease was clearly defined. Although *M. tuberculosis* is recognized as the human strain, *M. bovis* (the bovine strain) and *M. avium* (the avian strain) are also important causes of the disease in human beings.

The mycobacteria have very distinctive characteristics in respect to morphology, staining properties, and virulence mechanisms. In terms of structure, they are rod-shaped, but irregular branching is a typical characteristic of most species. All mycobacteria are difficult to stain by Gram's staining procedure. The tubercle bacilli and related species have cell walls that contain large quantities of complex lipid materials (approximately 20 to 40 per cent of cellular dry weight). As a result of difficulty encountered in staining these bacteria by Gram's methods, the mycobacteria are generally referred to as "acid-fast" bacilli. The latter designation refers to a specific staining procedure in which heat is used to promote the penetration of a red dye, carbol fuchsin, into the cells. Subsequently, the cells are decolorized with acid alcohol (ethyl alcohol + hydrochloric acid), a process that removes the carbol fuchsin from most bacteria within a few seconds. The mycobacteria retain the red color even after the application of a counterstain (methylene blue), therefore the name "acid-fast bacilli." On the basis of this unique staining property the mycobacteria can be differentiated easily from numerous kinds of non-acid-fast bacilli that may be present in sputum or other kinds of clinical materials. Virulence properties associated with the mycobacteria are not clearly defined. The tubercle bacilli are not known for their invasive powers, and they do not produce any of the classical exotoxins. To a degree, virulence may be associated with endotoxin-like posions that are released when the cells disintegrate.

Tuberculosis in human beings is primarily a disease of the lungs, but other organs and tissues of the body may become infected. The type of infection that may develop is related to the individual's prior exposure to the causative agent and general state of health. *Primary tuberculosis (child-type)* develops during childhood or in adults when exposed to the tubercle bacillus for the first time. The characteristic pathological lesion that develops during the initial infection is called the *tubercle,* and such lesions are self-limiting in most individuals. During the initial infection the body's defense mechanisms respond with the proliferation of polymorphonuclear leukocytes, which migrate to the site of infection. Subsequently, the bacilli become surrounded and engulfed by the phagocytes, a process that results in the enlargement and modification of the polymorphonuclear leukocytes into a tubercle. Dead tissues in this area may eventually become calcified. Thus, the initial infection is usually localized in the lungs and/or regional lymph nodes. Accompanying this phenomenon is the development of an immunity, generally referred to as a *hypersensitivity* to the tubercle bacilli. This type of hypersensitivity generally develops within a month of the initial infection and persists throughout one's life. This peculiarity is the basis for the tuberculin skin test, to be discussed later.

Chronic pulmonary tuberculosis (adult-type) develops from either the progression of the primary infection after a period of latency, or from a reinfection with pathogens from an exogenous reservoir. Regional lymph nodes are seldom involved in reinfection. Once these organisms are coughed up from a primary lesion in the lungs and swallowed, or ingested from an exogenous source, the organisms pass into the gastrointestinal tract and produce lesions in that area of the body.

The tuberculin skin test is a useful epidemiological tool. The test is based on the principle that the body develops a specific type of long-lasting hypersensitivity or allergy during the course of the initial primary infection with pathogenic tubercle bacilli. This phenomenon was first recognized by Robert Koch in the latter part of the nineteenth century. He demonstrated that guinea pigs infected with tubercle bacilli reacted to a subsequent subcutaneous injection of the same pathogens by developing an inflammatory reaction at the site of injection, while noninfected guinea pigs showed no reaction. Similar results were obtained when heat-killed tubercle bacilli or their culture filtrates were used.

Although the human tuberculin skin test evolved from the peculiar phenomenon that Koch observed, the actual procedures that are used today are considerably different. Two kinds of antigenic substances are widely used: old tuberculin (OT) and purified protein derivatives (PPD). The former is a crude extract prepared from a broth culture of the tubercle bacilli, and the latter is purified protein derived from the chemical fractionation of OT. A number of techniques have been developed for introducing either of the antigens into human skin. In general, PPD is preferred, because it is void of nonspecific antigenic materials that may be present in OT. When the test is administered to individuals that have been previously infected, a positive delayed type of hypersensitivity develops (indicated by a hardened or inflammed area of the skin at the test site, which develops within 24 to 48 hours). Hypersensitive individuals are called *reactors,* and the size of the zone is generally related to the risk of an active infection: the larger the zone, the greater is the risk. Such individuals should be further evaluated by x-rays and by a bacteriological culture of tissues (usually sputum) to ascertain the clinical status of the disease. Individuals that show negative reactions to the tuberculin skin test either have not had a prior infection with tubercle bacilli, or they may be undergoing the preallergic stage of a recent infection. In either case, the information obtained from a negative test is epidemiologically important, because timing, relative to exposure, is important in the decision to employ appropriate therapeutic or prophylactic measures.

Current therapeutics for active cases consist of administering one or a combination of the following antimycobacterial drugs: streptomycin, isoniazid (INH), para-aminosalicylic acid (PAS), rifamycin, or ethambutol.

Prevention and/or control consist of a number of measures: (1) immunization with the bacillus of Calmette and Guerin (BCG) vaccine, derived from a strain of *M. bovis,* provides some protection, and is recommended for skin-test-negative individuals who are repeatedly exposed to active cases of tuberculosis; (2) active cases should be isolated and treated immediately with chemotherapy; (3) each community should have a continuing surveillance program that employs x-rays and tuberculin skin tests. By exercising such measures, active cases of tuberculosis will continue to be reduced. During 1975, more than 33,500 cases of tuberculosis were reported in the United States. The manner in which the annual morbidity data were reported in 1975 is significant, because that was the first time information on the bacteriological status of all reported cases was available. Bacteriological analyses were performed on 82.3 per cent of cases.

Prior to 1975, a case was defined as a person who had developed active tuberculosis during the year and had never before been reported as active. *Reactivated* cases (previously reported active) were excluded from the new case count. Beginning in 1975, tuberculosis cases reported to the Federal Center for Disease Control, Atlanta, Georgia were defined as persons (1) having tissues or body exudates that are bacteriologically positive; and/or (2) who are being treated with two or more antituberculosis drugs.

Typhoid fever Typhoid fever, recognized since ancient times as an important communicable disease, occurs in all countries of the world. The disease is more prevalent, however, in communities where environmental sanitation is poor. The relationship between the incidence of the disease and the state of sanitation is especially evident in areas where sewage-treatment and water-purification facilities are inadequate.

Typhoid fever is caused by *Salmonella typhi,* and the classical disease develops only in human beings. Even when a susceptible individual is exposed, the causative bacterium *(S. typhi)* must be swallowed to produce the natural disease. If the pathogen gains entry to the body by other than the oral route, the natural disease does not occur. After being swallowed, the bacilli establish loci of infections within the gastrointestinal tract. There the pathogens multiply rapidly, penetrate the intestinal mucosa, invade regional lymph nodes, and gain entry to the bloodstream. This infection is both invasive and progressive and culminates with a fulminating systemic disease. During severe infections peritonitis may develop from typhoid bacilli that have entered the peritoneal cavity through perforations in the intestinal wall.

During the septicemia, the typhoid bacilli become widely disseminated in the body: some are spread to the spleen, kidney, and bone marrow, and others tend to become localized in the gallbladder. While traversing the body in the bloodstream, many of the pathogens undergo lysis, a phenomenon that results in the release of endotoxins. Thus, typhoid fever is a toxemia, and the complications experienced during the disease result from the release of endotoxins. Concomitantly, the infected individual exhibits overt symptoms. Following an incubation period of from 10 to 14 days, a fever develops and rises in increments to approximately 104°F. Accompanying the fever are abdominal pain, bowel disturbances, and a tendency to become constipated. Simultaneously, lesions appear on the skin and remain for several days.

By utilizing appropriate bacteriological techniques, *S. typhi* can be recovered from the blood, feces, and urine during the course of the disease. From the onset of the disease through the convalescence period, the pathogens can be recovered from the feces. Within the first three weeks after ingesting the pathogens, bacteria can be recovered from the blood, but the number of viable cells in the blood continues to decline. Since the body's immune system is provoked by the infection, pathogens present in the blood are destroyed progressively as antibodies continue to be synthesized. When the kidneys become infected, typhoid bacilli can be recovered from the urine after the second week of the infection and during the convalescent period. After recovery, most individuals have a lifelong immunity to reinfection.

Of particular importance to the control of typhoid fever are the large numbers of individuals who become carriers. Approximately 40 to 50 per cent of all infected persons become convalescent carriers and continue to excrete pathogens from their bodies in the feces and urine. From 2 to 5 per cent of the fecal carriers remain in the carrier state for their entire life or until their gallbladder is removed surgically. Once the gallbladder has become infected, organisms continue to multiply and are released into the intestinal tract. A major contributor to the spread of typhoid fever are carriers among food handlers in public eating establishments, although the number of individuals that may be infected from a carrier depends upon the nature of contacts and the degree of personal hygiene exercised by the carrier. However, the potential for spread of the disease by a carrier is almost unlimited. "Typhoid Mary" is the classic example of a typhoid carrier. During the 1920s she worked as a cook for several families and in public eating establishments. Over a 10-year period, a number of typhoid outbreaks, which included 51 cases and 3 deaths, were traced to her.

Since this disease does not occur in nonhuman hosts, environmental reservoirs are: individuals with active infections, carriers, and inanimate materials contaminated with human excreta (water and food items). From such reservoirs, the pathogens (S. typhi) are transmitted directly and/or indirectly to susceptible individuals. In 1900 typhoid fever was a major cause of death in the United States, but the number of cases and deaths have declined continuously. The significant and progressive decline of this disease among Americans can be attributed to the widespread application of modern methods for the treatment of sewage and for purification of drinking water (see Chapter 11). However, sporadic episodes of the disease do occur. The largest recent outbreak of typhoid fever in the United States occurred in 1973 at a migrant farm labor camp in Homestead, Florida. Among the number of individuals that exhibited symptoms of typhoid fever, 150 suspected cases were hospitalized, and of that group, 63 cases were confirmed by bacteriological cultures. Subsequent epidemiological studies revealed that the camp's drinking water came from two wells, and both were being contaminated with groundwater from a faulty drain. Although S. typhi was not isolated from the water, large numbers of fecal coliform bacteria were present. It is important to note that S. typhi lives only about one week in sewage-contaminated water, but remains viable in fecal matter for a longer period.

To control the spread of typhoid fever, a number of measures must be utilized: (1) Infected individuals should be treated therapeutically with broad-spectrum antibiotics; (2) when the carrier state can be related to the release of pathogens multiplying in the gallbladder, it should be removed surgically; (3) an active surveillance program should be implemented by appropriate officials to detect and remove carriers from positions of food handlers in public establishments; (4) shellfish from polluted waters should not be eaten raw; and (5) prophylactic immunizations should be administered to those working in high-risk exposure areas.

Other complications in humans that are caused by *Salmonella* bacteria are paratyphoid fever and a localized gastroenteritis. Paratyphoid fever resembles typhoid fever in symptomology but is a much milder disease and is caused by different *Salmonella* species, most often *S. paratyphi* A and *S. schottmulleri*. The localized gastroenteritis is a foodborne illness that occurs after the ingestion of a foodstuff contaminated with any of several different *Salmonella* species. The foodborne illness can generally be differentiated from the systemic fever by the sudden onset of abdominal pain and diarrhea within 48 hours after ingesting the pathogens.

Cholera   John Snow, a British physician, conducted one of the classic experiments in epidemiology during the *cholera* epidemics of 1853–1854 in London. During a very short period, a large number of individuals became ill with cholera and more than 500 died of the disease. Snow made a thorough investigation of the outbreak, and formulated some fundamental concepts about the transmission of the disease. This was a laudable accomplishment at that period when one recognizes that the microbial causation or "germ theory" of diseases had not been established. Assumptions that guided Snow's investigations were as follows. Since the disease appeared to involve only the gastrointestinal tract, he believed that systemic manifestations were secondary to the initial disturbance in the alimentary canal. On this basis, he postulated that the causative agent would leave the body of cholera patients in their feces. He postulated that if the causative agent was a chemical substance, it would soon become too dilute, in its passage from person-to-person, to exert an effect. Therefore, an incubation period would allow time for a living causative agent to multiply. He knew that cholera could be transmitted from the sick to the healthy, probably by fecal contamination on the hands of those attending the sick, because he had observed the progressive development of the disease among individuals that had contact with cholera patients. And, finally, to explain the more widespread outbreaks of cholera, he postulated that the causative agent might be transmitted through drinking water that had been contaminated with human feces.

During the cholera epidemic, the affected individuals were concentrated in one small area of the city. Further investigations revealed that practically all of the ill obtained their drinking water from a well located on Broad Street. Snow observed that other individuals, who lived in the same neighborhood but obtained their drinking water from another source, were not ill with cholera. Continued investigation also revealed that sewage from a cesspool at a nearby house was seeping into the Broad Street well. Furthermore, prior to the cholera outbreak, an individual at the house with the faulty cesspool had been ill with a gastrointestinal disorder. These findings strongly supported Snow's hypothesis that the causative agent of cholera was present in sewage that contaminated the Broad Street well.

Approximately 30 years after Snow's brilliant observations, the causative agent of cholera was characterized by Robert Koch. In his investigation of cholera epidemics during 1883 in India and Egypt, a comma-shaped bacterium called the cholera vibrio, was isolated and demonstrated to be consistently present in fecal matter and in tissues taken during autopsies on those that had died of the disease. Vibrios are curved rod-shaped bacteria, rather than straight rods. Currently, the causative agent of cholera is called *Vibrio cholerae*. This species, and the El Tor biotype, are considered to be the most important pathogens.

Like typhoid fever, naturally occurring cholera develops only in human beings. The disease is characterized by its rapid onset within 2 to 5 days after ingesting the pathogens, most often in sewage-polluted water or in contaminated food. The pathogens multiply rapidly in the gastrointestinal tract, and the infected individuals experience abdominal cramps, nausea, and diarrhea, as evident by the passage of "rice water" feces. In severe cases, the loss of 10 to 15 liters of fluid per day may lead to dehydration, shock, and death. Manifestations of the disease are believed to result from the excretion of a potent exotoxin by the rapidly proliferating pathogens in the gastrointestinal tract. The toxin is sometimes referred to as an enterotoxin because it exerts its action on mucosal cells in the lumen of the small intestine.

Cholera occurs most often in individuals who are in a debilitated condition and in areas where environmental sanitation is poor. Epidemics are generally associated with sewage-contaminated drinking water supplies, because *V. cholerae* remains viable in water for only a short time. In the early 1970s, cholera outbreaks reached pandemic proportions and involved nearly 40 countries. A recent *Morbidity and Mortality Weekly Report* shows a general decline in worldwide epidemics during 1975. Although there were decreases in the number of countries involved and in the total number of cases, more than 87,000 cases of cholera were reported to the World Health Organization in that year.

Untreated cholera is a serious disease and may have 60 to 70 percent mortality among those infected. However, it can be treated effectively with broad-spectrum antibiotics. After recovery, immunity is of short duration. A cholera vaccine is available, but it only provides about 50 per cent protection and is also of short duration. Thus, prevention lies in the establishment and maintenance of facilities for the sanitary treatment of sewage and for the purification of drinking water.

Selected viral diseases

Diseases caused by viruses vary considerably in terms of their effects on susceptible (nonimmune) individuals: mild or moderately damaging in some cases, and extremely devastating in others. In Chapter 9 several of them will be characterized as viral zoonoses. In this section our discussion will be restricted to certain kinds of viral diseases that occur only in human beings. Although lower animals may become infected with human viral pathogens, the agents do not replicate and produce symptoms in these animals that are characteristic of the respective human diseases.

In many aspects, epidemics caused by viruses present formidable challenges to public health officials charged with the responsibility of preventing and controlling them.

Smallpox Smallpox, an ancient disease, was endemic in many countries for centuries. However, tremendous progress has been made toward its eradication in the twentieth century. When considering the merits of measures that are being used to prevent and control viral diseases, the eradication of smallpox from within the United States and Europe can be recognized as a hallmark of success in public health, obtained by vaccinating the masses of susceptible individuals with a smallpox vaccine, and by implementing vigorous programs for the sanitary handling and disposal of human infectious materials. However, a few cases of smallpox do occur sporadically, but most of them result from contact with international travelers who have visited areas where the disease is endemic. An example of a current endemic area for smallpox is Ethiopia. The fact that smallpox vaccination requirements for all age groups that reside within the United States have been canceled attest to the rarity of the disease. However, vaccination is required for travelers going to Ethiopia and for travelers returning to the United States from countries where there has been a recent outbreak of smallpox.

Smallpox is often referred to as *variola,* and the causative agent belongs to the poxvirus group. Currently, two clinical forms of smallpox are recognized: variola major (the severe type) and variola minor (the mild type). When a susceptible individual (nonimmune) becomes infected, the onset of symptoms is rapid, usually within two weeks. Symptoms include fever, headache, pains in the back and limbs, and, within a few days, the appearance of characteristic pustules on the skin. The pustules usually erupt into ulcerative "pox" lesions, which become dried and leave permanent and very distinct scars. Death from smallpox is extremely common and probably results from hemorrhages in various parts of the body. Survivors have a lifelong immunity that protects them from reinfection.

Smallpox, a disease that affects only human beings, is usually transmitted directly through contact with infected persons or indirectly by contact with fomites or droplets. Within the body, the variola viruses spread from the upper respiratory tract to lymphatic tissues, in which they multiply. From the lymphatic system, the virus gains entry to the bloodstream. A viremia develops, which leads to a systemic disease with the symptoms we have described.

There is no effective treatment for smallpox. Prevention and control rely primarily on vaccinating susceptible individuals and implementing rigorous sanitary measures. The latter includes the exercise of strict personal hygienic practices for those that care for the infected, and thorough terminal disinfection of the surrounding environment after the death of smallpox victims.

Although artificial immunization with the smallpox vaccine is highly effective, it subjects people to the possibility of certain complications. For this reason it is inadvisable to vaccinate persons with the vaccine while they are experiencing skin disorders, such as impetigo or dermatitis; severe burn injuries; pregnancy; or conditions in which the normal immune status may be altered (e.g., leukemia or therapy with immunosuppressants).

Polio   Poliomyelitis is an acute disease caused by viruses that belong to the picornavirus group. Three antigenic types of the polio virus (designated as types I, II, and III) have been characterized as the pathogenic agents responsible for producing the disease. The disease has three forms: (1) a mild gastrointestinal disorder of only 2 or 3 days' duration; (2) a moderately mild aseptic meningitis, accompanied by pains in the lower back and neck regions; and (3) a severe paralysis, called *paralytic poliomyelitis* or *spinal meningitis.*

Human beings are the only natural reservoirs of the polio virus, and all age groups of susceptible (nonimmune) individuals may become infected. Water and food may become contaminated with human excreta (sewage) and serve as environmental reservoirs from which the pathogens may be dispersed, or consumed. The polio virus gains entry to the body through the mouth, and replicates within the gastrointestinal tract. After an incubation period of a few days, or in some cases weeks, the virus may escape from that region and spread to the central nervous system. Multiplication of the polio virus within the spinal cord and brain produces damage which manifests itself as a paralysis of various muscles. The paralysis may be of varying degrees of severity, ranging from transient weakness to permanent impairment. Often there is a loss of muscular functions in the legs. When the infections produce widespread damage to cells in the spinal cord and to nerves that control the diaphragm, the individual experiences difficulty with breathing and swallowing. At that point, mechanical devices must be used to support respiration, without which death occurs rapidly.

Prevention and control of poliomyelitis is based primarily on the development in susceptible individuals of an effective immunity to each of the antigenic types of the polio virus. Most individuals develop a natural active immunity to the polio virus from exposure early in life. However, immunity can be induced in susceptible individuals by vaccinating them with either of two types of approved poliomyelitis vaccines: *inactivated poliovirus vaccine* (IPV), and *trivalent oral poliovirus vaccine* (TOPV).

The IPV was developed by Jonas Salk and made available worldwide in 1955. IPV is prepared by cultivating each of the three antigenic types of polio virus separately in rhesus monkey kidney cells in tissue culture. After being produced in tissue culture, each virus type is inactivated by treatment with formaldehyde or ultraviolet light. Current practice is to mix the three antigenic types into a single trivalent preparation. For effective immunization, an individual is administered IPV by an initial injection, followed by booster injections at prescribed intervals. When the series is completed, the individual is protected by circulating antibodies which the body produces in response to vaccinations. The vaccine is highly effective against paralytic poliomyelitis, but it does not prevent intestinal carriage of the polio virus. Thus, the polio virus can continue to be excreted in the feces by IPV-immunized individuals.

TOPV was developed by Albert Sabin and made available in 1962. It is prepared by cultivating each of the antigenic types of the poliovirus separately in human cells in tissue culture. After production in tissue culture, each of the antigenic types is attenuated (weakened, not inactivated) and mixed into a single trivalent preparation. Unlike

IPV, which is administered by injection; TOPV is administered orally. After an initial dose of TOPV is taken, booster doses are administered orally at prescribed intervals. To be effective, the series must be completed. By administering TOPV orally, the attenuated viruses replicate within the lymphatic tissue of the gastrointestinal tract and provoke the production of circulating antibodies that protect against paralytic poliomyelitis. While replicating within the gastrointestinal tract, the attentuated viruses prevent replication of wild or normal polio viruses that might be present in the gastrointestinal tract. Thus, the excretion of active polio viruses is reduced—an added benefit of this vaccine.

Both types of poliomyelitis vaccines are highly effective in protecting individuals against paralytic poliomyelitis. However, TOPV is currently more widely used. To provide maximum protection, it is advisable to administer the initial dose between 6 months and 12 months of age. This is extremely important when a new generation of infants is being born into an area where the majority of adults are effectively immunized through vaccination. In such instances the young remain highly susceptible to the virus because of its absence. Because this condition exists within most areas of the United States, vaccination of young susceptibles is highly recommended, as without it there is serious danger of the onset of an epidemic of paralytic poliomyelitis.

In an effort to prevent such a devastating epidemic, U.S. Public Health Service officials have defined "poliomyelitis epidemic" as two or more cases caused by the same polio virus antigenic type within a four-week period, and within a given population (e.g., city, county, or village). In general, poliomyelitis epidemics occur most often in susceptible (nonimmune) individuals who live in areas of poor sanitation and in warm climates. Furthermore, the probability of an epidemic in such areas is enhanced if methods for purification of drinking water and disposal of sewage are substandard (see Chapter 11). It might be of interest to note that the last major epidemic of paralytic polio in the United States occurred in the early 1950s, during the prevaccine era. As a result of the effectiveness of vaccines, paralytic poliomyelitis is now preventable. During the first six months of 1976, only five cases of paralytic poliomyelitis were reported to the Federal Center for Disease Control.

Measles   Measles is an extremely common acute and self-limiting viral disease that spreads rapidly to susceptible (nonimmune) individuals. The causative viral agents, members of the paramyxoviruses group, are commonly referred to as *rubeola*. The rubeola virus is widely dispersed in the environment among human hosts, and the disease is endemic in all countries. Nonimmune individuals of all ages are susceptible to rubeola virus infections, but measles is still considered to be primarily a childhood disease. Throughout the world the incidence of the disease is highest among the preschool group and children between 5 and 7 years old. Many adults are not susceptible to measles, because they have developed a natural immunity to the rubeola virus as a result of being infected during childhood.

Measles is one of the most contagious infectious diseases. The rubeola virus is spread by direct contact with the secretions of infected persons, and indirectly by fomites and aerosols. The virus gains entry to the body through the respiratory tract and replicates within lymphatic tissues of that area. Actually, measles is a disease of the upper respiratory tract from which the virus enters the bloodstream and spreads to lymphoid tissues in the skin. The onset of the disease is rapid and symptoms appear within 3 to 4 days of initial infection. The first symptoms are general and include headache and fever, but later symptoms are the characteristic rash and lesions that appear on the skin and within the mouth region. In most cases diagnosis can be made on the basis of clinical symptoms alone, and the complications usually subside within about 10 days without evidence of serious damage. However, in rare instances the virus gains entry to the nervous system and produces a severe encephalomyelitis, a concurrent infection of the spinal cord and brain.

Usually, an individual develops a lifelong immunity as a result of first infection with rubeola virus. Therefore, the disease may periodically disappear in small communities owing to the absence of enough susceptible individuals to continue transmission. For the disease to persist in a community, two conditions are necessary: (1) the presence of human reservoirs, and (2) a continuous supply of susceptible persons. Both conditions exist in undeveloped countries, where overcrowding and poor sanitation are common. In well-developed countries, measles epidemics often occur in young children in cycles every 3 to 5 years as a result of the emergence of a new population of nonimmune individuals.

Prevention and control of epidemics within a community depend primarily on the maintenance of a high level of induced immunity among its individuals. Although passive immunizations with hyperimmune serum (human immunoglobulins) will limit the severity of the damage if administered early after the onset. In the early 1960s an attenuated measles vaccine was developed and is used widely to induce an immunity in large numbers of individuals in areas of an epidemic. Mass-immunization programs have proved to be extremely beneficial in reducing the incidence of measles. Such programs can be managed logistically with little difficulty, because only a single injection of the attenuated virus is required to provoke the production of antibodies sufficient to provide an effective immunity.

German measles or *rubella* is caused by a different virus, but is also an extremely communicable childhood disease. Usually, rubella virus infections in small children occur without complications and a lifelong immunity develops after a single attack. Symptoms are similar to those described above, but the incubation period is longer (2 to 3 weeks).

The importance of rubella virus infections relates primarily to its ability to produce congenital defects in children born to women who became infected during the first trimester of pregnancy. During that period the rubella virus can penetrate the placental barrier and infect the fetus. Consequences of such infections are infants born with brain damage, heart defects, and a host of other abnormalities.

There is no specific treatment for rubella. Prevention and control are based upon the development of an effective immunity in susceptible individuals, especially females during the prepuberty years. In 1969 an attenuated rubella vaccine was licensed for use within the United States, and extensive immunization programs were initiated in an attempt to control epidemics and prevent the occurrence of congenital defects. If possible, all susceptible (nonimmune) individuals in the following categories should be vaccinated: (1) children between 1 and 12 years of age, and (2) all susceptible (nonimmune) nonpregnant women of childbearing age. In the second group, effective methods of birth control must be used for at least 3 months following vaccination.

The effectiveness of the widespread use of the rubella vaccine reflects itself in the following data, which pertain to the occurrence of rubella cases within the United States: 16,343 cases of rubella were reported for the year 1975. That represents a 51 per cent decrease in the average number of cases per year (33,375) reported for the period 1970–1974.

Hepatitis Traditionally, two different forms of viral hepatitis are recognized: *hepatitis A* (formerly *infectious hepatitis*); and *hepatitis B* (formerly *serum hepatitis*). The types are distinguished primarily on the basis of differences in modes of transmission, severity, and immunological features. After the onset of the disease, individual cases are difficult to differentiate on a clinical basis.

The causative agent of hepatitis is a virus or viruslike particle that has not been fully characterized. Human beings are the only known natural reservoirs from which the viruslike agent can be dispersed into the environment.

Hepatitis A infections appear to be spread primarily by the ingestion of the causative agent in fecal-contaminated water or food. After gaining entry to the body, the normal incubation period is 2 to 14 days, during which the agent replicates within the gastrointestinal tract, produces a viremia, and spreads to the liver. Simultaneously, the pathogens are being released from the body in the feces.

Hepatitis B infections are transmitted directly by physical contact with an infected person, and indirectly by contact with contaminated blood or materials. The agents enter the susceptible individuals orally or through the skin. Most often, though, hepatitis B infections are transmitted indirectly by one or a combination of the following mechanisms: blood transfusions, sharing contaminated needles while injecting drugs, ear piercing, tattooing, and the handling of contaminated blood products. Other mechanisms (airborne and vectorborne) for spread have been postulated but not proved.

During severe infections of either type, the classical symptoms include yellowing of the skin (jaundice), the excretion of dark urine, and light stools, all consequences of liver damage.

Hepatitis-associated antigens have been detected consistently in the blood of individuals during acute hepatitis B infection. Such antigens have not been found in the blood of individuals infected with hepatitis A pathogens, however, so the presence of such antigens in the blood can serve as a marker for differentiating the two types of infection. Such antigens appear in the blood during the incubation period and often remain for as long as 8 weeks after the initial infection. Highly sensitive serological tests are used to detect them. In those tests, commercially prepared antiserum (hepatitis B antibodies) is reacted with the blood from infected persons. If specific hepatitis-associated antigens are present in the blood, the test gives a positive reaction. Individuals that show positive reactions are in a state of infectivity, and are living reservoirs from which hepatitis B pathogens can be dispersed.

Hepatitis B infections are emerging as a major public health problem in the United States. More than 55,000 cases were reported to the Center for Disease Control during 1975. The risk of becoming infected is greater among individuals that handle blood or infected persons in their occupations (e.g., professional personnel in surgery units, blood banks, hematology—oncology units, and plasma fractionation facilities) than among individuals in the general population.

Some general preventive measures are as follows:

1. Attempt to minimize contact with infected persons
2. Members of a family or household among which an acute- or asymptomatic hepatitis B-infected person resides should be in-

formed of the mechanisms through which the infection can be spread.

3. Parents and others responsible for the care of the institutionalized or the mentally retarded persons should be made aware of the necessity for exercising good hygienic practices and of the possibility of contributing to the spread of the pathogens.

4. Health-care professionals should exercise precautions, establish rigid guidelines, and adhere to those guidelines when working in high-risk areas with potentially contaminated blood, or with hepatitis B-infected individuals

Influenza Sporadic episodes of influenza occur within the United States each year, and serious influenza pandemics occur periodically on a worldwide basis. Historically, the 1918–1919 influenza pandemic is recognized as one of the most serious and devastating outbreaks of an infectious disease known to mankind. That outbreak alone is believed to have contributed to the death of more than 20 million persons. Even within the United States, thousands have died of influenza during the past 20 years: nearly 70,000 deaths were associated with the Asian influenza epidemic of 1957–1958, and more than 30,000 deaths resulted from the Hong Kong influenza epidemic of 1968–1969.

Influenza is an extremely communicable acute respiratory illness that is caused primarily by three types of myxoviruses (influenza A, influenza B, and influenza C). The major influenza epidemics of the past have been caused by viruses designated as type A and type B. Apparently, type C viruses are low-virulent and produce human influenza only on rare occasions. It also appears that type A viruses are more virulent than type B, because most of the severe influenza epidemics have been caused by various strains of type A viruses. Within type A and type B viruses, there are a number of subtypes or strains, each of which is differentiated by major antigenic determinants associated with the respective virus particles.

An important characteristic of influenza viruses is their unusual degree of genetic variability. As a result of this property, influenza viruses periodically undergo major antigenic shifts with the emergence of new variants (i. e., strains with different antigenic determinants). This phenomenon of antigenic shift in influenza viruses is of considerable practical importance, because each major shift in antigenic determinants renders a large segment of the world population susceptible to infection by the newly emerged strain. Thus, individuals with an immunity to a prior strain as a result of a natural infection, or as a result of a vaccination, have no protection to the antigens of the newly emerged strain. The sudden appearance of such strains and their rapid spread (by secretions and aerosols) among large numbers of susceptible individuals account for the explosive epidemics of influenza that occur periodically.

During an epidemic, the infective strain enters the body through the mouth and nose and replicates in the upper respiratory tract. Complications result from the extensive virus multiplication in cells of that region. Normally, influenza is a self-limiting disease and clinical complications last only for a few days. Thus, epidemics subside quickly. Most deaths that occur during an influenza epidemic result from viral destruction of cells in the respiratory tract plus complications caused by secondary bacterial pneumonia. Observations from past epidemics show that most deaths occur among elderly (over 65 years) and chronically ill individuals.

Owing to its extreme communicability, the short incubation period, and the short duration of epidemics, the control of influenza is a formidable problem. Efforts to prevent or control influenza in the United States have generally been aimed at annual immunization of individuals considered to be in the high-risk group, which includes those with the following conditions:

1. Heart diseases of any kind
2. Chronic diseases of the respiratory tract (asthma, tuberculosis, emphysema, bronchitis)
3. Chronic renal disease
4. Diabetes and other metabolic disorders

Prior to 1976, influenza control through widespread vaccination programs was not a public health objective, for the following reasons:

1. Limited duration of the immunity
2. Variable effectiveness due to lack of specificity to the causative strain
3. Availability of influenza vaccines
4. The low frequency of serious complications from the disease among healthy people in the general population

In February 1976 a new strain of human influenza A virus (A/New Jersey/76) was isolated in an outbreak of influenza among United States Army recruits at Fort Dix, New Jersey. The new influenza virus strain was found to be related antigenically to the virus strain that is believed to have caused the devastating 1918–1919 pandemic. The emergent strain represented a major antigenic shift from the influenza A viruses that caused the 1968 epidemic (A/Hong Kong/68), and the 1975 epidemic (A/Victoria/75). Therefore, a major segment of the population would have no immunity to the newly emerged strain. Based on the prospect that the A/New Jersey/76 strain would persist and spread rapidly, as have other viruses in past epidemics, federal health officials initiated a bold and unprecedented nationwide campaign to vaccinate the general public within the United States. The National Influenza Immunization Program for 1976–1977 provided for the development and use of two vaccine formulations: (1) a bivalent vaccine that contained both A/Victoria/75 and A/New Jersey/76 viruses, for immunizing the traditionally identified high-risk groups; and (2) a monovalent vaccine that contained only the new virus (A/New Jersey/76), for immunizing the rest of the population.

Subsequently, the National Influenza Immunization Program became plagued with a sequence of problems, which included:

1. A delay in the production of the vaccines while Congress debated the merits of enacting laws that would provide insurance to protect the vaccine manufacturers from liability claims that might be filed against them by persons who experienced complications after being vaccinated
2. Inefficient operation of many immunization centers because of logistical problems encountered by manufacturers in making and distributing the vaccines
3. The sudden death of a number of elderly persons who had received swine flu vaccinations
4. Widening of the credibility gap when a large number of individuals who had been immunized with swine flu vaccines developed a paralysis called *Guillain-Barre syndrome* (French polio)

Finally, Federal health officials imposed a moratorium on the nationwide immunization program, because only a few cases of influenza infections caused by A/New Jersey/76 virus had been confirmed, and the risk of developing the Guillain-Barre syndrome from the vaccinations appeared to be considerable.

The 1976–1977 flu season was interesting in another respect. There was a major outbreak of A/Victoria influenza among residents of a nursing home in Miami, Florida. Although a few other isolated cases had been reported, no A/Victoria vaccines had been produced during the year and targeted for protecting high risk groups against A/Victoria influenza. All of the A/Victoria/75 vaccines that were manufactured in 1976 were incorporated in the bivalent vaccine. Consequently, Federal officials had no choice but to use the bivalent vaccine for protecting high risk groups against A/Victoria influenza. In February, 1977, the moratorium on the bivalent vaccine (A/Victoria/75 and A/New Jersey/76) was lifted. However, the moratorium on the monovalent vaccine remained in effect.

Although the 1976–1977 nationwide immunization program will probably remain a controversial issue, it demonstrates a general lack of understanding of influenza epidemiology among health officials.

Selected fungal (mycotic) diseases

Fungi are widely dispersed in the environment as inhabitants of soil, water, vegetation, and the bodies of human beings and other animals. Thus, infinite opportunities exist for direct or indirect contact with various kinds of fungi, many of which are pathogens.

Human diseases caused by fungi are commonly referred to as *mycoses,* and complications range from superficial infections of the skin (dermatomycoses) to extremely severe diseases that involve internal tissues and organs of the body (systemic mycoses).

Dermatomycoses   Fungal infections that only involve the superficial layers of the skin and skin appendages (hairs and nails) are called *dermatomycoses,* and the pathogens that cause them are called *dermatophytes.* Fungi known to be dermatophytes comprise an extremely heterogeneous group of organisms, and the clinical syndromes they produce are also diverse. Some of the commonly occurring superficial infections will be summarized in this section.

Ringworms (also called *tinea*) are among the most common and widely distributed types of infections known. Clinically, ringworms are characterized by circular lesions in the epidermal layer of the skin that tend to spread equally in all directions. Most often ringworms are classified on the basis of the specific area of the body that is affected rather than by the taxonomic name of the fungus that produces them: on the scalp (tinea capitis), on the beard (tinea barbee or "barber's itch"), on the body (tinea corporis), and on the feet (tinea pedis or "athlete's foot"). Those designations are less confusing than a system based on the names of specific dermatophytes that produce them. For example, various species within the genus *Microsporium* (Figure 8.4) may cause ringworms of the scalp, beard, or other area of the body. Furthermore, dermatophytes that belong to other genera, such as *Trichophyton* and *Epidermophyton,* may also cause ringworms on any of various regions of the body.

Other kinds of dermatomycoses include infections that are caused by a number of different kinds of fungi that produce infections within structures or regions of the skin: (1) tinea versicolor is an infection of the horny layer of the epidermis, and is recognized as scaly patches on the skin of the back and chest regions of the body, (2) tinea nigra is a

Figure 8.4.   Light microscope photo showing characteristic spores (conidia) of *Microsporium canis,* a common cause of tinea corporis. (Courtesy of T. J. Kloeckl.)

clinical syndrome characterized by the development of lesions in the epidermal layer within the palm of the hands, and (3) piedra is an infection that involves only the hairs of the scalp (called black piedra), and hairs of the beard or mustache (called white piedra).

The important fact to remember about the dermatomycoses is that the pathogens that produce them are not invasive. Therefore, complications, although troublesome, are not life-threatening. But if the conditions are allowed to persist and the area becomes infected with secondary bacterial pathogens, serious complications may develop. Therefore, prompt treatment of dermatomycoses should be initiated with griseofulvin, the drug of choice for dermatophytes. During a persistent infection prolonged treatment with that orally administered drug may be required.

Systemic mycoses   Fungal diseases that involve subcutaneous tissues and internal organs of the body are called *systemic mycoses.* Most systemic infections are caused by fungi that live normally in the soil, from which fungal spores are aerosolized. Human beings become infected by inhaling the pathogens and/or their spores. Within the respiratory tract, the spores develop vegetatively and initiate a localized infection which in most cases is mild and self-limiting. In other instances, the pathogens invade other tissues and spread progressively to various organs of the body.

It is of interest to note that systemic mycoses are caused most often by those fungi that exhibit *dimorphism,* the ability to grow as yeastlike cells in body tissues and in culture at 37°C and grow as filamentous molds in culture at room temperature. Representative kinds of systemic mycoses and their causative agents are listed in Table 8.3.

Three of the selected systemic mycoses (Histoplasmosis, Cryptococcosis, and Aspergillosis) are characterized in Chapter 11 in the discussion that deals with the potential health hazards of nonindustrial microbial aerosols. Here we shall present some general aspects of Candidiasis, Blastomycosis, and Sporotrichosis.

Candidiasis is an extremely common clinical syndrome in which complications range from mild superficial infections to severe systemic involvement. *Candida albicans* is the most common pathogen, although other *Candida* species are sometimes involved. *Candida albicans* is often referred to as an "opportunistic" pathogen, because it is found normally within the gastrointestinal tract of asymptomatic persons. This pathogen assumes an invasive role only under certain conditions, such as those which occur when an individual is compromised by a chronic illness, is undergoing prolonged treatment with broad-spectrum antibiotics, or is in a generally debilitated state. A fulminating systemic disease often develops under those conditions. Infants often become infected with *C. albicans* during birth, acquiring the organisms from the mucous membranes of the vaginal tract. In such instances an infection within the mouth region, called *Thrush,* is the most common syndrome the infant experiences. Superficial *C. albicans* infections can often be managed effectively with nystatin, a topically administered drug. Orally administered amphotericin B is usually preferred when a person is experiencing systemic candidiasis.

Blastomycosis is caused by two different kinds of fungi, depending upon whether the infection is acquired in South America or North America. Infections acquired in North America are caused by *Blastomyces dermatitidis*, a dimorphic fungus (Figure 8.5) that inhabits the soils of the United States and Canada. Apparently, the fungi gain en-

Table 8.3. Selected Types of Systemic Mycoses That Occur Within the United States

Disease	Causative Agent	Dimorphism	Habitat
*Histoplasmosis	*Histoplasma capsulatum*	Yes	Soil, bird droppings
*Cryptococcosis	*Cryptococcus neoformans*	No	Soil, bird droppings
*Aspergillosis	*Aspergillus fumigatus*	No	Soil, compost piles
Candidiasis	*Candida albicans*	Yes	Human mucous membranes
Blastomycosis	*Blastomyces dermatitidis*	Yes	Soil
Sporotrichosis	*Sporotrichum schenckii*	Yes	Soil, vegetation

*See Chapter 11 for characterization and a discussion of their epidemiology.

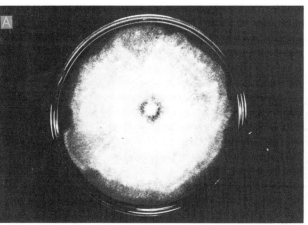
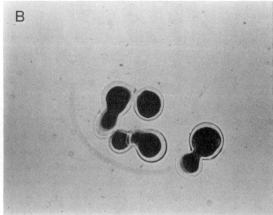

Figure 8.5. (A) Fluffy white colony of *Blastomyces dermatitidis* growing on Sabouraud dextrose agar at room temperature. (B) Light microscope photo of *Blastomyces dermatiditis,* showing large oval budding yeastlike cells when cultivated at 37° C. (Courtesy of T. J. Kloeckl.)

try to the body through the respiratory tract, after which they invade other internal organs and produce the characteristic granulomatous disease. Tissue destruction is extensive. Skin lesions are also common, suggesting that the pathogens may gain entry to the body through wounds in the skin. If such conditions are not treated promptly, they may be fatal. Chemotherapy with amphotericin B is the recommended treatment.

Sporotrichosis is commonly referred to as a subcutaneous mycosis, because tissue involvement is localized within the subcutaneous layer of the skin. The causative agent, *Sporotrichum schenckii,* is a dimorphic fungus that inhabits the soil and various kinds of vegetation (Figure 8.6). Most infections occur when contaminated thorns, splinters, bristles, or other woody materials penetrate the skin. This condition is now recognized as an occupational disease among forestry workers, especially those who handle sphagnum moss. The first symptom that one should be alerted to is a splinter prick that does not heal. If not treated, the infection spreads to the lymph nodes and produces ulcerative lesions. Sometimes surgical drainage is required in addition to chemotherapy with amphotericin B to supress the infection.

The following case demonstrates the manner in which sporotrichosis is acquired. Thirteen Mississippi forestry workers developed sporotrichosis during a 2-month period (December 1975 to February 1976). Epidemiological investigations revealed that all had planted pine seedlings supplied by a particular nursery. Sphagnum moss was packed around bundles of pine seedlings to provide moisture for the

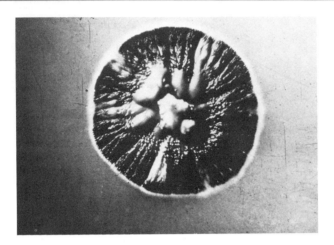

Figure 8.6. Colony of *Sporotrichum schenckii* growing on Sabouraud dextrose agar at room temperature, showing the characteristic black, wrinkled surface. When cultivated at 37° C, this fungus grows as yeastlike colonies with oval budding cells. (Courtesy of T. J. Kloeckl.)

roots during shipment. At the planting site, seedlings were separated from the moss by the forestry workers, the process through which the infections were probably acquired. *Sporotrichum schenckii* was cultured from samples of the moss at the planting site, and also recovered from the moss remaining in the nursery. There is no established method for eradicating *S. schenckii* from contaminated moss.

Selected protozoan diseases

The protozoans pathogenic for human beings are a diverse group of unicellular organisms. Most of the pathogenic types live as parasites in or on the bodies of humans or other animals, and some species must spend a portion of their life cycle in different kinds of animal hosts to complete their development (see Chapter 9). Other kinds of protozoans live primarily as parasites in the gastrointestinal tract of humans and, when released to the external environment in human excreta, exist primarily in a dormant state called a *cyst*. Soil and water contaminated with human excreta are the common mechanisms through which the infected cysts are transmitted.

Amebiasis   Entamoeba histolytica is the causative agent of *amebiasis,* a condition commonly referred to as *amebic dysentery* to distinguish this protozoan disease from bacillary dysentery, a diarrheal infection caused by the bacterium *Shigella dysentery.*

The protozoan *Entamoeba histolytica* lives normally as a parasite in the large intestine of humans. During acute infections, ulcerative lesions are produced in the intestinal wall, through which the protozoans escape and penetrate the liver, lungs, and other organs. Characteristic symptoms of the acute infection are abdominal pain and a bloody diarrhea.

When the parasites are released to the external environment in human excreta, they do not multiply but form cysts that can survive for long periods. Human beings become infected by ingesting the cyst in contaminated water or food. As people are the only animal reservoir, the spread of the infection is related to the status of environmental sanitation in a given area. The disease occurs in all parts of the world but is most common in areas that have inadequate water- and sewage-treatment facilities, and in warm climates in areas where the overall level of sanitation is poor.

Infected individuals should receive chemotherapeutic treatment with chloroquine to inhibit the protozoans and with tetracycline to prevent secondary bacterial infections. Furthermore, infected persons should be prohibited from working as food handlers. Other general preventive measures consist of providing adequate facilities for purifying drinking water, and for disposing of sewage in a sanitary manner.

Giardiasis The protozoan *Giardia lamblia* is the causative agent of the clinical syndrome called *giardiasis*. Normally, the protozoan lives as a commensal in the human gastrointestinal tract. The carrier rate is believed to be between 1.5 and 2.0 per cent in the United States. Periodically, the protozoan replicates to the extent that huge numbers accumulate and cause blockage of fat absorption, referred to as severe malabsorption syndrome. Characteristic symptoms are abdominal cramps, dehydration and diarrhea with pale stools. Transmission of the infection is by the ingestion of water and/or food contaminated with protozoan cysts. Treatment and prevention is the same as was described above for amebiasis.

The largest outbreak of giardiasis in the United States occurred in Rome, New York, during 1974–1975. Out of a population of 46,000, more than 4,800 individuals experienced symptoms of giardiasis. Epidemiological investigations revealed that the city's drinking water was contaminated with *G. lamblia* cysts. Those laboratory data were the first to demonstrate the presence of *G. lamblia* cysts in water, and also the first to confirm the infectivity of those cysts. Infectivity was established in pathogen-free beagle puppies after they were allowed to drink samples of the contaminated water.

A more recent outbreak of giardiasis occurred between June 1 and July 28, 1976, in Estes Park, Colorado. Giardiasis was confirmed in several individuals from among two groups who had stayed on separate occasions at a particular cabin that was supplied with water from a small reservoir. The water in that reservoir had been chlorinated but not filtered. Further laboratory investigations confirmed the presence of *G. lamblia* cysts in water samples taken from that reservoir.

These case reports demonstrate the importance of access to a drinking water supply that is adequately purified—one that has been both chlorinated and filtered. (These processes are discussed in Chapter 11.) *G. lamblia* cysts are not destroyed by chlorination at dosages and contact times commonly used in municipal water-purification facilities.

Selected venereal diseases

Venereal disease (VD) is a general term used to designate infections transmitted from one person to another primarily by sexual contact. One can, however, contract VD through intimate contact with the mucous membranes of an infected person in ways that do not involve sexual intercourse.

A variety of types of venereal diseases have been characterized, but our discussion will be limited to the two that are the most serious public health problems in the United States, gonorrhea and syphilis.

Gonorrhea   In terms of frequency of occurrence or total number of infected persons, *gonorrhea* ranks as a major public health problem in the United States. During 1975, nearly 1 million cases were reported to the Center for Disease Control, and those reported cases are believed to represent only a fraction of the actual number of cases.

Neisseria gonorrhoeae, the causative agent (Figure 8.7), is a gram-negative bacterium normally found in human beings only during an active infection; it does not have a nonhuman host. In most cases the bacterium grows within the mucous membranes of the genitourinary tract, from which it is transmitted from person to person by direct contact with mucous membranes. Symptoms usually develop within 4 to 7 days and consist of a urethral discharge accompanied in the male by a burning sensation when urinating, and in the female by vaginal secretions. Unfortunately, a large percentage of females remain asymptomatic during active infection. Serious complications, such as endocarditis, or arthritis, may develop in both sexes if infections are not treated. The infections do not confer immunity, so repeated infections are possible.

Newborn infants may contract a serious gonococcal infection of the eyes (neonatal ophthalmia) during birth if the mother has an active case of gonorrhea. To prevent that condition, a 1 per cent solution of

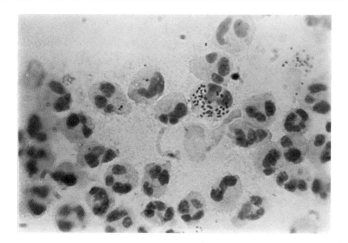

Figure 8.7. Light microscope photo of *Neisseria gonorrhoeae* cells. Note the bacterial cells within the polymorphonuclear leukocytes. (Courtesy of T. J. Kloeckl.)

silver nitrate, or an antibiotic such as tetracycline, is routinely administered to the eyes of infants immediately following birth.

Treatment of gonorrhea can usually be accomplished without difficulty if the infected individual consults a physician or public health clinic. The present drug of choice is aqueous penicillin, but tetracycline is administered to persons allergic to penicillin. Prevention is difficult, because it depends on each infected individual accepting responsibility for his/her sex partner, which in turn requires informing the health authorities of sexual contacts. Although such information is strictly confidential, many infected persons refuse to cooperate with officials. Thus, gonorrhea continues to be spread. Furthermore, the problem of treatment is becoming more complicated with the emergence of *Nesseria gonorrhoeae* strains that are resistant to penicillin. *N. gonorrhoeae* strains that produce penicillinase have been identified in 11 countries. Many physicians screen for gonorrhea by taking cultures from females during routine pelvic examinations. Since prevention and control consist of both moral and microbiological aspects, education as to the nature of the disease seems to be the ultimate answer.

Syphilis Unlike gonorrhea, *syphilis* infections do not involve large numbers of individuals. During 1975, less than 26,000 cases of the disease were reported to the Center for Disease Control. In terms of severity, however, syphilis is an extremely important disease and a major public health problem.

Treponema pallidum, a spiral-shaped bacterium (Figure 8.8), is

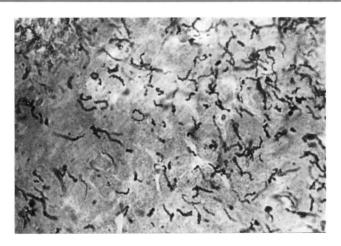

Figure 8.8. Light microscope photo of *Treponema pallidum* cells. (Courtesy of T. J. Kloeckl.)

the causative agent of syphilis. Like other venereal diseases, syphilis is transmitted primarily by sexual contact; the bacterium does not have a nonhuman host. Unlike gonorrhea, syphilis is a chronic disease and, if untreated, three stages are recognized: primary, secondary, and tertiary.

Primary syphilis is characterized by the development of lesions, called *chancres,* on the penis or vaginal wall or in other areas of the genitourinary tract. The bacteria grow in the lesions and often spread to the lymph nodes. The chancre heals spontaneously and may be mistaken for a mild sore or even go unrecognized.

Secondary syphilis gives rise to characteristic symptoms from 2 weeks to 3 months after the appearance of the chancre. Symptoms at this stage consist of the development of a skin rash or lesions on hands and around the mouth, and may involve bones and the nervous system. Direct contact with the skin lesions can cause transmission of the infection.

Tertiary syphilis is the late or terminal stage, and is characterized by the development of granulomatous lesions throughout the body. The person is recognized to be in a serious state of chronic illness. Infections produce antibodies, but not lasting immunity.

Treatment can be accomplished without difficulty if an infected person consults a physician during the early stages. By administering penicillin, a person can be rendered noninfective in 24 hours. Penicillin chemotherapy is the treatment of choice for all three stages of the disease. Tetracycline is used for persons allergic to penicillin.

Prevention consist of identifying and treating persons who have been in contact with infected individuals. Tracing contacts is a difficult

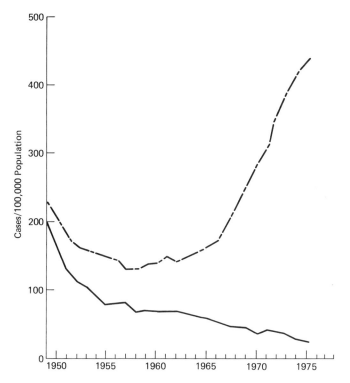

Figure 8.9. Reported cases of gonorrhea, broken line, and syphilis (all stages), solid line, within the United States by Year, 1950–1975. (Center for Disease Control, *Morbidity and Mortality Weekly Report,* **24** (No. 54): **August 1976; Annual Summary for 1975.)**

task for officials, owing to the common unwillingness of infected individuals to reveal their contact. All persons should be treated who have contacted a case of primary syphilis at any time during the previous 3 months. Those with secondary and tertiary infections, then, may have contacts going back several years, who should be treated. Prevention depends upon educating all age groups to the nature of the disease. The trends of reported cases of venereal diseases in the United States is shown in Figure 8.9.

Prevention and control of infectious diseases

The ultimate aim of epidemiology is to understand the web of causation for a specific disease and to establish methods for its control. The specific approach to the establishment of controls is determined by the type of infective agent, geographical location, health status of the population, overall level of environmental sanitation, and a host of other factors. More specifically, control measures can be considered in terms of three objectives: to eliminate the reservoir, to interrupt the transmission of the infective agent, and to increase host resistance.

In a broad sense, the control or eradication of communicable diseases is a major public health problem of concern to the general public throughout the world. Many countries work cooperatively on an international scale through the World Health Organization (WHO) in the development and implementation of programs to prevent the spread of pathogens that produce devastating diseases. Furthermore, separate official agencies within many of the developed countries are charged with protecting the general public from threatening diseases. Within the United States, the broad responsibilities for control of communicable diseases lie within the Federal Department of Health, Education, and Welfare. Its separate programs are implemented through a diverse group of federal, state, and local agencies.

In general, all aspects of prevention and control of communicable diseases are implemented through programs that involve one or a combination of the following: chemotherapy, prophylactic immunizations, and environmental sanitations.

Chemotherapy

When specific chemical substances are administered to the human body (externally or internally) for the purpose of interrupting the growth (multiplication) of microorganisms, the process is called *chemotherapy*. The kinds of chemical substances that can be used in chemotherapy are called *chemotherapeutic agents*. When applied externally, they are called *topical agents*, and when taken orally or by injection, they are called *systemic agents*. Chemotherapeutic agents are defined or characterized by their unique properties: the ability to inhibit the growth (multiplication) of microorganisms at concentrations that can be administered to the human body preferably without producing complications (side effects) to the host that are harmful. Chemotherapeutic agents exhibit the phenomenon of *selective toxicity* by exerting their effects on microbial cells without damaging human cells.

Numerous kinds of chemotherapeutic substances have been characterized, among them are the antibiotics (discussed in Chapter 4). Here we shall present an overview of chemotherapy and discuss some problems associated with the broad-scale use of antibiotics.

Antibiotics vary considerably in terms of the type of microorganism they will inhibit, and they also vary in terms of their mode of action (i.e., specific mechanisms through which microbial cells are altered). Furthermore, the altered cell may be affected only temporarily by a chemotherapeutic agent or it may be damaged permanently. Those that exert their effects on bacterial cells only while present in the body are referred to as *bacteriostatic substances,* and those that exert a lethal effect on bacteria are referred to as *bacteriocidal agents.* It is also important to remember that some antibiotics act preferentially on a particular kind of bacteria, while others exert equal effects on bacteria of various kinds. For example, antibiotics that will inhibit only gram-negative or only gram-positive bacteria are referred to as having a narrow range of activity, whereas types of antibiotics that will inhibit both gram-positive and gram-negative bacteria are referred to as broad-spectrum antibiotics. The successful use of chemotherapy in the control of bacterial infections is sometimes aborted by the emergence of antibiotic-resistant strains. In such instance combined therapy with two or more antibiotics is useful in preventing the emergence of resistant mutants.

One of the most interesting and perplexing problems that physicians must deal with is the management of infections that are caused by multiple drug-resistant microorganisms. In general, the emergence of drug-resistant strains of bacteria accompanied the introduction and broad-scale use of antimicrobial agents, and the number of resistant strains has increased accordingly. In 1940 most strains of *Staphylococcus aureus* were susceptible to penicillin, but within a short period after the widespread use of this antibiotic, penicillin-resistant strains of *S. aureus* emerged. A similar phenomenon has accompanied the introduction and broad-scale use of most, if not all, of the chemotherapeutic drugs that are currently being used in modern medical practice. The increased number of drug-resistant pathogens was attributed for many years to the broad-scale use of antimicrobial agents, which created environments that favored the growth and development of normally occurring antibiotic-resistant mutants. As the problem continued to expand, research investigations soon revealed that drug-resistant mutants were only one aspect of the problem. The other aspect, and indeed the most serious, relates to R-factor-mediated drug resistance.

An *R-factor* is a genetic element (a fragment of DNA that contains the gene for drug resistance) that may or may not be present in the cytoplasm of bacterial cells; when present, it may exist autonomously or become incorporated into the bacterial chromosome. An R-factor is neither essential to a cell nor damaging to it. Cells that carry R-factors are designated as F^+, and cells that do not contain R-factors are designated as F^-. An F^+ cell is the donor or male, and can transfer its R-factor to a susceptible F⁻ recipient by conjugation or by transduction. It is important to remember that R-factor-mediated drug resistance in bacteria is an infectious process and differs from the drug resistance that develops from a spontaneous mutation. Like mutants, F⁺ cells have a competitive advantage over F⁻ cells, especially when subjected to the selective pressures of chemotherapeutic drugs in hospital environments.

Hospital environments, or certain reservoirs within them, are well-known sources of F⁺ cells. Furthermore, certain type of resistant F⁺ cells are endemic in hospitals. As opportunist pathogens, the frequency of hospital-acquired infections due to F⁺ cells is relatively high when compared with the frequency of similar infections caused by F⁻ cells. Even when such infections have been rendered clinically safe, there remains the possibility that the person will be a carrier. Thus, F⁺ pathogens have ample opportunity for widespread dispersal in the ecosystem outside the hospital environment. Therefore, R-factor-mediated drug-resistant cells are of practical significance not only from the standpoint of making treatment of an immediate infection more difficult, but from the standpoint of conferring resistance to commensals that are widely dispersed in the biosphere. In laboratory experiments, transfer of R-factors among bacteria that inhabit the intestinal tract of laboratory animals have been detected after feeding the animals F⁺ cells.

The problem of managing infections caused by R-factor-mediated drug-resistant pathogens becomes more difficult when one considers the diversity of bacterial genera that may contain susceptible F⁻ cells.

Another factor of importance in the use of chemotherapy is a change in the normal flora that may result from prolonged administration of broad-spectrum antibiotics. In such instances, an ecological imbalance develops among pathogens and commensals. Those types that are normally present in or on tissues of the body in low numbers such as *Candida albicans* and "opportunists" such as *Pseudomonas aeruginosa,* often increase rapidly and produce superinfections while commensals are surpressed. In such instances more rigid measure are often required to manage the newly emerged infections. In this regard it is important for us to recognize that infections caused by fungi, protozoans, and viruses are not manageable with antibiotics. Amphotericin B and griseofulvin are two widely used antifungal chemotherapeutic agents; and quinine derivatives such as chloroquine are widely used in the management of protozoan infections. Virus infections seldom respond to specific treatment. Therefore, prevention of viral epidemics is dependent primarily upon increasing the level of host resistance through prophylactic immunizations.

Prophylactic immunization

Vaccinations have been used for many years to induce an immunity in individuals who are susceptible to particular kinds of diseases. For viral diseases, mass-immunization programs have proved to be highly effective and the most rewarding kind of control measure. For example, small pox is becoming an obsolete disease largely as a result of an effective vaccination program that was implemented on a worldwide basis. In a similar manner, paralytic poliomyelitis has been reduced to an extremely low level as a result of the widespread use of mass immunizations. Many kinds of bacterial infections have also been controlled by mass immunizations: diphtheria and whooping cough are examples of such diseases. In a large number of other diseases, prophylactic immunizations with hyperimmune sera (preformed antibodies) have proved to be beneficial. Yet, with the exception of smallpox, mass immunizations have not succeeded in the prevention of sporadic outbreaks. Therefore, continuous surveillance and follow-up immunizations are required at periodic intervals to maintain a high level of protection in a given population.

The effectiveness of any immunization program is dependent upon:

1. The nature of the antigens in the vaccines in active immunizations,
2. The potency of the antibodies in passive immunizations,
3. The body's ability to tolerate the vaccine preparations,
4. The body's ability to synthesize antibodies

Prophylactic immunizations have proved to be an invaluable life-supporting measure for certain individuals: the very young, the very old, and those whose health has been compromised by traumatic injury or chronic disease.

Environmental sanitation

It is important to remember that many nonliving reservoirs of microorganisms exist within the environment. Those within the following categories are potential sources of infectious organisms: (1) food and food products, (2) drinking water, and (3) recreational waters. Therefore, various official agencies within the United States and other developed countries have established standards and regulations that govern the manner in which these items are used by people, in an effort to prevent the transmission of infectious diseases. Another major concern of environmental sanitation programs is the disposal of liquid and solid waste (see Chapter 11).

Key Words

acid-fast   Refers to the ability of certain kinds of bacteria (mycobacteria and related types) to retain the red-colored dye (carbol fuschin) after being subjected to the acid-fast staining procedure.

bacteremia   Refers to bacteria present in the bloodstream but not actually multiplying.

endocarditis   An inflamation of the mucous membranes of the heart.

immunosuppressant   A chemical compound or substance that retards the functioning of the immune system in animals.

osteomyelitis   An infectious disease that involves the destruction of bone tissue.

pus   Exudates formed during an infection, consisting of bacteria, white blood cells, dead tissue cells, and other foreign materials from the bloodstream.

septicemia   Refers to the invasion and multiplication of microorganisms within the bloodstream.

Selected Readings

1. Anderson, F. M., N. Data, and E. J. Shaw. 1972. R-Factors in Hospital Infection. British Med. J. 3:82 – 85.
2. Center for Disease Control. "Collected Recommendations of the Public Health Service Advisory Committee on Immunization Practices." *Morbidity and Mortality Weekly Report* 21 (No. 25):June 1972.
3. *Morbidity and Mortality Weekly Report* 24 (No. 23):June 1975.
4. *Morbidity and Mortality Weekly Report* 24 (No. 43):October 1975.
5. *Morbidity and Mortality Weekly Report* 25 (No. 21):June, 1976.
6. *Morbidity and Mortality Weekly Report* 25 (No. 28):July 1976.
7. *Morbidity and Mortality Weekly Report* 26 (No. 7):February 1977.
8. *Morbidity and Mortality Weekly Report* 26 (No. 5):February 1977.
9. Rosebury, Theodor, 1971. *Microbes and Morals* New York: The Viking Press.

CHAPTER 9

Epidemiology of zoonotic diseases

- **Selected types of zoonoses**
    Bacterial zoonoses
        Anthrax
        Brucellosis
        Bubonic plague
        Salmonellosis
    Viral zoonoses
        Encephalitides
        Yellow fever
        Rabies
    Protozoan zoonoses
        Malaria
        Toxoplasmosis
- **Prevention and control of zoonoses**
    Prophylactic immunization
    Environmental sanitation
- **Key words**
- **Selected readings**

The epidemiological principles discussed in Chapter 8 are equally applicable to microbial diseases that occur among lower animals. In spite of the fact that natural immunity protects human beings from most animal diseases, the protection is not absolute. Thus, we must consider two kinds of hosts, human beings and other animals (both wild and domestic), and also consider the manner in which infectious agents can be exchanged between them. Diseases of animals that are transmissible to people, are called *zoonoses,* and are most prevalent during periods when there is a sudden increase in the number of animals that develop a specific disease within a community or other geographical region. Instances of such occurrences are referred to as *epizootics.* The terms "epizootic" and "epidemic" are used to refer to synonymous conditions that may occur in animal and human populations, respectively.

Selected types of zoonoses

The large populations of wild and domesticated animals that live in our immediate communities constitute reservoirs from which agents that cause zoonotic diseases can be dispersed. Furthermore, dispersal is facilitated by unsanitary conditions that often exist in areas that are overpopulated with both people and animals. Some of the best studied, and most commonly found types of zoonoses will be discussed in the following section. Collectively, they are caused by environmental pathogens that must be controlled.

Bacterial zoonoses

In terms of historical significance and in terms of severity, anthrax, bubonic plague, and brucellosis rank among the classical diseases that occur in human beings and animals. Some bacterial zoonoses that still threaten us today are listed in Table 9.1.

Anthrax The historical significance of the anthrax bacillus *(Bacillus anthracis)* was noted in Chapter 1. Robert Koch and Louis Pasteur studied this bacterium extensively during the formative stages of microbiology as a science.

Anthrax is primarily a disease of lower animals (horses, cattle, sheep, goats) and is caused by *Bacillus anthracis*. This bacterium has the ability to maintain itself either vegetatively or in the spore state in many kinds of soils for years. Animals become infected from contact with contaminated soils and/or by ingesting vegetation contaminated with this pathogen. Then, infected animals become living reservoirs from which the pathogens are dispersed. Anthrax, a serious disease, is responsible for heavy losses in livestock in Asia, Africa, and certain countries in the Middle East.

During the period of July 22 to September 10, 1976, an outbreak of animal anthrax occurred in Texas. One hundred and sixty cattle and horses died of this disease during that epizootic. Thirteen sites were affected (11 ranches, a feedlot, and one isolated pasture), and all but three were geographically contiguous. One ranch was located approximately 10 miles from the majority of affected sites, but the feed lot and

Table 9.1. Selected Types of Bacterial Zoonoses

Disease	Animal Reservoir	Agent	Mode of Transmission
Anthrax	Horses, other animals	*Bacillus anthracis*	Hair, infected tissue, dust
Brucellosis	Cows, swine, goats, sheep	*Brucella* species	Milk, infected meat products
Bubonic plague	Rodents	*Yersinia pestis*	Fleas
Salmonellosis	Poultry, rodents, pet turtles	*Salmonella* species	Flesh, eggs, excreta from turtles

the isolated pasture contained cattle that had been transferred from the epizootic area. No human anthrax cases were reported in connection with the outbreak.

The Texas Animal Health Commission reacted quickly. After bacteriological confirmation of the disease, the Commission instituted a program of quarantine and vaccination of animals on affected and adjacent ranches. As a result of this vigorous control program, the epizootic subsided quickly and the last animal death from *B. anthracis* occurred eight days after the vaccination program was completed. This case report emphasizes the importance of a quick response by health officials to unexplained sudden deaths of animals in an anthrax endemic area. Such prompt action not only reduces the financial loss from animal deaths, but also decreases the risk of human infections.

Human anthrax is primarily an occupational disease, most prevalent among individuals that have contact with infected animals or their contaminated products (hides, meat, hair, etc.). Two clinical forms of human anthrax are generally recognized: (1) cutaneous anthrax, and (2) inhalation anthrax (wool-sorters' disease).

Cutaneous anthrax is characterized by the development of pustules on the skin after this pathogen (vegetative cells or spores) gains entry to the body through scratches or abrasions. When the infection remains localized, the disease can often be managed effectively by prompt treatment with antibiotics (penicillin or tetracyclines). In some cases the bacilli do not remain localized but spread from the infected pustules into regional lymph nodes, from which they gain entry to the circulatory system. Then, the conditions progress rapidly into a fulminating systemic disease that is often fatal.

Inhalation anthrax is the clinical form of the disease that develops after aerosolized vegetative cells and/or spores gain entry to the body through the respiratory system. This form of anthrax is usually more severe than the cutaneous form. In many cases a severe bacteremia has developed by the time symptoms appear, and in these cases chemotherapy is ineffective.

Virulence of *B. anthracis* is now known to be associated with the production of an exotoxin. The exact site or tissues on which the toxin acts is not known. However, toxins spread rapidly through the circulatory system, and death usually occurs suddenly as a result of cardiac failure.

Human anthrax is not an obsolete disease. Cases still appear sporadically in all parts of the world. In January 1976, a 32-year-old California craftsman who operated a weaving business in his home died of inhalation anthrax. Epidemiological investigations showed that the most probable scource of the pathogens was his working materials (yarns that consisted of imported camel hair, goat hair, or sheep wool in varying combinations). Further laboratory tests showed that *B. anthracis* was present in his clinical and autopsy specimens, and also in a variety of the yarns that were recovered from the distributors.

Prevention and control consist of prompt therapeutic treatment with penicillin G or other broad-spectrum antibiotics. Chemotherapy is usually ineffective unless initiated early. Survivors of an attack of this disease appear to develop an immunity that protects them to some degree from subsequent attacks. Yet, the human vaccine (a toxoid) appears to offer only marginal protection against the lethal anthrax toxin. It appears that the animal vaccine that is used to protect livestock is more effective. Once the animals die of the disease their carcasses should be cremated. Strict adherence to sanitary measures when handling infected materials seems to be the most effective mode of prevention.

Brucellosis   Brucellosis is caused most often by one of three different species of bacteria, each named for the kind of animal that it was originally associated with: *Brucella abortus* (cattle), *Brucella suis* (swine), and *Brucella melitensis* (goats). Infected animals are the reservoirs from which humans acquire the disease. This type of zoonosis occurs most often in individuals who handle meat and/or meat products in the meat-processing industry.

In the United States, 2,047 confirmed cases of brucellosis in human beings were reported to the Center for Disease Control during the 10-year period 1965–1974. From case reports, the most probable sources of infection were determined. Persons working in various occupations within the meat-processing industry (slaughterhouses and

Table 9.2.   Brucellosis Cases in Employees of the Meat-Processing Industry in the United States, 1965–1974

| Year | Most Probable Source of Infection | | | | | Total |
	Swine	Cattle	Swine or Cattle	Sheep or Goats	Unspecified Farm Animals	
1965	51	16	30	—	6	103
1966	42	6	27	—	10	85
1967	52	14	35	1	7	109
1968	80	9	15	1	11	116
1969	113	10	12	3	11	149
1970	79	9	21	1	8	118
1971	82	4	7	—	2	95
1972	80	16	11	1	1	109
1973	49	17	13	—	1	80
1974	74	20	15	—	—	109
Total	702	121	186	7	57	1,073
Percent	65.4	11.3	17.3	0.7	5.3	100.0

Source: CDC Brucellosis Surveillance, Annual Summary for 1975, issued July 1976.

rendering plants) accounted for 1,073, or 52 percent, of the confirmed cases (Table 9.2).

The 1975 surveillance data (not included in the table) show that human cases of brucellosis continue to increase in the United States. A total of 309 confirmed cases were reported in 1975, and 60 percent of them were among employees of the meat-processing industry. It is important to note that the general decline in human cases since 1947 was reversed in 1973, and the increasing trend has continued from year to year (Figure 9.1). The increase appears to be related to an increase in infections resulting from contact with infected cattle. Such

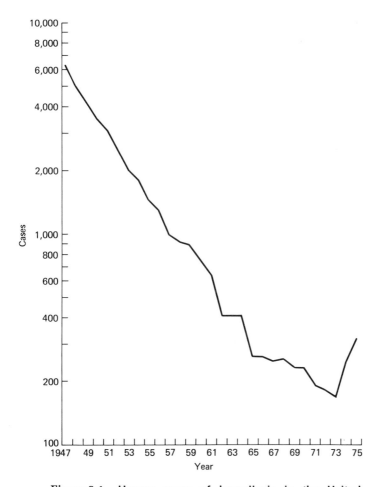

Figure 9.1. Human cases of brucellosis in the United States, 1947–1975. [Data from *HEW Publication No. (CDC) 76-8086,* **Brucellosis Annual Summary for 1975, issued July 1976.]**

infections accounted for 104, or 34 percent, of the 309 cases, as compared to 94, or 30 percent, of the cases attributed to contact with swine or swine products, the most common source of human infection during the preceding 10 years.

Brucellosis is not readily transmitted from person to person, but the infectious agents spread rapidly in animal populations. Those diseased animals that recover often become carriers (i. e., harbor virulent organisms with no apparent symptoms). For this reason infected animals are usually destroyed and cremated. Other preventive measures consist of pasteurizing milk, and the establishment of measures to improve environmental sanitation in all areas of the meat-processing industry.

In human beings the infectious bacteria tend to become established in the lymphatic system and multiply as intracellular parasites. Virulence is believed to be associated with endotoxin poisoning. Symptoms that develop during the course of the infection are nonspecific and may involve enlargement of the lymph nodes, bacteremia with intermittent fevers, and a variety of muscular pains. In most cases the disease is not fatal, and the infection can be managed effectively with antibiotics, tetracycline or tetracycline in combination with streptomycin.

Bubonic plague Bubonic plague, the "Black Death," took the lives of more than 100 million people in Europe and Asia during the Middle Ages. The name was coined from observations of infected persons. Characteristically, their skin turned purple as a result of loss of oxygen from the blood (cyanosis) during the terminal stages of the disease.

The killer disease remained a mystery for many years, but through diligent investigation by several individuals, the causative agent, *Yersinia pestis,* was identified and its association with rats was explained. Bubonic plague is primarily a disease of rats, and the infectious agent usually remains within rodent populations until epizootics occur. During such episodes, the disease is called *sylvatic plague.* Transmission of the disease occurs through the bite of a flea. While feeding on an infected rat, the flea's intestinal tract becomes colonized with the infectious agent. Subsequently, when the infected flea feeds on another host, it regurgitates the bacteria into the wound. Human beings become infected indirectly after fleas from wild rodents infect urban rats. Then infected fleas from urban rats bite human beings, and bacteria from their intestinal tract enter the wound during the feeding process.

Bubonic plague is the clinical form of the disease in which the bacteria gain entrance to the lymph nodes, which become the enlarged "bubos," that gave the disease its name. From the lymph nodes, the bacteria enter the blood and circulate through the body. Death usually follows within hours or a few days. Usually there is no direct transfer of this disease from person to person. However, if the respiratory tract becomes infected with the bacterium, the clinical disease is called *pneumonic plague.* This condition is transmissible from person to person through aerosols or respiratory droplets. Death from this form of plague is also rapid.

Prevention is based upon a thorough understanding of the epidemiological factors that relate to the disease. The interacting factors are human beings, a bacterium, rat, flea, and environment. The application of drastic measures to eliminate rodents might actually enhance the epidemic, because fleas will abandon dying rats and seek other hosts, which might be human beings if other living animals are not available. Artificial immunization with killed or attenuated vaccines will provide protection for about six months. Other measures could be directed toward destruction of fleas and toward the overall upgrading of environmental sanitation.

Plague is enzootic (i. e., always present within populations of wild animals) in certain parts of Arizona, California, and New Mexico. Confirmed cases of human plague among individuals that frequent those enzootic areas appear to be increasing. As of this writing, two fatalities and a total of 13 cases of human plague have occurred in the United States during 1976. Therefore, travelers, campers, and others who visit wilderness areas are warned to exercise precautions and avoid contact with rodents, prairie dogs, and ground squirrels. Furthermore, sick or dead animals in those areas should be reported to health authorities or other officials.

Salmonellosis　*Salmonellosis* is caused by many species of bacteria in the genus *Salmonella.* Organisms within this genus are widely dispersed in nature. Birds, lower animals, and human beings harbor them in their gastrointestinal tract and are natural reservoirs from which the bacteria can be dispersed. One species *(Salmonella typhi)* lacks the ability to live in nonhuman organisms. For this reason typhoid fever develops only in human beings. This classical disease was discribed in Chapter 8.

Other types of *Salmonella* are usually associated with a low-grade fever, less severe than the classical disease, or with food poisoning. The latter can usually be differentiated from the other two conditions by the sudden onset of abdominal pain, vomiting, and diarrhea within a few hours after ingesting the organisms. Of significance is the fact that these organisms remain localized in the gastrointestinal tract. The severe pain is believed to be caused by endotoxin poisoning.

Food poisoning due to *Salmonella* is actually a foodborne gastrointestinal infection. The organisms gain entry into foodstuffs by contamination from a food handler or from lower animals. The food items most commonly involved are meat and meat products (e. g., meat pies, sausage, cured meats), poultry, eggs, and dairy products. It is not uncommon to trace *Salmonella* to food products in which uncooked eggs were incorporated (e. g., custards, cream pies, cakes, eggnog). Large numbers of organisms must be present in such items to produce infection.

Symptoms usually develop 12 to 24 hours after the ingestion of *Salmonella*-contaminated food. The longer incubation time is usually required for the bacteria to mutliply in the gastrointestinal tract. The symptoms are characterized by nausea, vomiting, severe abdominal pain, and diarrhea.

The disease is rarely fatal, and recovery usually occurs within 2 or 3 days without treatment. Prevention can easily be effected by eating only properly cooked and/or properly processed foods. The latter requires good hygienic practices and proper refrigeration and storage of food.

Recently, turtle-associated salmonellosis in children has emerged as a significant public health problem. It has been estimated that during 1970 and 1971 more than 200,000 cases of turtle-associated salmonellosis occurred in the United States, with associated medical expenses that ranged in millions of dollars. Most cases were found in children who had small green pet turtles (Figure 9.2).

The underlying basis of the problem is the difficult, and most likely impossible, task of producing *Salmonella*-free turtles. Turtles sold commercially are normally produced on turtle farms, some of which are located in Mississippi and Louisiana. Such farms have separate ponds in which pet turtles breed. Although artificially constructed, turtle ponds must have dirt banks for egg laying and dirt bottoms where the breeding stock can hibernate. Such requirements, plus the fact that ponds hold large volumes of water, make it highly impractical to rear turtles in a sanitary environment. Furthermore, *Salmonella* species are widely associated with other types of organisms (animal and bird species) that inhabit the area. Consequently, the *Salmonella* species are constantly entering ponds from other environmental reservoirs. As a result, most shipments of turtles certified as *Salmonella*-free in accord with federal regulations are in fact contaminated.

The clinical picture of turtle-associated salmonellosis indicated that turtles harbor the bacteria in their gastrointestinal tract, from which they are excreted into the water of their aquaria. Subsequently, while handling the turtles or while changing the water, children become contaminated with the organisms.

On May 20, 1975, the U. S. Food and Drug Administration banned

Figure 9.2. Small pet green turtle. (Courtesy of R. Seidler.)

the interstate shipment of pet turtles with shells that measure less than 4 inches across. Although the hazards will probably be lessened by such federal and state regulations, it seems reasonable that children should be advised to select other kinds of pets, and adults should be educated of the seriousness of the problem.

Viral zoonoses

As obligate intracellular parasites, viruses cannot reproduce themselves in nature as free-living entities. This is a universal feature of all viruses. Viral agents that cause zoonoses maintain themselves in lower animals (mammals, birds, and arthropods) in the form of a latent infection or as an overt disease. In either case infected animals become reservoirs from which the viruses are dispersed. Normally, dispersal is believed to occur through animal-to-animal contact, or through animal–arthropod–animal contacts. Human infections are incidental to the natural development of those viruses. However, infections in human beings are often severe. Zoonoses that we shall discuss are listed in Table 9.3.

Table 9.3.   Selected Types of Viral Zoonoses

Disease	Animal Reservoir	Agent	Mode of Transmission
Encephalidites			
Eastern equine encephalitis (EEE)	Wild birds	Arbovirus group A	Mosquito (*Culex* species)
Western equine encephalitis (WEE)	Wild birds	Arbovirus group A	Mosquito (*Culex* species)
Venezuelan equine encephalitis (VEE)	Wild burds	Arbovirus group A	Mosquito (*Culex* species)
St. Louis encephalitis (SLE)	Wild birds	Arbovirus group B	Mosquito (*Culex* species)
Yellow fever (jungle)	Monkey	Arbovirus group B	Mosquito (*Aedes aegypti*)
Rabies	Bats, other wild animals	Rabies virus	Animal bites, infected tissue

Encephalitides   In this section we shall present an overview of some related diseases that are caused by arboviruses. This group of viruses has the ability to replicate themselves within tissues of certain lower animals and within the gut of certain arthropods. The former serve as reservoirs and the latter as vectors. Mosquitoes in the genus *Culex* (Figure 9.3) are the most active vectors in the transmission of the encephalitides. As blood-sucking arthropods, they acquire the viral agent while feeding on an infected host (wild bird). Then the virus replicates itself in the gut of the mosquito. Subsequently, when that mosquito feeds on a different animal, viruses are injected into the wound. A generalized cycle for the transmission of encephalitis is shown in Figure 9.4. Many different bird species are susceptible to arbovirus infections. Research has shown that those viruses multiply faster in the blood of wild birds than in the blood of domesticated birds, thus supporting the role of wild birds as primary reservoirs.

Three factors seem to account for differences in names that have been given to the encephalitides: (1) the geographical region in which the disease was first recognized, (2) the difference in strains of arboviruses that cause them, and (3) distinct differences in host susceptibilities to them.

Three of the encephalitides (EEE, WEE, and VEE) are caused by group A arboviruses. In human beings the disease resulting from either type is characterized by a viremia (virus multiplication in the blood) and subsequent infection of the central nervous system. The virus that causes EEE is the most virulent for human beings. Mortality is high, and those victims that survive experience neurological problems. In contrast, the virus that causes WEE seems to be of low virulence, because individuals recover with little difficulty. The third

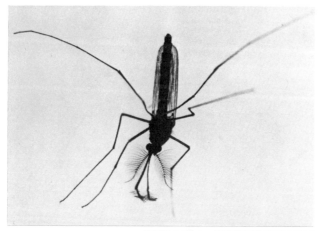

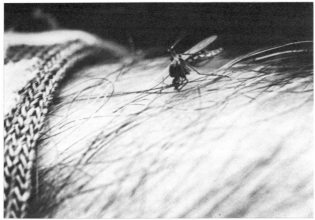

Figure 9.3.   Top: A mosquito (*Culex* species); bottom:
a *Culex* mosquito on the human arm. (*The Milwaukee*
Journal: a "Milwaukee Journal Photo.")

type (VEE) is rarely found in people, but the infection rate is high in horses. In 1972 VEE in horses reached epizootic proportions in Mexico. During the period May–July of that year, 1,248 horses died of the disease. Yet, horses are not significant reservoirs. They appear to be dead-end hosts.

St. Louis encephalitis (SLE) is caused by group B arboviruses. This type of disease is rarely found in horses, but human beings are highly susceptible. The disease is characterized by a central nervous system disorder, but mortality is usually low. In the summer of 1975 there was an epidemic of St. Louis encephalitis among people in the United States. Confirmed cases of SLE were reported from 14 states, with the highest numbers occurring in Mississippi and Illinois, respec-

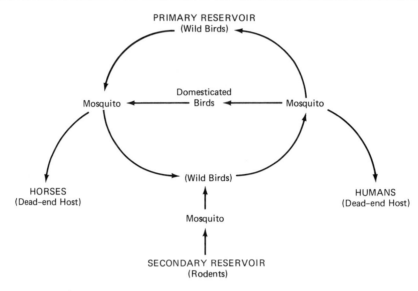

Figure 9.4. Generalized cycle for the transmission of encephalitis.

tively. In the transmission cycle human beings appear to be dead-end hosts, wild birds the primary reservoirs, and *Culex* mosquitos the vectors.

Yellow fever Among group B arboviruses are the agents that cause yellow fever, which occurs in two epidemiological patterns: jungle and urban. The jungle cycle has reservoirs in monkeys, and the vectors are jungle mosquitoes. Human beings in the forests of South America, Central America, and Africa can become infected with the agent through the bites of mosquitoes. Otherwise, people play no role in the cycle. On the other hand, the urban yellow fever cycle involves only *Aedes aegypti* and human beings. Man is the reservoir for this type of yellow fever. Therefore, control of the vector allows control of the disease. The eradication of yellow fever from the United States can be attributed to highly effective mosquito-control programs and to the absence of human reservoirs. Furthermore, prophylactic vaccination with the yellow fever vaccine is recommended for individuals who work in laboratories, where the risk of exposure to the virus is high, and is mandatory for travelers to high-risk countries.

It is significant to note that the yellow fever virus was the first arbovirus to be discovered. Furthermore, the role of the mosquito in the transmission of the yellow fever virus was definitively explained in 1901. During the Spanish–American War, Major Walter Reed and a team of scientists had conducted unprecedented experiments with human volunteers in which mosquitoes that had bitten diseased individuals were allowed to feed on healthy nonimmune volunteers. Results from those experiments proved that mosquitoes were the agents responsible for transmitting the yellow fever virus. However, Major Reed never learned the significance of that research. He died of appendicitis in 1902. The Walter Reed Army Medical Center, Washington, D. C., was named to honor his work.

Rabies   Historical records show that rabies was present in early civilizations, and it ranks today among the most serious infectious diseases. All mammals are susceptible to the rabies virus, but it is found primarily in wild animals (e.g., skunks, foxes, bats). Wildlife accounted for 84 per cent of 2,674 confirmed cases of animal rabies in the U.S. in 1975. Such animals are reservoirs from which the virus is spread to domesticated animals and to people. Untreated rabies is a highly fatal disease in all animals except bats. They often harbor the virus in the form of a latent infection, but many appear to be healthy carriers. For this reason, bats are considered to be the primary environmental reservoir of the rabies virus. The rabies virus is transmitted from animal to animal or from animal to person most commonly through bites of infected animals. In most cases human beings are infected by domestic pets (dogs and cats), but the disease can also be acquired from direct contact with infected tissues (especially if the skin is scratched), and indirectly by inhalation of infected secretions (aerosols).

Clinically, the disease is characterized by a variable incubation period (6 days to 1 year), after which there is a series of progressive events beginning with headache and spasmatic contraction of muscles. Often the individual swallows liquids only with difficulty, because the sight of water provokes muscle spasms (hydrophobia). Unless treated prior to the onset of symptoms, the disease is usually fatal. In fact, only one person with a confirmed infection is known to have survived.

The disease can be treated prophylactically with a postexposure vaccine. Several types are available, but all are versions of the original vaccine prepared by Louis Pasteur in the nineteenth century. The treatment consists of subcutaneous inoculations daily for a period of 14 to 21 days, depending on the type of vaccine. The individual then develops an active immunity to the rabies virus. One problem that physicians find to be perflexing is deciding whether or not to immunize those that have been exposed, because adverse reactions (side effects) may follow vaccination. In some instances, the risk from vaccination is greater than the risk of becoming infected with rabies virus from an animal scratch. For this reason every possible exposure to rabies must be evaluated individually. In the United States, physicians consider the following factors prior to treatment with antirabies vaccines.

1. The species of the biting animal
2. The circumstances of the biting incident
3. The presence of rabies in the region
4. The vaccination status of the biting animal
5. The type of exposure to infectious materials (tissues, fluids, etc.)

Often passive immunization to the rabies virus with preformed antibodies (hyperimmune horse serum) is used to treat high-risk exposures. Another preventive measure is the preexposure immunization. It is recommended for individuals employed in positions where rabies is a threat: veternarians, animal handlers, laboratory workers, and children living in areas where rabies is a constant threat.

These immunization practices, plus rigorous laws that require rabies vaccinations for pets, have contributed significantly to the reduction of rabies in the United States. In 1946, there were more than 8,000 cases of rabies in dogs, compared to 129 in 1975. Rabies in humans has decreased from an average of 22 cases per year during the period between 1946 and 1950 to an average of one to two cases per year since 1963. It is significant to note that one of the human cases that occurred in 1975 was in a 60-year-old farmer and part-time trapper in Minnesota who had been bitten on the finger by an unprovoked stray cat. This was the first human case of rabies from a cat bite in the continental United States since 1960. Also pertinent to this case is the fact that an epizootic of both wild and domestic animal rabies has been present in Minnesota and neighboring states since 1970.

Protozoan zoonoses

Some of the historical diseases of human beings and animals are caused by protozoa. This group of microorganisms are extremely complex, and some of the parasitic types require different hosts to complete their development. In such instances the parasite may have an obligatory requirement for one type of an animal host but not for the other. As a consequence of this versatility, many different animals serve as reservoirs for protozoan parasites. The zoonoses that we shall examine are malaria and toxoplasmosis (Table 9.4).

Table 9.4.   Selected Protozoan Zoonoses

Disease	Animal Reservoir	Agent	Mode of Transportation
Malaria	Human beings	*Plasmodium* species	Mosquito (*Anopheles* species)
Toxoplasmosis	Wild and domestic cats	*Toxoplasma gondii*	Aerosols, cat feces

Malaria   In terms of total human deaths that have been attributed to infectious diseases, malaria ranks among the greatest killers.
Although eradicated from many parts of the world, malaria is still
endemic in certain tropical regions of Central America, South America, Africa, and Asia. Countries in which the disease has been eradicated are still confronted with the problem of controlling imported malaria (i.e., the entry of infected individuals into a malaria-free region). An
example of such a country is the United States. Malaria had been eradicated from the United States and effectively controlled until 1965,
when military personnel began to return from Vietnam, an area where
malaria is endemic. Between 1965 and 1971, the United States experienced a dramatic increase in cases of malaria (Figure 9.5). These cases
in military personnel reflect the incidence of imported malaria, and

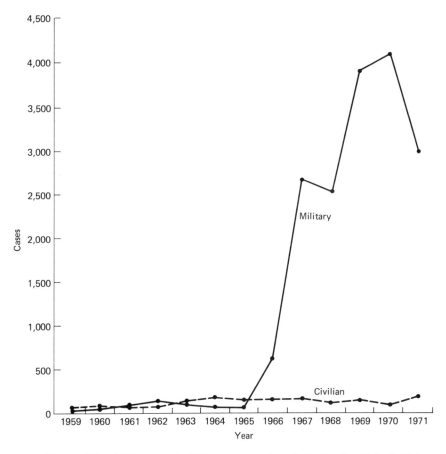

Figure 9.5. Military and civilian cases of malaria in the United States
(1959–1971). (Data from Center for Disease Control, 1971 Annual Report
on Malaria Surveillance, issued June 1972.)

cases in the civilian population reflect the incidence of indigenous malaria (i.e., malaria acquired by mosquito transmission within the United States).

Malaria is caused by protozoans within the genus *Plasmoduim,* and the most commonly found species are *P. vivax, P. falciparum, P. malariae,* and *P. ovale.* However, the protozoa do not act alone. In order for these parasites to survive, certain stages of their development must take place within the gut of female *Anopheles* mosquitoes. Consequently, as blood-sucking arthropods, infected mosquitoes transmit the protozoa to human beings through bite wounds while feeding. For this reason, malaria is called an arthropod-borne disease.

The life cycle of the plasmodia is extremely complex but directly related to the development of the disease in people. Specifically, the cycle of development consists of a human phase and a mosquito phase. Infected human beings are the environmental reservoirs from which the mosquito acquires the parasites. An abbreviated description of the developmental processes will be presented.

The mosquito phase of development begins when the mosquito bites an infected person. Immediately after being ingested by the mosquito, the protozoan parasites undergo certain modifications and attach themselves to the intestinal wall of the mosquito. While located at that site in a stationary form, the parasite undergoes a sexual reproduction process, after which the stationary form ruptures and releases motile (asexual) infective forms. These infective forms migrate throughout the mosquito's body and enter the mosquito's salivary glands, from which they are released when the mosquito bites another person.

In the new human host, the parasite within the blood is carried to the liver and undergoes repeated divisions within liver cells. Some may continue to replicate within liver cells, but many are released and enter into red blood cells (erythrocytes). These forms then reproduce themselves (asexually) through the erythrocyte cycle, a developmental process that terminates with the rupture of the erythrocyte and release of progeny that repeats the cycle. When this occurs, toxic end products from the parasites' development enter the human bloodstream and produce a fever, characteristic of the malarial syndrome. The intermittent high fever with normal periods of well-being in persons with malaria correspond with the abovementioned rupture of erythrocytes.

It is significant to remember that the mosquito phase appears to be commensalistic, because the mosquito is not harmed by the parasite's development, a fact that may contribute to the *Anopheles* effectiveness as a vector in regions where malaria is endemic.

Control measures are directed primarily toward eradication of

mosquitoes, and toward treatment of infected individuals with chemotherapeutic drugs. Even when drug therapy appears to be effective,
individuals harbor resistent types of protozoa which tend to persist
in the liver. Such individuals often experience repeated blood infections (relapse of the disease syndrome).

Toxoplasmosis The protozoan parasite that causes toxoplasmosis *(Toxoplasma gondii)* is an obligate parasite of cats, but a variety of
organisms can serve as reservoirs after becoming infected.

The domestic cat and related felines appear to be the only natural
host for *Toxoplasma gondii*. The parasite develops itself through a cyclical process in the intestinal epithelial lining of the cats, after which
infective forms of the parasite are excreted in the feces. The cat, noted
for its fastidious habits, buries its feces in the soil or litter box, from
which the infective material can be dispersed (Figure 9.6). Thus, in the
home the litter box is an important environmental reservoir from
which the protozoan parasites are dispersed. Other animals become
contaminated by ingesting food contaminated with the parasites or by
inhalation of aerosols containing them. In the body of nonfeline hosts,
the parasite persists in the tissue in an infective state. Consequently,
carnivores and human beings that eat uncooked meat can acquire the
parasites indirectly. Thus, cats are believed to acquire their initial
infection by eating wild birds, raw meat, and so on. Thus, through any
of the above processes, a wide variety of animals may actually harbor
infective *Toxoplasma*.

**Figure 9.6. Domestic cat and its litter box, both po-
tential reservoirs for** *Toxoplasma gondii.* **(Courtesy of
D. Amo.)**

The human disease can be of two types: acquired and congenital. The *acquired form* in adults is usually a mild disease, but the *congenital disease* produces serious complications. If an uninfected woman acquires the parasite during the first trimester of pregnancy, the fetus may become infected, because the parasite can pass through the placental wall. The child may be born with active toxoplasmosis, or with a number of serious birth defects, such as brain damage and/or blindness. As a result of such, it is believed that the number of birth defects from toxoplasmosis is equal to, or greater than, the number of birth defects from any one of the following: rubella (measles), cystic fibrosis, phenylketonuria (PKU), or congenital syphilis.

As a result of the effects that might result from an infection with toxoplasmosis, several preventive measures have been proposed.

1. Feed domestic cats only dried, canned, or cooked meat.
2. Change litter boxes daily and disinfect them with boiling water.
3. Use work gloves when working with soil contaminated with cat feces.
4. Cover children's sandboxes when not in use.
5. Avoid eating raw meat; heating to 66°C (150°F) will kill *Toxoplasma*.
6. Pregnant women should let someone else take care of the cat and contact a physician concerning a screening test for the presence of anti-*Toxoplasma* antibodies.
7. Exercise good sanitary practices around animals and rodents.

We can conclude by stating that toxoplasmosis is a type of zoonosis that individuals can help to control by modifying behavior.

Prevention and control of zoonoses

In general, the measures described in Chapter 8 for the control of infectious diseases in human beings also apply to the control of zoonotic diseases. Yet, there are some basic differences in the approaches that are sometimes used very effectively. For example, the control of brucellosis in the United States relies heavily on measures aimed at eliminating the reservoir. When cattle become infected with *Brucella abortus,* diseased animals, which sometimes include an entire herd, are isolated and destroyed, an approach that cannot be used in the control of human diseases.

Prophylactic immunizations

Both active and passive immunizations are used widely in the control of zoonotic diseases. Vaccination methods for animals are not vastly different from those used for human beings, except that the vaccine formulations (actual content of vaccine preparations) are often different. In either case the ultimate aim is to enhance the animal's resistance to various kinds of infectious pathogens.

Perhaps the significance of prophylactic immunizations can be visualized if we consider the manner in which they are used to control rabies. Prophylactic immunizations are absolutely essential to the prevention of rabies in both animals and people. The natural reservoirs of the rabies virus are rodents and other kinds of wild animals, especially bats. To prevent the transmission of the infectious agent from those living reservoirs to pets (dogs and cats), preexposure vaccinations are required for the pets. Preexposure immunizations are also recommended for human beings employed in high-risk occupations. However, postexposure immunizations have proved to be highly effective as a prophylactic tool in the prevention of human rabies and in the control of this devastating disease, which is almost always fatal after the symptoms develop.

Environmental sanitation

All effective environmental sanitation programs contribute, directly or indirectly, to the control of zoonotic diseases. Such programs encompass a number of diverse measures, aimed primarily at disrupting the transmission cycle through which various kinds of pathogens are dispersed from a reservoir to a susceptible human or animal host. One important measure involves rodent-control programs, which aim to eliminate rodent breeding places, to rat-proof buildings, to destroy rodents by the use of appropriate poisons, and so on. These programs are directed toward the elimination of plague.

A second group of programs attempts to destroy mosquito breeding places, to protect individuals and animals from mosquito bites, and to destroy mosquitos by the appropriate use of insecticides. Such measures contribute to the control of encephalitides, malaria, and yellow fever.

Sanitary control within the food-processing industry would ensure that food for human consumption is processed and handled in a manner that will not allow the transmission of infectious agents, and programs that provide for the sanitary purification of drinking water and the sanitary disposal of human and animal waste (see Chapter 11) would extend the protection of human beings.

Key Words

attenuate Pertains to pathogens that have been treated in a manner to render them less virulent but to retain their antigenicity.

carrier An individual that harbors an infectious agent in or on tissues of its body but exhibits no apparent symptoms of the disease the agent may cause.

endemic A disease, illness, or agent that is always present within a population or a given geographical area.

epidemic A sudden increase in frequency of a disease, above the normal expectancy, in a population of human beings.

epidemiology The field of science that concerns itself with the determinants of a disease or condition and with factors that influence its distribution.

episome An extrachromosomal genetic element (piece of DNA) that may exist autonomously in the cytoplasm of a bacterial cell or may become integrated into the chromosome.

epizootic A sudden increase in the frequency of a disease above the normal expectancy in a population of lower animals.

lymph node A lymph gland occurring in lymphoid tissue.

pandemic A disease that occurs in worldwide proportions, not restricted by geographical or other boundaries.

pasteurization The process of using heat under controlled conditions of time and temperature to reduce the number of microorganisms that may be present in liquids or other food substances (this is not a sterilization process).

vehicle Any contaminated inanimate material (e.g., water and food) from which microorganisms can be transported or dispersed to a susceptible host.

zoonoses The group of diseases primarily confined to lower animals (wild and/or domestic) but transmissible to human beings directly or indirectly.

Selected Readings

1. Altman, R., et al. 1972. Turtle-associated salmonellosis. II. The relationship of pet turtles to salmonellosis in children in New Jersey. *Am. J. Epidemiol.* 95 (No. 6):518–520.
2. Frenkel, J. K. 1974. Breaking the transmission chain of *toxo-*

plasma: a program for the prevention of human toxoplasmosis. *Bull. N.Y. Acad. Med.* 50 (No. 2):228–235.

3. Greenberg, B. 1965. Flies and disease. *Scientific American* 213 (No. 1): 92–99.

4. Kaufman, A. F., et al. 1972. Turtle-associated salmonellosis. III. The effect of environmental salmonellae in commercial turtle breeding ponds. *Am. J. Epidemiol.* 95 (No. 6):521–528.

5. Lamm, S. H., et al. 1972. Turtle-associated salmonellosis. I. An estimation of the magnitude of the problem in the United States, 1970–1971. *Am. J. Epidemiol.* 95 (No. 6):511–517.

6. Mc Coy, R. H., and R. J. Seidler. 1973. Potential pathogens in the environment: isolation, enumeration, and identification of seven genera of intestinal bacteria associated with small green pet turtles. *Appl. Microbiol.* 25:534–538.

7. *Morbidity and Mortality Weekly Report.* 25 (No. 37), September, 1976.

8. *Morbidity and Mortality Weekly Report.* 25 (No. 40), October, 1976.

9. Ruben, R. H., et al. 1970. St. Louis encephalitis in Saline County, Illinois, 1968. *J. Inf. Dis.* 122 (No. 4):347–353.

CHAPTER 10

Microbial toxins in the environment

- **Types of microbial toxins**
  - *Bacterial toxins*
    - *Exotoxins*
    - *Endotoxins*
  - *Algal toxins*
    - *Blue-green toxigenic algae*
    - *Eucaryotic toxigenic algae*
  - *Fungal toxins (aflatoxins)*
- **Ecological consequences of microbial toxins**
  - *Effect on human beings*
  - *Effect on food chains*
- **Microbial toxins as insectidical agents**
  - *Bacterial toxins harmful to insects*
  - *Fungal toxins harmful to insects*
- **Key words**
- **Selected readings**

Microbial toxins are organic poisons produced by microorganisms. When such poisons are ingested, absorbed, or otherwise introduced into the body of a different living organism, they cause damage to tissues and/or interfere with normal physiological functions. Microbial toxins are complex in terms of structure and chemical composition, and they possess antigenic properties, features that differentiate them from simple poisons such as arsenic and cyanide.

In this chapter we shall present an overview of toxigenic microorganisms from an ecological perspective. Our discussion will focus on toxin producers selected from among the various microbial groups, with emphasis on their distribution in the biosphere and the manner in which they interact with susceptible hosts.

Types of microbial toxins

Numerous types of microbial toxins have been described, and the literature abounds with names that are somewhat confusing. Therefore, some familiarity with toxin nomenclature is essential for subsequent discussions. Criteria that have been used most often for classifying the various toxins include:

1. The manner in which the toxin is associated with anatomical sites within the producer cell
2. The mode of action or tissue affinity within susceptible hosts
3. The chemical structure of the toxin molecule
4. The name of the phylogenetic microbial group to which the toxin producer belongs

In some instances only one of these criteria has been used to describe a microbial toxin, and in other circumstances the criteria have been used collectively. Such usages will become obvious as we proceed with our discussion of toxic substances produced by representative organisms from within the following groups.

Bacterial toxins

Toxins produced by bacteria have been studied extensively since the latter part of the nineteenth century, when diphtheria, tetanus, and botulinum toxins were discovered. Today, these three diseases are classical examples of the harmful affects that bacterial toxins can produce. Potent toxins produced by *Corynebacterium diphtheriae, Clostridium tetani,* and *Clostridium botulinum* are entirely responsible for specific clinical symptoms that characterize the respective diseases. Knowledge gained from the study of those diseases and their causative organisms provided a basis for subsequent work which led to the discovery of toxins that many other kinds of microorganisms produce. Consequently, a great deal of detailed information is now available on microbial toxins.

In Chapter 8 we briefly discussed bacterial toxins as a mechanism of pathogenicity and stated that they could be broadly divided into two categories: exotoxins and endotoxins. Now, we shall discuss these groups of toxins in greater detail. The basis for the categories is the manner in which the two groups of toxic substances are associated with cells that produce them. *Exotoxins* are soluble substances produced within cells but secreted to the cell's exterior environment during periods of active growth. *Endotoxins* are bound to the bacterial cell wall as a structural component — or, in certain types of microbes, contained within the cell's cytoplasm — but they are not released until the cell disintegrates. In general, exotoxins are produced by gram-positive bacteria, and endotoxins are produced by gram-negative bacteria. However, two gram-negative organisms are noteworthy exceptions *(Shigella dysenteriae* and *Vibrio cholerae)* because both of them produce exotoxins in addition to their cell-bound endotoxins. Other differences, between exotoxins and endotoxins, which are probably more significant than the location of the toxin within the cell, will be discussed below.

Exotoxins   In addition to being excreted from the producer cell to its surrounding environment during periods of active growth, exotoxins have many other interesting features. They are proteinaceous substances, specifically characteristized by their lack of chemical association with other macromolecules. As such, they can be referred to as toxic antigens. Since antigen–antibody reactions are highly specific, exotoxins can be neutralized by their homologous antibodies. In terms of toxemia and other harmful effects, each kind of exotoxin exhibits a characteristic and specific type of action on a susceptible host. Thus, specific diseases are attributed to them (Table 10.1).

Furthermore, exotoxins tend to behave as enzymes both in terms of sensitivity to denaturation by heat and other agents and in terms of their catalytic action. Although the toxicity of most exotoxins can be destroyed or markedly reduced by exposure to heat, strong acids, or proteolytic enzymes, there are a few exceptions. For example, the

Table 10.1.   Some Exotoxins and the Diseases They Cause

Bacterial Producer	Habitat	Representative Toxin	Disease (Common Host)
Bacillus anthracis	Soil	Complex	Anthrax (human beings and horses)
Clostridium botulinum	Soil	Neurotoxins	Botulism (human beings and other animals*)
Staphylococcus aureus	Human skin	Enterotoxin	Emetic (human beings)
Corynebacterium diphtheriae	Human	Diphtheria toxin	Diphtheria (human beings)
Yersinia pestis	Rat flea	Plague toxin	Plague (human beings and rodents)
Vibrio cholerae	Human gastrointestinal tract	Enterotoxin	Cholera (human beings)
Clostridium tetani	Soil	Neurotoxin	Tetanus (human beings)
Shigella dysenteriae	Human gastrointestinal tract	Neurotoxin, enterotoxin	Dysentery (human beings)

*Host susceptibility varies with specific antigenic type (see Table 10.3).

Staphylococcus aureus enterotoxin is heat-stable, and the potency of botulinum type E exotoxin appears to be enhanced by the action of proteolytic enzymes present in the human gastrointestinal tract. In terms of potency or toxicity, bacterial exotoxins rank among the most poisonous substances that have been described. Toxicity is usually measured in terms of the quantity of an exotoxin (called LD_{50}) required to kill 50 per cent of susceptible laboratory animals (mouse, rat, rabbit, guinea pig, etc.) within a stated period after administering the toxin, or by the quantity of homologous antibody required to neutralize it. Utilizing the former method for measuring toxicity, it has been calculated that 1 milligram of botulinum toxin is enough to kill more than 1 million guinea pigs.

Although the above characterization of exotoxins is abbreviated, it will suffice to show that exotoxins as a group are vastly different from endotoxins, which will be described next.

Endotoxins   The classical or true endotoxin is a complex lipopolysaccharide–protein component of the bacterial cell wall. However, a few proteinaceous substances contained within the cytoplasm of certain bacteria during periods of active growth have also acquired the name "endotoxin." In either case endotoxins are not released until the producer cell is disrupted or until it disintegrates spontaneously by autolysis.

The lipopolysaccharide–protein endotoxin is relatively heat-stable, and each component within the complex plays a specific role in the activity of the molecule. Antigenic determinants are associated with the protein and polysaccharide components, and toxicity is associated with the lipid component. As a result of the specificity that is associated with each component of the molecule, toxicity from endotoxins is not neutralized by their homologous antibodies, because they interact with the nontoxic antigenic determinants (protein and polysaccharide components). Furthermore, endotoxins are less potent and less specific in their mode of action than the exotoxins are. Endotoxins characteristically produce a generalized syndrome in susceptible hosts that is common to all endotoxins regardless of the bacterial species from which the endotoxin was obtained. As a result of this generalized syndrome associated with endotoxins, there are no classical endotoxin diseases. However, the biological effects that can be attributed to endotoxins range from a mild fever to shock and even death. Potency of endotoxins is also measured in terms of the LD_{50} dose.

The preceding discussion pertains to the true or classical endotoxins. Another kind of endotoxin that has been studied extensively is the proteinaceous crystalline parasporal body of *Bacillus thuringiensis*. This is another example that adds to the confusion of the endo–

exotoxin nomenclature. The bacterium *B. thuringiensis* is gram-positive and the crystalline substance is protein in composition. Yet, by being produced within the cell's cytoplasm and contained there until the cell disintegrates, it has acquired the name "endotoxin." This toxin does not cause diseases in human beings or higher animals. Of particular interest is the fact that it causes diseases in insects. At present, studies of this toxin are centered around its use as a microbial insecticide. These aspects will be discussed in a later section.

Algal toxins

The algae are one of the most conspicuous and widely distributed groups of microorganisms known to mankind. They are natural inhabitants of fresh and marine bodies of water throughout the world, and periodic episodes of excessive algal growth have for many years been recognized as algal blooms. Yet, this ubiquitous group of microbes do not cause any of the classical human or zoonotic diseases. Consequently, toxigenic algae have been studed less than toxigenic bacteria. However, some species of algae produce toxic substances that have caused fatalities in human beings and other animals. People generally encounter such toxins indirectly as a result of eating shellfish, such as mussels and clams, that have been collected from toxin-polluted waters. In a similar manner, livestock become poisoned by drinking water that is contaminated with algal toxins.

Research on toxigenic algae has been intensified in recent years. Consequently, we are becoming more aware of such organisms and of the poisonous substances they produce. Species that produce toxins are found both within the procaryotic blue-green algae and within other eucaryotic algal groups (Table 10.2).

Blue-green toxigenic algae The blue-green algae are abundant in both running and stagnant bodies of water, and their blooms are pronounced in such habitats, especially those that are polluted. As a result of their numerous gas vacuoles, blue-green species float at the water surface and are easily visible. Thus, it might often be a misconception that they are the dominant species, unless biomass is measured. However, blue-greens are generally more abundant in freshwater habitats than in marine waters.

Toxins of blue-green algae that have been studied most extensively are from three freshwater genera: *Microcystis aeruginosa, Anabaena flos-aquae,* and *Aphanizomenon flos-aquae.* For each of the three species, toxin production in culture under laboratory conditions occurs best when environmental conditions (light, temperature, aeration) are carefully controlled. It must be remembered that most of the early evidence, which tended to incriminate blue-green algal toxicity with ani-

Table 10.2.   Some Toxigenic Algae

Algal Producer	Toxin	Susceptible Host and/or Method of Acquisition
Blue-green algae		
Microcystis aeruginosa	Microcystis-FDF	Livestock; death to mice 30 minutes after injection
Aphanizomenon flos-aquae	Aphanizomenon toxin	Livestock and fish.
Anabaena flos-aquae	Anabaena-VFDF	Livestock, waterfowl, and fish; death to mice 2 to 10 minutes after injection
Lyngbya majuscula	Dermatitis toxin	Human dermititis (Hawaiian swimmers' itch)
Eucaryotic algae		
Gonyaulax catenella	Saxitoxin	Human paralytic shellfish poisoning
Gonyaulax monilata	Not characterized	Poison to fish (but not to warm-blooded animals)
Peridinium polonicum	Glenodine toxin	Poison to fish
Gymidinum breve	Neurotoxin	A type of shellfish poison*

*Symptoms are similar to those of ciguatera (see *Morbidity and Mortality Weekly Report* 23 (No. 23):June 8, 1974): a type of fish toxicosis (ichthyosarcotoxism) seen in human beings who eat raw fish that inhabit certain marine waters.

mal poisoning, came from field observations. Many of the early reports from widely different geographical areas and countries describe the death of animals as being within a short time (few hours) after drinking water from ponds and lakes that were densely populated with algae.

Microcystis aeruginosa toxins have been isolated and described as two distinct toxins. The first, called the *fast death factor* (FDF), is an endotoxin and is released from laboratory-grown cells during the early stages of growth — an observation that suggested leakage or autolysis of some cells while conditions for growth were optimal. FDF isolated from cells that were artifically disrupted killed mice within 30 to 60 minutes after an intraperitoneal (IP) injection. The second toxin required 24 to 48 hours to kill mice and was therefore called the *slow death factor* (SDF). Further research on these two toxic fractions revealed that SDF was associated with the bacterial partner, which grows in symbiotic association with the blue-green algae. The FDF has been chemically characterized as a cyclic polypeptide.

Anabaena flos-aquae is another blue-green toxigenic genus that has been extensively studied. The toxin from this species is more potent than the previously described *Microcystis*-FDF. In terms of chemical composition, the *Anabaena* toxin is an alkaloid that will kill laboratory mice within 2 to 10 minutes after an IP injection of the purified toxin. Thus, it is a very fast death factor (VFDF), and can now be visualized as the toxin that was probably responsible for some of the most dramatic incidents of wildlife and waterfowl poisonings associated with *Anabaena* blooms.

The third genus *(Aphanizomenon)* of toxic blue-green algae that has been extensively analyzed has a toxic principle (uncharacterized) that appears to be different from those that were described for *Microcystis* and *Anabaena*. It also is a fast-acting toxin. Field observations suggest that *Aphanizomenon flos-aquae* has been responsible for many deaths of livestock and for many fish kills. Although not purified, some reports indicate that one of the toxin fractions from *Aphanizomenon flos-aquae* appears to be similar to the "saxitoxin" of *Gonyaulax catenella,* a toxic eucaryotic algae that will be described in our subsequent discussions.

Another blue-green toxigenic alga *(Lyngbya majuscula)* exerts its toxicity in a manner that is different from either of the toxigenic species mentioned above. This alga has been associated with "Hawaiian swimmers itch," a dermatitis that develops in swimmers after they contact blooms of *Lyngbya majuscula*. The toxic substance has been partially characterized as a lipid substance that causes death in rats when administered intravenously. It also lysed protozoans and exhibited antibacterial properties when tested against some species of mycobacteria.

Eucaryotic toxigenic algae   The toxigenic species that we shall discuss in this section have often been associated with human cases of shellfish poisoning. They represent types of eucaryotic algae that grow abundantly in marine waters. Some members do grow in fresh waters, but usually to a lesser extent. Many of the toxic species belong to the algae called dinoflagellates.

Gonyaulax catenella was the first species to be unequivocally characterized as the causative agent of paralytic shellfish poisoning. The condition occurs in human beings after they eat shellfish (usually mussels and clams) that have been feeding on poisonous dinoflagellates. Symptoms begin with a numbness of lips, tongue, and fingers within minutes after consuming the poisonous shellfish. Progressively, the legs, arms and neck become numb. Then there is a feeling of dizziness, drowsiness, incoherence, and loss of muscular coordination, which culminates with a paralysis of the diaphragm and death due to respiratory failure, usually within 2 to 12 hours of eating the poisonous shellfish. If one survives the first 24 hours, the prognosis is good. There is no effective antidote. Unlike many bacterial toxins, it is non-antigenic.

Under optimum growth conditions, *Gonyaulax* grows rapidly, with a generation time of 1 to 2 days. Maximum growth is usually attained within 2 to 3 weeks, a population density sufficient to cause the water to turn red or brown (commonly called "red tide"). Clams or mussels in such waters are not affected by the poisonous species. The shellfish feed on the poisonous algae and filter the water through their bodies. In the process, the toxin becomes concentrated in the shellfish, which people consume as food. In this manner the toxins are acquired indirectly by a secondary consumer in the food chain.

The toxin isolated from *Gonyaulax catenella* has been described as a heat-stable nonprotein substance called *saxitoxin*. The amount of this toxin required to cause death in humans is not known, but its potency is measured in terms of *mouse units*, a numerical value that has been calculated for use in the assay procedure, and takes into consideration extraction methods, purity of extract, and other factors. Human beings appear to be more sensitive to saxitoxin than other animal species are.

It is important to note that poisonous and nonpoisonous species of *Gonyaulax* have identical characteristics. Therefore, human protection is totally dependent on educating the general public to refrain from gathering shellfish for food, and by requiring commercial producers to screen harvested clams and mussels by the standardized mouse assay technique. Very little is known about the toxicity of other *Gonyaulax* species. Toxins isolated from *G. monilata* only affect fish.

Gymnodinium breve is a dinoflagellate that inhabits the coastal waters in the Gulf of Mexico and is responsible for shellfish poisoning in human beings that mimics ciguatera poisoning, a neurological syndrome (ichthyosarcotoxism) that is caused by eating raw fish in tropical and subtropical regions. Fish most commonly associated with ciguatera toxin(s) are barracuda, amberjack, and red snapper. The toxin is heat-stable, nonpoisonous to fish, and tends to concentrate in their tissues with the highest concentration in the liver and decreasing amounts in intestines, ovaries, testes, and muscles. However, muscles of large fish can contain enough to produce the characteristic symptoms in humans. In February and March 1974 a large number of persons in Hawaii developed ciguatera poisoning after eating Kahala sashimi (raw, sliced amberjack) at a wedding party.

Another interesting eucaryotic alga is *Prymnesium parvum*. This alga produces several closely related toxins that have been characterized as proteolipids. All exhibit toxicity to fish (in different degrees) and are called ichthyotoxins, some of which cause respiratory paralysis. They are exotoxins that this alga excretes into its surroundings. This organism has been incriminated as the causative agent for massive fish kills, and it may be damaging to other gill-breathing animals. In laboratory experiments, the toxins have been shown to have different modes of action: lysis of red blood cells (carp, chicken, cattle, rabbit, human), increase in membrane permeability, and lysis of *Mycoplasma* and cell-wall-less forms of bacteria. Other aspects of these toxins will be discussed in a later section.

Fungal toxins (aflatoxins)

Fungal toxins, like algal toxins, have not been associated with any of the classical diseases in human beings or other animals. This does not imply that fungi are harmless. On the contrary, some are serious pathogens. Selected types of pathogenic fungi and the diseases they cause are discussed in other sections (see Chapters 8 and 11).

Fungal toxins (also called mycotoxins) encompass a wide variety of poisonous substances, but this discussion will be limited to that special group of fungal metabolites that have been characterized as *aflatoxins*. Aflatoxins were discovered in 1960–1961 following a mysterious outbreak of "turkey X" disease in England. During that year more than 100,000 young turkeys died suddenly after eating contaminated feed that had been produced at a particular London mill. Intensive investigations revealed that the dead turkeys had consumed contaminated peanut meal. Through subsequent chemical analyses, the contaminant was isolated and characterized. During the investigations, Brazilian peanuts, from which the feed had been prepared, were also shown to be contaminated with a fungus *(Aspergillus flavus)*. When a toxin, produced by that fungus *in vitro*, was fed to young turkeys, they developed symptoms of turkey X disease. Subsequently, the substance that was isolated from the contaminated peanut meal and the toxic substance that the fungus produced were shown to be identical. Thus, it was named "aflatoxin."

Shortly after aflatoxin was demonstrated to be a toxin of *Aspergillus flavus*, many scientists directed their research efforts toward the characterization of that poisonous substance. By using various sophisticated chemical analyses, the poisonous substance was found to be a family of closely related toxins, designated as aflatoxin B_1, B_2, G_1, G_2, M_1, M_2, B_{2a}, and G_{2a}. All the toxins except M_1 and M_2 were initially identified as components of the fungus toxin. M_1 and M_2 were used to designate toxic substances isolated from milk and other tissues of cows and sheep after being fed aflatoxin-containing rations.

Since those initial studies, which were aimed primarily at characterizing the substance, aflatoxin research has expanded in many directions. In addition to *Aspergillus flavus*, other *Aspergillus* species and several *Penicillium* species are now known to produce aflatoxins under laboratory conditions. Such fungi are common members of the soil microflora, but as a result of the role they play in the spoilage of stored cereal crops and/or their products (wheat, oats, soybeans, barley, corn, flour, peanuts, etc.), they have become known as storage fungi. Large grain elevators, where such products are stored for long periods, can be considered as important reservoirs of aflatoxin-producing fungi, because environmental conditions (optimal substrates, moisture, temperature) favor their proliferation. Although soils throughout the world contain types of fungi that produce aflatoxins, field conditions must be considered as a source of inoculating the grain, but not as a suitable environment for aflatoxin production. The latter occurs in bulk grain elevators.

In terms of toxicity of aflatoxins in natural materials, components B_{2a} and G_{2a} appear to be of less importance than the other types. B_1, B_2, G_1, and G_2 are the most active products. However, it must be remembered that concentrations of aflatoxin in natural materials are determined by many variables — species of fungi, type of cereal, moisture, temperature, and a host of other environmental factors. However, some general considerations relative to mycotoxicosis from aflatoxins (aflatoxicosis) can be made: (1) the condition is most likely to occur in young birds (especially ducklings and turkey poults), young farm animals, and fish (rainbow trout in hatcheries) after eating aflatoxin-contaminated rations (mixed feed); (2) it is noninfectious and is most damaging to the liver, but effects vary with species of animal and type of toxin; (3) therapeutic treatment with drugs is not effective; and (4) natural cases of human aflatoxicosis remain speculative.

On the basis of these facts, several precautions have been recommended as a safeguard against aflatoxicosis. Of course, the first would be to harvest, store, and process feeds and foodstuff in a manner that would be unfavorable to the growth and development of toxigenic molds.

Biological activity of aflatoxins in suspected products is monitored by feeding aflatoxin B_1 to 1-day-old ducklings (called the day-old duckling assay), because ducklings are the most sensitive animal to aflatoxicosis. Actual concentration of aflatoxin in grain or other products is measured quantitatively by extremely sensitive chemical tests. Both the biological assay and the chemical test must be used to monitor grains and grain products for the presence of aflatoxins. Heating is usually ineffective, because aflatoxins are stable up to their melting points (approximately 250°C). However, some reports have indicated that autoclaving at 120°C for 4 hours could significantly reduce the potency of aflatoxin B_1.

From the above we can see that aflatoxins continue to represent a potential economic and environmental problem that must be monitored. Further ecological considerations that pertain to them will be discussed in the section to follow.

Ecological consequences of microbial toxins

In general, microbial toxins must be considered as natural poisons that are potentially present in the environment at all times. Toxigenic species from among all the microbial groups live in aquatic, terrestial, and biological habitats throughout the world. Furthermore, their detrimental effects are often potentiated by human-related activities. Therefore, knowledge of environmental conditions that favor the growth of toxigenic microorganisms, and knowledge of methods and/or processes that can be used to control them or neutralize the activity of their toxic products, are important ecological concerns.

As the world population continues to increase, interactions among various population groups have been intensified. Thus, opportunities for human contact with certain kinds of toxigenic microorganisms have increased. Furthermore, natural aquatic and terrestrial ecosystems are continuing to be modified in an unprecedented manner by activities associated with urbanization and industrialization. For example, those activities that are associated with the production and processing of foodstuffs, and those that are associated with the disposal of domestic and industrial waste often contribute to the growth and dispersal of toxigenic microorganisms.

Effect on human beings

People in modern societies often interact with toxins of organisms that cause the classical toxigenic diseases (diptheria, tetanus, and botulism) without adverse effects. The harmlessness of such encounters is directly related to the use of prophylactic immunizations, therapeutic treatments, and other public health measures. The successful development and use of such measures are related to the nature of those toxins. As mentioned previously, those classical diseases are caused by proteinaceous toxins that are highly antigenic. Consequently, their activity can be neutralized by their respective homologous antibodies. Active immunization with toxoids (i. e., detoxified toxins), or passive immunization with antitoxins (i. e., preformed antibodies to a toxin) have proved to be highly effective means of protecting human beings from diptheria, tetanus, and botulism. Yet, those practices are not absolute. In order to be more effective at controlling such organisms, all ecological and environmental aspects that pertain to the growth of toxigenic organisms and to the production and dispersal of their toxins must be considered. For example, the severity of botulism in human beings and other animals is directly related to the antigenic specificity of the exotoxin (Table 10.3).

Table 10.3.   Antigenic Specificity of Clostridium botulinum Toxins

Antigen (Immunological) Type*	Common Hosts
A	Human beings and chickens
B	Human beings and chickens
C (alpha)	Chickens and ducks
C (beta)	Cattle
D	Cattle
E	Human beings
F	Human beings

*Antigenic types cannot be differentiated on the basis of history or symptoms. Laboratory tests are required to determine which type of toxin is responsible. Prophylactic antitoxins (vaccines) have been developed for each type.

While human cases of botulism can often be traced to improperly processed canned food items such as beans, corn, and beets, it is important to note that several cases of botulism have been traced to canned mushrooms and to smoked fish. The anaerobic bacterium *(Clostridium botulinum)* inhabits the soil and forms spores that are widely dispersed in soil and water habitats. Consequently, many types of food items can become contaminated with the various antigenic types. Types A and B are the most prevalent causes of human botulism. However, any of the four human types (A, B, E, F) can be fatal. In terms of total numbers of cases, botulism occurs in human beings at an extremely low rate as compared to other types of food poisoning (Figure 10.1), but it is the most important type in terms of severity. Without proper processing of certain food items, fatalities do occur (see Table 3-3). It is important to remember that all types of botulism toxins are heat-labile and can be destroyed by boiling food items for at least 10 minutes. Note that *Clostridium perfringens* causes food poisoning that is less severe.

We should remember that the enterotoxin of *Staphylococcus aureus* is heat-stable. This toxin is commonly found in picnic-type foods (ham salad, chicken salad, cream pie, etc.). The organisms usually gain entrance to such items from people who have handled the food. Although *Staphylococcus* food poisoning accounted for 28.8 per cent of the total reported outbreaks of foodborne illnesses in 1971, it seldom causes fatalities. In 1973 a total of 307 foodborne outbreaks were reported to the Center for Disease Control, in which 12,447 persons were affected. Fifteen of the cases were fatal. Four of the deaths were caused by botulinum toxins, one by *Clostridium perfringens*, and the other by nontoxigenic organisms.

In general, bacterial toxins that are present in the environment seem to play at least two important roles, both of which have health-related implications. One of the roles relates to their ability to produce severe diseases in human beings and other animals; the other role

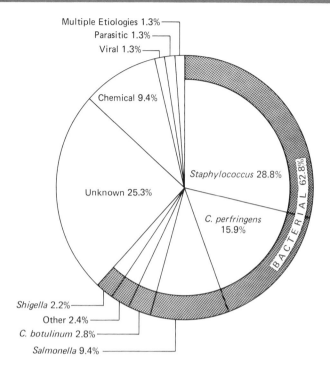

Figure 10.1. Foodborne disease outbreaks in the United States in 1971. [Data from *HEW Publication No. (HSM) 73-8185*, **CDC Foodborne Outbreaks, Annual Summary 1971, issued October 1972.**]

relates to indirect diseases or illnesses that may occur in human beings and/or animals as a result of consuming them as contaminants in water or food items. Other ecological roles that bacterial toxins may play will be discussed in the next section along with the roles that fungal and algal toxins play in food chains.

Effect on food chains

The role that food chains play in ecosystems was discussed in Chapter 6. Although many different food chains exist in nature, all of them are interconnected and they provide for the continual movement of energy and materials in the biosphere. In other words, food chains are vital to the self-perpetuating nature of ecosystems. We characterize food chains as being composed of different nutritional or physiological types of organisms. The primary producers (photoautotrophs) were described as organisms that could satisfy their nutritional needs by utilizing radiant energy and carbon dioxide for the synthesis of their cellular constituents (biomass). Subsequently, energy and materials stored in that biomass are utlized by heterotrophic organisms (primary consumers), which produce more biomass for use by subsequent organisms. Thus, microbial toxins can be transferred through food chains as an integral part of the biomass.

Since all algae are photoautotrophs, they are primary producers of biomass in food chains. In this regard endotoxins that they produce can become poisonous food for subsequent consumers. However, the primary consumer may remain unaffected by the toxin but transfer the toxin (in the form of biomass) to a subsequent consumer. A mechanism of this nature is characterized by the manner in which human beings are affected by paralytic shellfish poisoning. *Gonyaulax catanella,* a major food source for shellfish (mussels and clams), produces an endotoxin. These shellfish are not affected by the endotoxin of *Gonyaulax,* but it can be stored in their tissues. Subsequently, when people eat toxin-containing shellfish, the outcome may be shellfish poisoning, a severe toxicosis.

It is important to note that exotoxins (and endotoxins if released in waters) exert their effect on food chains in a different manner. Organisms that consume toxin-containing water are often poisoned. Massive fish kills, livestock, and waterfowl poisoning that have been traced to waters that contain toxin-producing species of *Microcystis, Aphanizomenon,* and *Anabaena* are typical examples. We consider such events as phenomena that disrupt food chains. Furthermore, such episodes usually occur in polluted waters when toxigenic algal populations become excessive, and human-related activities are often the source of the pollutants. These examples will suffice to demonstrate the complex nature of interrelated factors that influence the manner in which algal toxins can interfere with food chains.

It is reasonable to assume that algal toxins can interact with other organisms in various ways. For example, toxigenic species may interact antagonistically toward other microorganisms that are present in their microhabitats, thus indirectly altering populations that might be vital to a particular food chain.

Aflatoxicosis may be viewed as a phenomenon that may disrupt or modify food chains, especially when large numbers of animals of a single type are fatally poisoned. In addition to the loss of animals in food chains, it has been speculated that M_1 and M_2 aflatoxins could be passed to human beings from cows through milk and milk products if the cows had eaten aflatoxin-contaminated rations. In view of this possibility, further research on M_1 and M_2 aflatoxins is warranted to clarify their status in milk and milk products. Furthermore, aflatoxins must be regarded as environmental contaminants that must be controlled. Thus, the monitoring program that is required by federal agencies must be continued and its effectiveness evaluated periodically to ensure that cereal grains and their products are safe for animal and human consumption.

At this point we shall briefly describe another type of mycotoxicosis that may occur in human beings and domesticated animals, *ergotism*. The fungus *Claviceps purpurea* causes a specific infection in cereal crops, especially rye. The infected kernels do not develop normally but produce large darkened structures called *sclerotia* (Figure 10.2). The sclerotia contain toxic fungal metabolites. When people consume large quantities of bread that has been prepared from grain that was heavily contaminated with sclerotia, they may develop ergotism. A single dose usually produces no adverse affects, but repeated ingestion of small quantities of contaminated bread may prove fatal. The classical symptoms of ergotism are convulsions, hallucinations, and, in pregnant females, spontaneous abortion. In some instances, gangrene develops. Abortion is a common symptom of ergotism in pregnant cows.

Ergotism is an ancient disease, but periodic episodes do occur in this age. In the early 1950s, several human deaths and a large number of illnesses that occurred in a small village in France were attributed to ergotism.

Another interesting aspect that relates to ergotism is the type of

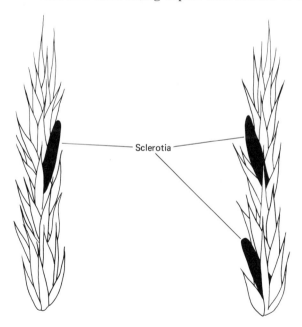

Figure 10.2. Schematic version of rye grains infected with a fungus, *Claviceps purpurea.* **The large darkened structures (sclerotia) contain the toxic alkaloids.**

toxic metabolites that are found in sclerotia. They are several different alkaloids, among which is lysergic acid, the active compound from which lysergic acid diethylamide (LSD) is synthesized. LSD is the well-known hallucinogenic (or psychedelic) drug that was extensively abused during the 1960s.

In many ways *Claviceps purpurea* is unique. It is a soil inhabitant that produces a family of toxins, among which is the etiogical agent of ergotism and the active compound from which a hallucinogenic drug is synthesized. Thus, the environmental effects of this fungus are both economic and health-related.

Microbial toxins as insecticidal agents

Following the much publicized adverse environmental effects that resulted from the widespread and often indiscriminate use of chemical insecticides, especially DDT, the concept of biological control of insect pests has become more attractive to an increasing number of individuals. Biological control simply means the intentional use of natural enemies (predators, parasites, and pathogenic microorganisms) to lower the population density of insect pests. The concept of biological control of insect pests was coined over 50 years ago by H. S. Smith of the University of California at Riverside. In spite of many promoters among entomologists and ecologists, the potentialities of biological control were not fully explored, and the concept was overshadowed by the emergence of the highly effective chemical insecticides during the decades of 1950 and 1960.

In a broad sense, biological control encompasses the use of many kinds of natural enemies to destroy or cause disease in insect pests. The term *insect pest* is used to designate those arthropods that destroy agricultural and/or ornamental plants and to those that transmit or cause diseases in human beings and other domestic animals. Thus, any agent (animate or inanimate) that will kill insect pests is called an *insecticide*. Such agents are intentionally introduced to an area or applied to individual plants and/or animals for the purpose of reducing the number of insect pests to a lower level than would occur otherwise.

The basic goals or aims of biological control are to reestablish a steady-state condition in ecosystems. Insects are an integral part of food chains. Therefore, their populations are regulated and held within limits by a combination of interacting forces in ecosystems. The need for using insecticides arises when the steady-state condition of an ecosystem becomes imbalanced. For example, imbalances often occur when a pest is accidentally imported into an ecosystem where it has no natural enemies or when natural enemies of a specific pest are suddently destroyed. Instances of the latter were observed frequently after extensive applications of DDT. Concomitantly, many kinds of insect pests developed resistance to that chemical insecticide. Those facts, plus the enumerable detrimental affects that DDT had on ecosystems, are largely responsible for the current interest in biological control.

Biological control is an integral part of the overall phenomenon of natural control, those biotic and abiotic forces that function to regulate population density of all organisms within ecosystems. In this regard many of the factors and processes that have been discussed in previous chapters also apply to interactions among insect pests. Biological control actually involves two separate, but interrelated aspects: macrobial control of insect pests and microbial control of insect pests. The former deals with predators and parasites of insects, and the latter is concerned with the activities of microorganisms that cause diseases in insects.

Pathogenic microorganisms of insects have been divided into three categories: (1) obligate pathogens are those microbes that cause a specific insect disease and in nature reproduce themselves only within tissues of some insect host; (2) facultative pathogens are those microbes that possess some mechanism for damaging or invading a susceptible insect, but they can be cultivated in artificial media; and (3) potential pathogens are those microbes that normally do not multiply within the gut of insects but can grow in the hemocoele (body cavity) once they gain access to it. The latter group can be easily cultivated in laboratory media, and they are not associated with a specific disease in insects. Although many kinds of pathogenic microorganisms fall within one or another of the categories above, they are not equally suitable for use as a microbial agent for controlling insect pests. The effectiveness of microbial control is directly related to the virulence of the microbe, the susceptibility of the insect host, and the manner in which the host- – parasite interaction is influenced by environmental conditions.

In view of the considerations above, microbial control agents must be selected with great care. Some desirable attributes that the selected pathogens should possess are listed in Table 10.4.

The effectiveness of microbial control agents is also influenced by the method of application. In some instances it is advantageous to introduce a new pathogen into a given area in an attempt to effect rapid colonization of the area by the pathogen and to initiate an epizootic among the pests. Under different circumstances, it might be more advantageous to apply a toxic product from the pathogen instead of introducing the pathogen itself. The latter is the most commonly used method of applying microbial toxins.

When considering the merits of using microbial insecticides, the importance of environmental safety cannot be overemphasized. In

Table 10.4.   Desirable Attributes of Microbial Agents

Attribute	Comments
Host specificity	Highly virulent for the insect pests, but harmless to human beings and other organisms
Stable (hardy) in the environment	Able to withstand desiccation, exposure to sunlight, and be compatible with other abiotic and biotic forces
Stable (hardy) during preparation	Able to withstand emulsification and other formulation procedures
Easily dispersed	Spread easily, actively or passively
Reproduce rapidly	Able to colonize and replicate in the given ecosystem

other words, microbial insecticides (insect pathogens and/or their toxic products) must be innocuous or harmless to other kinds of organisms. Most countries that use large quantities of pesticides have established laws to regulate the use of microbial insecticides in order to safeguard beneficial insects and to protect the health of human beings and domestic animals. Embedded within those laws are detailed requirements of safety that candidate microbial insecticides must meet before they are certified for use. In the United States such regulations are enforced by the U. S. Department of Agriculture (USDA) and the Food and Drug Administration (FDA). In other countries pesticide regulations are enforced by comparable agencies. Although numerous kinds of microorganisms are known to be harmful to insects, most of them fail to meet the requirements for environmental safety. Therefore, they are immediately disqualified for use as insecticides. In this category are all microorganisms that produce diseases in human beings and domestic animals. Consequently, the number of microbial insecticides that have been certified for use is relatively small. Some of those that have been approved contain viable organisms, but others contain only their toxic products.

Bacterial toxins harmful to insects

Most research that is currently being conducted on microbial insecticides deals with various aspects of the endotoxin that is produced by *Bacillus thuringiensis*. The endotoxin of this bacterium is a proteinaceous crystal, and many insect larvae are highly susceptible to it. Currently, microbial insecticides that contain the crystalline toxin of *Bacillus thuringiensis* have been certified for use in many countries to control insect pests that are harmful to agricultural crops, forest trees, and ornamental plants. The toxin has been shown to be highly effective against more than 100 different larvae of the Lepidoptera (butterflies and moths).

Among the various types of toxigenic bacteria that have been considered for use as insecticidal agents, *Bacillus thuringiensis* have proved to be the most effective and the most interesting. The ability to produce the crystalline proteinaceous endotoxin is a unique feature of *Bacillus thuringiensis* and its 11 varietal types or subspecies. With the exception of the ability to produce the endotoxin, called a "crystalline body," these bacteria are closely related to other aerobic gram-positive spore-forming bacteria that are common inhabitants of the soil.

Detailed research studies have shown that the crystalline substance is formed within cells of *Bacillus thuringiensis* during the process of sporulating. In fact, the protein crystal forms simultaneously within the cell while the endospore is developing. The bacterial cell disintegrates during the last stage of sporulation, and both the endospore and the endotoxin are released. In most varieties the endospore and the crystalline body are distinct entities within the cell's cyctoplasm. The endospore of *B. thuringiensis* is nontoxic to insects and the endotoxin has been characterized as the mechanism responsible for pathogenicity that these bacteria exhibit toward insects. Knowledge of the fact that the endotoxin is released from the cell simultaneously with the mature endospore is of practical importance. Commercial establishment currently grow *B. thuringiensis* in large quantities under laboratory conditions and allow them to sporulate, after which the endotoxin is recovered from the growth medium. It is then mixed with some type of inert carrier, from which it can be applied as a dust or spray. When the endotoxin-containing material is ingested by susceptible insects, it dissolves in their gut. Then the gut becomes paralyzed and nonfunctional within 24 to 48 hours and the insect dies.

From both laboratory and field tests, microbial insecticides (sold under many tradenames) that contain *B. thuringiensis* endotoxin have proved to be both safe and effective against pests it is said to destroy. Furthermore, knowledge gained from research on the nature of the crystalline body may enable future researchers to synthesize a broad-spectrum insecticide with similar characteristics.

It is important to note that two other species within the genus *Bacillus* are currently being used as microbial insecticides. Although their pathogenicity for insects is not based upon toxins, *B. popilliae* and *B. lentimorbus* cause the "milky disease" in the Japanese beetle. The mechanism of pathogenicity for both species is a fulminating septicemia in the beetle's hemocoel (cavity through which the blood, or hemocyanin flows), and culminates with a massive development of endospores which gives the beetles a whitish appearance, thus the name.

These bacteria were demonstrated to be an effective means of controlling the Japanese beetle long before the elucidation of the *B. thuringiensis* endotoxin. In fact, demonstrations that the milky disease bacteria could be used to control the Japanese beetle can be considered as the classical example of using microbial agents for the control of insect pests.

Fungal toxins harmful to insects

In general, investigations of the fungal toxins for possible use as insecticides have been approached with less vigor than those that were concerned with the bacterial toxins. Of course, the fungi are notorious for their ability to produce antibiotics (see Chapter 4), but only a few of their metabolites have been shown to exhibit toxicity toward insects.

Since the aflatoxins are historically associated with liver damage in many kinds of vertebrates and are known as hepatotoxins, they are immediately disqualified from use as insecticidal agents. Such is the case with most other fungal toxins that have been purified. However, a study of their toxicity in insects might yield valuable information relative to the mechanisms of toxin action in insects and contribute to better understanding of the manner in which biological materials could be used to control insect pests.

Numerous fungi are known to be pathogenic for insects by mechanisms that are not known to involve toxins. In this category a few types have been developed for use as insecticidal agents. *Beauveria bassiana* and *Metarrhizium anisopliae* are the most widely used. The latter has been shown to be effective in a very wide host range of insects. However, both have been shown to cause minor pathological conditions in mammals. Therefore, their future use as insecticidal agents is still being evaluated. Thus, further research in the development of microbial insecticides is warranted, especially in view of the possibilities that through genetic manipulations, desirable traits may be obtained in a pathogen or in its toxic products.

Key Words

dinoflagellate   The group of algae that belong to the class Dinophyceae and have cells that are naked (unarmored) or covered with a layer of cellulose, sculptured in definite patterns (armored).

gram-negative bacteria   Pertains to that group of bacteria that do not retain the crystal violet stain in their cells when stained by the procedure described by Christian Gram, and thus appear pink or red when viewed under the light microscope.

gram-positive bacteria Pertains to that group of bacteria that retain the crystal violet stain in their cells when stained by the procedure described by Christian Gram, and thus appear blue or purple when viewed under the light microscope.

heat-labile Pertains to heat-sensitive materials or substances (i.e., chemical structures and/or properties that can be altered or destroyed by heat).

heat-stable Pertains to heat-resistant materials or substances (i.e., chemical structures and/or properties that are relatively resistant to alteration or destruction by heat).

LD$_{50}$ The concentration (lethal dose) of microorganisms or other substances that is required to kill 50 per cent of the test animals to which the substance or organism has been administered.

toxemia Pertains to a toxin in the blood or circulatory system.

toxigenic Pertains to the ability of an organism to produce a toxin.

Selected Readings

1. Aziz, K. M. S. 1974. Diarrhea toxin obtained from a water-bloom-producing species, *Microcystis aerugenosa* Kutzing. *Science* 183:1206–1207.

2. Bernheimer, A. W. 1968. Cytolytic toxins of bacterial origin. *Science* 159:847–850.

3. Debach, P. 1974. *Biological Control by Natural Enemies*. London: Cambridge University Press, p. 323.

4. Echlin, P. 1966. The blue-green algae. *Scientific American* 214 (No. 6):75–81.

5. Hughes, J. M., et al. 1975. Food-borne disease outbreaks in the United States, 1973. *J. Inf. Dis.* 132 (No. 2):224–228.

6. Irving, G. W., Jr. 1970. Argricultural pest control and the environment. *Science* 168:1419–1424.

7. Moikeha, S. N., and G. W. Chu. 1971. Dermatitis-producing alga *Lyngbya majuscula* Gomont in Hawaii. II. Biological properties of the toxic factor. *J. Phycol.* 7:8–13.

8. Moikeha, S. N., G. W. Chu, and L. R. Berger. 1971 Dermatitis-producing alga *Lyngbya majuscula* Gomont in Hawaii. I. Isolation and chemical characterization of the toxic factor. *J. Phycol.* 7:4–8.

CHAPTER 11

The role of microorganisms in polluted environments

- **Microbiological aspects of air pollution**
    Non-industrial microbial aerosols
    Industrial microbial aerosols
- **Microbiological aspects of water pollution**
    Drinking water — problems of purification
    Sewage — problems of treatment
    Solid wastes — problems of disposal
- **Microorganisms as tools for detecting specific pollutants**
    Bacterial and protozoan assays
    Tissue culture assays
- **Key words**
- **Selected readings**

Throughout this book we have emphasized the functional aspects of microorganisms in ecosystems and shall continue that theme in this final chapter by discussing several interrelated aspects of pollution. Before discussing the role that microorganisms play in polluted environments, it is necessary for us to define pollution, an exceedingly difficult term when one considers the complexity of ecosystems. A specific material may be a pollutant when present in one ecosystem under a given set of conditions, but the same material may be a nonpollutant when present in another ecosystem where environmental variables are different.

For example, hydrogen sulfide (H_2S) is a toxic pollutant for human beings when present in high concentrations. Yet, certain microorganisms utilize H_2S as an important substrate for their growth and metabolism. Therefore, a precise definition for pollution can be stated only in reference to a specific kind of material and a given habitat. For this reason various specific definitions for pollution are present in the literature. Furthermore, broad definitions of pollution are sometimes vague and meaningless. Nevertheless, a broad definition of the term will provide us with a framework for these discussions. Thus, an environment is considered to be polluted when it receives by-products from the activities of human beings and/or other organisms (materials and/or effects), directly or indirectly, in concentrations that are harmful to the ecosystem. This definition is subjective, but it emphasizes our two major concerns for pollutants in ecosystems: (1) the kind and quantity of materials, and (2) the effects of such materials (i. e., the manner in which they interact with biotic and abiotic components).

Pollutants in ecosystems may be specific chemicals such as DDT, PCBs, lead, mercury, hydrogen sulfide, sulfur dioxide, or a combination of materials, such as microorganisms and their toxins, oxides of nitrogen, radiations, and sewage. We readily see from these examples the variability of pollutants that may be present in ecosystems. It is also important to recognize that pollutants may produce effects in ecosystems that are also extremely variable, but, in general, excessive pollution of any kind tends to create imbalanced ecosystems.

Microbiological aspects of air pollution

One does not need to be trained in science, or to be unusually concerned about the quality of the environment, to recognize that the air we breathe is polluted. In the vicinity of large urban centers, indicators of excessive air pollution are obvious: eye and throat irritations, decreased visibility, and large amounts of orange, brown, or black material being emitted from smokestacks. Not so obvious are the serious health effects that may result from the accumulation of certain air pollutants in ecosystems.

A great deal of information pertinent to the health effects of air pollutants has been obtained from research studies in which laboratory animals housed in closed chambers were allowed to breathe air polluted with various concentrations of chemical substances. Data from exposure studies have shown that high concentrations of certain air pollutants tend to increase susceptibility to respiratory infection by impairing such normal body defense mechanisms as ciliary action in the upper respiratory tract and phagocytosis. Furthermore, normal microbiological processes, under certain conditions, may produce air pollutants that are hazardous to human health.

Evidence in support of the contention that imbalanced ecosystems can cause normal microbiological processes to be hazardous to human health can be seen in the following case history. On June 30, 1975, six men died in an animal-waste-products rendering plant in Franklin County, Ohio. The probable cause of death was hydrogen sulfide, alone or in combination with methane. The routine procedure required trucks to dump the waste products into a holding pit, after which the trucks were washed out. The drainage entered an adjoining pit through a 6-inch drain pipe. The pipe had been clogged for several days. In an attempt to unclog the drain, one of the men descended into the pit, stayed for approximately 20 minutes, and returned to the surface, with no apparent ill effects. Later, he reentered the pit to turn off a sump pump, and collapsed. Three of his partners went to aid him and collapsed. The two-man rescue squad arrived and assumed the problem to be electrical difficulties. They entered the pit and collapsed almost immediately. At that point, additional help was obtained from the fire department, and all wore supportive air packs while removing the six men. Four of the men were dead, one died enroute to the hospital, and one died four days later. Subsequent epidemiological investigations revealed that the animal waste products that had accumulated while the drain pipe was clogged were being degraded anaerobically, releasing large quantities of hydrogen sulfide and methane gas from sulfur-containing amino acids (see Figure 6.9). Under normal conditions, organic waste materials would be degraded by microorganisms and their constituent elements recycled through the biosphere without adverse effects. In this instance human activities associated with modern technology modified the ecosystem, creating a health hazard from air pollutants that involved the indirect action of microorganisms.

Nonindustrial microbial aerosols

Technically, an *aerosol* is a gaseous medium that contains suspended particles. Since microbial cells are particulate matter, all studies that deal with airborne microorganisms are concerned with aerosols. Air does not contain an autochthonous or indigenous microflora. Therefore, microorganisms that are present within the atmosphere represent types that have been introduced to it from other sources.

Aerosols are generally recognized as being produced by shear forces (mixing and grinding operations), splashing of liquids, or turbulent dispersal of dry materials by air currents. Those aerosols that are of greatest concern to human health contain pathogens or other particles small enough to traverse the barriers of the respiratory tract and be retained within the lung alveoli, air sacs where gas is exchanged with the blood. Our discussion here is concerned with aerosols that contain live microorganisms and/or their viable spores.

Numerous kinds of aerosols are formed during activities not associated with industry. For example, routine human activities such as eating, talking, coughing, and sneezing often generate microbial aerosols. Likewise, aerosols are generated by normal activities associated with animals. Furthermore, the nonindustrial environment is laden with reservoirs from which pathogenic aerosols can be generated: soil, stockyards, domestic animals, old buildings where birds roost. Aerosols from such sources often contain pathogens that are responsible for serious human diseases, among them histoplasmosis, cryptococcosis, and aspergillosis.

Histoplasmosis is caused by a fungus *(Histoplasma capsulatum)* that inhabits the soil. This fungus is *dimorphic:* at room temperature, it grows characteristically as aerial mycelia colonies, but in infected tissue it grows characteristically as oval yeast cells. Human beings become infected by inhaling the aerosolized fungal spores, called *conidia* (Figure 11.1). Often the infection remains localized within the lungs and is self-limiting, but it may spread throughout the body tissues and produce a systemic disease. The disease is not transmitted directly from one human being to another or from an animal to a person. Aerosols represent the mechanism of infection. Thus, histoplasmosis is an airborne disease. The organism is found in large numbers in soils enriched with chicken and/or wild bird droppings, and are especially abundant in chicken houses, bird nesting areas, and in soils under trees in areas that are heavily populated with birds. For this reason wild birds are associated with the dispersal of the pathogen (Figure 11.2). The role that wild birds play in dispersal was dramatized in 1975 when millions of birds settled near Hopkinsville, Kentucky. In an effort to destroy the birds, the U. S. Army's 101st Airborne division sprayed them with a biodegradable detergent (Tergitol S-9) to wash the protective oils from their feathers, without which the birds froze to death. National attention was attracted to the event by opposition from environmentalists and bird lovers. The excessive number of wild birds in that geographical area can be viewed as a detrimental force

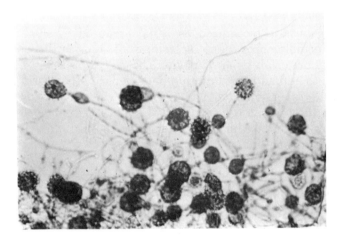

Figure 11.1. Aerial spores (conidia) of *Histoplasma capsulatum* **when grown on Sabouraud dextrose agar at room temperature. (Courtesy of T. J. Kloeckl.)**

Figure 11.2. (A) Large flock of birds (mostly blackbirds and starlings) flying over Fort Campbell, Kentucky, in the winter of 1974–1975. (B) Birds hanging on tree limbs after being sprayed with a detergent (Tergitol S-9), (C) Frozen Tergitol-sprayed birds that have fallen to the ground. (Courtesy of the *Kentucky New Era,* Hopkinsville, Kentucky.)

operating against a balanced ecosystem. Not only did they pose a threat to human health by dispersing *Histoplasma* spores, but economic damage was estimated to be in the millions of dollars.

To further emphasize the important role that nonindustrial aerosols play in the transmission of pathogens, we shall describe an outbreak of histoplasmosis that occurred in Delaware, Ohio. During Earth Day activities, April 22, 1970, pupils at a junior high school participated in a clean-up program whereby the school's courtyard was raked and swept (a beautiful way to generate aerosols!). The courtyard had previously been an old roosting area for starlings and blackbirds. Subsequently, clinical histoplasmosis developed in 384 (40 per cent) of the students and faculty. Epidemiological investigations revealed that *Histoplasma capsulatum* was abundant in the courtyard. In September 1970 the courtyard was decontaminated by three separate treatments with 3 per cent formalin. Following the decontamination program, the soil in the courtyard was sampled for *H. capsulatum* in September 1971, and again in September 1972, and found to be negative for the fungus on both occasions.

Cryptococcosis is another airborne disease caused by a fungus (*Cryptococcus neoformans*). It differs from *Histoplasma* in not being dimorphic. The cells of *C. neoformans* are yeastlike and can usually be recognized by their distinct capsule (Figure 11.3). This fungus is found in all parts of the world and is extremely abundant in soils enriched by pigeon droppings. Common reservoirs from which the yeast cells are aerosolized are pigeon roosts in barns and window ledges (Figure

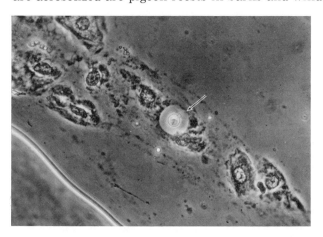

Figure 11.3. Phase-contrast microscope photo of *Cryptococcus neoformans* cell in sputum. Note the large capsule that surrounds the cell (arrow) (×1,000). (Courtesy of G. D. Roberts, *J. Clin. Microbiol.* **2:**261, 1975, with permission.)

Figure 11.4.   (A) Several domestic pigeons in an area heavily covered with pigeon droppings; (B) Pigeon droppings on a window ledge. (Courtesy of L. Simon.)

11.4). People inhale the aerosolized cells, which usually produce a self-limited pulmonary infection of which the victim is unaware. In some cases however, the disease progresses into a severe systemic infection and causes brain tumors and/or meningitis.

It is important to note that in both histoplasmosis and crypto-coccosis, although birds are important agents of dispersal, they themselves do not become infected. Both of the causative organisms are sensitive to elevated temperatures. They grow well at human body temperature (37°C) but not at the body temperature of birds (40–42°C).

Aspergillosis is another disease that is transmitted by nonindustrial aerosols. The causative agent is also a fungus *(Aspergillus fumigatus)* that inhabits the soil. It is widely distributed in the soil, is commonly found in decaying vegetation, and is especially abundant in compost piles. Farmers and others who handle such materials inhale the aerosolized spores. Asthma and other allergies are the most common types of conditions caused by this fungus, but it produces a severe pulmonary disease in debilitated individuals. Cells of this fungus are shown in Figure 11.5. *A. fumigatus* grows well at temperatures up to 50°C, and produces infections in birds. Abortion in sheep and cows is also attributed to this fungus.

Industrial microbial aerosols

There are an infinite number of processes and activities in industry that may generate aerosols, but our concerns are with those types that are most likely to contain infectious organisms. Industrial facili-

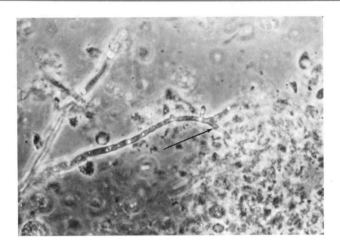

Figure 11.5. Phase-contrast microscope photo of *Aspergillus fumigatus* in sputum, showing distinct branching hyphae (arrow). (Courtesy of G. D. Roberts, *J. Clin. Microbiol.* 2:261, 1975, with permission.)

ties in which hazardous aerosols are commonly produced are slaughterhouses, animal-waste rendering plants, hospitals and nursing homes, and various research and diagnostic laboratories. An attempt to enumerate the processes and/or activities in such facilities would be futile. Therefore, we shall present only a few examples of industrially produced aerosols that are hazardous.

Long before the development of microbiology as a science, those working in hospitals recognized the important role that aerosols play in the transmission of diseases. Consequently, methods and procedures that generate aerosols have been dramatically reduced and/or rigidly controlled. For example, normal activities that are associated with patient care, such as bed making and the handling of soiled linen, often generate aerosols (see Table 3.1). These data show a considerable increase in airborne microorganisms after laundry was handled. The laundry chute has long been controversial: praised by some as an effective method for moving soiled linen, and condemned by others as an effective way to spread pathogens. Other common generators of aerosols within hospitals are dishwashing machines and garbage grinders. After such procedures, aerosolized pathogens behave in a dynamic manner and may infect persons directly, become distributed throughout the hospital by the ventilation system, or be deposited on other surfaces. In the latter case pathogens may become a reservoir for subsequent dispersal. Hospital-acquired infections are now considered to be one of the largest communicable-disease problems in the United States, and the role that aerosolized pathogens play in producing them is considerable. However, aerosols are not of equal importance to the transmission of human pathogens, owing to the inability of fastidious types (obligate parasites) to live for long periods when separated from host tissues.

Hazardous aerosols are often formed from activities that occur in textile mills and slaughterhouses. Individuals working in such industries often develop diseases from inhalation of pathogenic aerosols. "Wool-sorters' disease" is a clinical form of anthrax that is acquired in this manner. Previously, we characterized anthrax as being caused by *Bacillus anthracis,* a bacterium that forms spores and inhabits the soil. We also characterized anthrax as an epizootic disease. When infected hairs, hides, and wool are handled, the worker inhales the aerosolized spores, which germinate and produce a very severe and usually fatal disease called pulmonary anthrax. When spores enter a lesion in the skin, the individual develops a less severe form of the disease, called cutaneous anthrax.

In Chapter 9 brucellosis, another epizootic disease, was characterized as being caused by three different species of bacteria within the genus *Brucella*. The majority of human cases occurred in slaughterhouse workers, and the largest number of infected individuals usually have contact with infected swine or infected swine products (see Table 9.2). During the handling of the infected meat, the infecting organisms are acquired directly by contact or indirectly after the causative agents have been aerosolized.

Because of the activities essential to the normal operation of hospitals, textile mills, and slaughterhouses, aerosolization of microorganisms is unavoidable. However, the risk of acquiring infections from aerosolized pathogens can be reduced by upgrading the sanitation of the environment within such institutions, and by educating the workers to the health hazards. In this regard considerable attention is being attracted to research laboratories in which pathogenic aerosols are being generated by activities of professional scientists. Not only do such aerosols threaten the health of scientists themselves; but, if not controlled, they are potentially dangerous to the general population should the aerosolized pathogens escape the laboratory environment.

Microbiological aspects of water pollution

Unfortunately, most individuals do not regard water as a limited resource. Although abundant in quantity, it is seriously lacking in quality. Excessive water pollution, like air pollution, is the result of the rapid growth of the human population, increasing urbanization, and our expanding technology. Not only is water pollution excessive, but industry continues to dump an increasing amount of complex materials into a basically fixed water supply.

To safeguard human health, numerous municipalities process domestic and industrial waste through sewage-treatment facilities before dumping it into waterways, and process water to be used for human consumption through water-purification plants. Although such facilities are operated at the local level, they must function in accord with state and federal guidelines. In the United States, the Environmental Protection Agency (EPA) is the overseer of state and local programs charged with the responsibility of providing the public with a water supply that is safe for its intended use.

Drinking water — problems of purification

Contrary to the belief of some people, pure water does not exist in nature. Water in the biosphere moves in a cyclical manner and involves three phases: precipitation, runoff, and evaporation. Although the large bodies of stored water are obvious, atmospheric precipitation (rain, snow, and hail) can be considered as the primary source of water. After reaching the surface of the earth, precipitated forms of water become surface water (runoff). Some of the surface water penetrates the soil and becomes the underground water table. Both underground and surface waters move through streams that enter lakes, and eventually, oceans. During transport and while in storage, water is continually being evaporated and reenters the atmosphere, from which it will again be precipitated.

Although our description is oversimplified, it will help us to visualize the manner in which natural water becomes contaminated. During precipitation, rain, snow, and hail absorb natural atmospheric gases (and some air pollutants) and, after reaching the earth's surface, become exposed to an infinite number of other pollutants. These pollutants are so complex, so varied, and so numerous that natural processes (exposure to sunlight, filtration through the soil, and biological degradation) are inadequate to remove them.

In addition to those complex chemicals of industrial origin, runoff and storage water collects numerous kinds of natural materials present in ecosystems, much of which is in the form of excreta (urine and feces) from humans and other animals. Human feces are of special interest, because they contain nondigestible residues of the food we eat and large populations of microorganisms that live in the gastrointestinal tract. Thus, individuals who become ill with gastrointestinal diseases such as cholera, giardiasis, and dysentery also excrete the causative agents in their feces. When such fecal material gains entrance to water, the pathogens survive, and can produce disease in other people if they consume the fecal-contaminated water. Diseases transmitted in this manner are called waterborne. Virus diseases such as hepatitis and polio (see Chapter 9) and a number of bacterial and protozoan diseases can normally be transmitted through contaminated water. Cholera is the classical example of a waterborne disease, because water contaminated with the causative bacteria *(Vibrio cholerae)* is the primary mode for transmitting the pathogens from an infected person to a susceptible noninfected one. Water as a vehicle for transmitting pathogens has been known since the London cholera epidemic in the middle of the nineteenth century, when John Snow, an English physician, showed that the infected persons had consumed water from a sewage-contaminated water supply. The U.S. Public Health Service Drinking Water Standards outline specific methods for measuring water quality. The aims are to prevent the transmission of diseases to the general public through drinking water. In spite of such standards and the modified version of them that has been adopted by the World Health Organization, waterborne diseases continue to be a public health problem. In 1970–1973, there was a worldwide outbreak of cholera, with more than 60 countries being affected. Contrary to the belief of many, well water in the United States is not bacteriologically safe. During 1972 an outbreak of shigellosis occurred in a population of students at a junior high school in Stockport, Iowa. All the infected students (208 cases) had consumed water from a shallow well. Epidemological investigations found the well to be substandard and susceptible to contamination with surface and groundwater. It is not difficult, then, to recognize the need for a system of purifying water that is to be used for human consumption.

Water-purification systems in municipalities have been designed primarily for the purpose of rendering water microbiologically safe for human consumption, but they also improve the overall quality by reducing odors and eliminating undesirable chemicals. However, most systems are inefficient at removing complex chemical substances, many of which are potentially dangerous to human health.

The kind and degree of treatment that must be applied to a body of water to render it safe depends to a great extent on the source, a body of fresh water (see Chapter 3). Well water, water from reservoirs, and water from larger lakes may vary considerably in terms of pollutant composition, but no natural waters can be considered safe.

Municipal water treatment consists primarily of three processes: sedimentation, filtration, and chlorination. Water is first pumped from the source to large *sedimentation* basins. Such basins are designed to hold large quantities of water for several hours to allow large particles of sand, rocks, and other debris to settle out. As part of the sedimentation process, certain chemical substances (e.g., aluminum, calcium, iron) are added to force insoluble substances to coagulate and settle. Many of the materials that impart turbidity to water are removed through the coagulation process. The actual time required for completion of the sedimentation process varies from one facility to another.

The second process is *filtration*. Water is filtered through carefully constructed sand filters (various layers and size gradients in a depth of about 4 feet). This process is highly efficient and critical, because it traps and removes protozoan cysts, and approximately 99 per cent of bacteria from the water.

The final process is *chlorination*. Although most of the bacteria have been removed, chlorine (a highly effective microbiocide) is added to destroy any remaining bacterial pathogens. Automatic devices perform this function, but the process is complex because many factors can negate the intent of chlorination. For example, the active chlorine content of the different kinds of chlorine compounds varies considerably, and organic materials present in sewage react strongly with chlorine. Thus, waters with high organic contents have high chlorine demands. Consequently, the actual amount of chlorine added may vary. For this reason water-quality criteria specify that drinking water should contain a residual chlorine concentration of 0.2 part per million (ppm) to maintain its bacterial killing power.

From this point water is pumped to storage tanks, from which it flows by gravity to the consumer. It is important to note that purification plants in many municipalities incorporate fluoridation as the final step as an extra benefit to the consumer. Fluoride in water at a concentration of 1 ppm protects against tooth decay.

Unless operated properly, these processes may be highly ineffective. For this reason personnel who supervise such operations should have an understanding of the principles (chemical and microbiological) behind each process. Furthermore, to guarantee that water is microbiologically safe, samples from it must be analyzed periodically by standard microbiological tests. One important test, designed to detect certain kinds of bacteria that live in the colon region of the human large intestine (colon bacilli), is the *coliform test*. Coliform organisms are not present in natural waters in large numbers unless the waters are polluted with human excreta. Thus, the number of coliforms present in water serves as an indicator of water quality. The United States Water Quality Criteria specify frequency of sampling, the types of tests to be performed, and the number of coliform bacteria that classify a drinking water supply as unsafe for human consumption.

The consequences of inadequately operated facilities, or of no treatment, can be disastrous. Illnesses associated with various types of drinking water sources and their probable cause are shown in Tables 11.1 and 11.2.

Table 11.1. Waterborne Disease Outbreaks, by Type of System and Cause of System Deficiency, 1974

	Municipal		Semipublic		Individual		Total	
	Outbreaks	Cases	Outbreaks	Cases	Outbreaks	Cases	Outbreaks	Cases
Untreated surface water	4	4,930	1	18	3	59	8	5,007
Untreated groundwater	—	—	4	1,290	1	5	5	1,295
Treatment deficiences	3	1,609	4	404	—	—	7	2,013
Deficiencies in distribution system	4	58	—	—	—	—	4	58
Miscellaneous	1	12	1	6	2	22	4	40
	12	6,609	10	1,718	6	86	28	8,413

Source: HEW Publication No. (CDC) 76-8185, "Foodborne and Waterborne Disease Outbreaks, Annual Summary 1974, issued January 1976.

Table 11.2. Waterborne Disease Outbreaks, by Etiology and Type of Water System, 1974

	Municipal		Semipublic		Individual		Total	
	Outbreaks	Cases	Outbreaks	Cases	Outbreaks	Cases	Outbreaks	Cases
Acute gastro-intestinal illness	4	440	5	847	2	25	11	1,312
Chemical poisoning	3	39	1	213	1	17	5	269
Giardiasis	4	4,930	1	18	2	39	7	4,987
Shigellosis	1	1,200	2	606	—	—	3	1,806
Salmonellosis (nontyphoid)	—	—	1	34	—	—	1	34
Typhoid	—	—	—	—	1	5	1	5
	12	6,609	10	1,718	6	86	28	8,413

Source: HEW Publication No. (CDC) 76-8185, "Foodborne and Waterborne Disease Outbreaks, Annual Summary 1974, issued January 1976.

We mentioned previously that natural water supplies should not be considered safe. That is true, but it is not uncommon for persons to become isolated from a source of purified water. When water purity is not known, boiling for 5 minutes is the simplest way to render water microbiologically safe. However, if one plans a hiking or camping trip, iodine tablets and special preparations of chlorine can be purchased at drugstores for the purification of small quantities of water. Instructions for proper use are provided with such materials.

Sewage — problems of treatment

Sewage is human and other animal waste (mostly liquids) that accumulates in ecosystems. Under natural conditions, when ecosystems are balanced, sewage does not cause problems. It would be degraded through processes that were characterized in Chapter 6. Since natural ecosystems have been modified extensively by human-related activities, modern sewage-treatment systems have been developed to render domestic and industrial sewage safe for disposal in waterways.

We shall first describe a municipal sewage-collection system, and then the processes for sewage treatment. A sewer system usually consists of a group of pipes connected in a manner to collect and transport sewage and wastewater from homes and industrial facilities to a treatment plant. There are two kinds of sewers: sanitary and storm. *Sanitary sewers* collect only sewage from homes, offices, and other institutions, regardless of kind, that contain human excreta (also called *domestic sewage*). *Storm sewers* collect only rainwater, melting snow, and normal runoff water from streets and buildings. Storm sewers are constructed to carry extremely large volumes. The two systems of sewers may operate separately or they may be linked and operate as a combined sewer system. In the combined system, the pipes are connected in a manner that allows for some of the sewage to bypass the treatment plant and enter the receiving waterway without undergoing treatment. In such systems the bypass is used during a storm, when the amount of water is much greater than normal. It must be remembered that when the bypass is used, some of the sewage from the sanitary portion also enters the receiving waterway untreated.

Industrial waste, unlike human waste, may vary considerably in terms of composition from one industry to another. Yet, in many instances industrial sewage is collected by sanitary sewers and treated in the same manner as domestic waste, which often is inadequate for the toxic and/or other complex materials that it may contain.

Treatment of sewage consists of subjecting it to a series of rigidly controlled processes, referred to as primary treatment, secondary treatment, and tertiary treatment (Figure 11.6).

Primary treatment of sewage consists only of physical processes. As the sewage enters the treatment plant, it flows through a series of screens, in which openings vary from coarse to fine, for the purpose of removing floating objects. After the sewage has been screened, it flows at a decreased rate through a sedimentation chamber in order to allow sand and other suspended solids to settle. After the settling process, chlorine gas is purged into the liquid as it is released to the receiving waterway. The remaining solids that settle out during the passage of sewage through primary treatment is called *sludge*. Sludge can be disposed of at this point or subjected to anaerobic digestion, which is a phase of secondary treatment. It is important to note that the organic content of sewage is not decreased by primary treatment. Although the process is only physical, many municipalities in the United States use

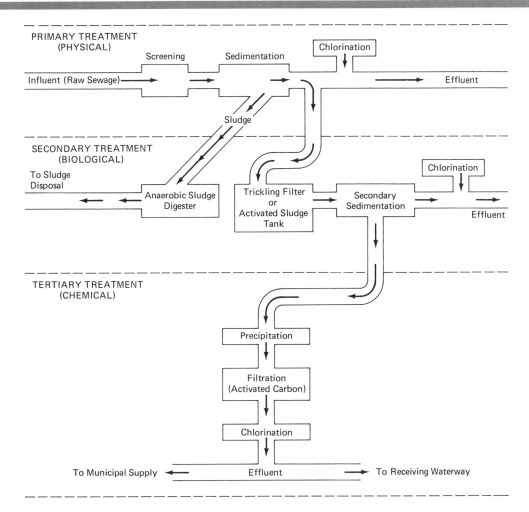

Figure 11.6. Municipal sewage-treatment facility.

only primary treatment. Such practices are highly undesirable, because they enhance eutrophication.

Secondary sewage treatment is entirely biological, and involves two basic phases: one for the treatment of the sludge (anaerobic sludge digestion) and the other for the treatment of the clarified effluent (oxidation in trickling filters or oxidation by activated sludge).

The *anaerobic sludge digester* is a tank designed in a specific manner to exclude oxygen, thus creating an anaerobic environment. After sludge enters the tank, it is vigorously attacked by a mixed population of anaerobic microbes. Through fermentation processes, various kinds of substrates are completely degraded. End products from the mixed fermentations are water and several gases (hydrogen, methane, and carbon dioxide). Some tanks are equipped with mechanisms for trapping the methane for subsequent use as a fuel. At this point it might be of interest to note the similarities between the gaseous end products from the fermentation of sludge and the gaseous end products from the fermentation of cellulose in the rumen (see Chapter 5).

A *trickling filter* is simply a bed of rocks (6 to 10 feet deep), over which the sewage is slowly sprayed. During the process the rocks become coated with a slimy layer of microorganisms. The mixed group of aerobic microbes multiply rapidly and degrade most of the organic materials present in the sewage. The last phase of treatment is to purge the liquid with chlorine gas while enroute to the receiving stream. Note that chlorine treatment at this stage is carried out primarily to remove odors and kill residual pathogens.

The *activated sludge process* is the other widely used type of aerobic treatment, and there are several different ways in which such units can be assembled (Figure 11.7). The process is rapid, because aeration is performed mechanically, thus enhancing microbial growth. We shall briefly review activated sludge treatment, making no attempt to detail the variations of the process.

Treatment begins when clarified sewage from the primary sedimentation is pumped into the aeration tank, the first tank in the activated sludge unit, so called because air flows into it from a series of small jets. Here many kinds of aerobic microorganisms multiply rapidly, forming aggregations or small clumps. During this period of rapid growth, most of the organic materials are degraded. Then the partially degraded sewage passes into a sedimentation tank, where the clumps of flocculent microbes settle out. Some of the flocculent clumps (activated sludge) are recycled into the aeration tank. Thus, the recycling of the activated sludge is the unique feature of the process because it keeps the microbes growing rapidly at all times. As a result of the vigorous oxidations and the constant inflow of sewage, treatment by activated sludge is extremely rapid. The clarified liquid is then purged with chlorine prior to its discharge into the receiving waterway.

Sewage that has undergone both primary and secondary treatment emerges as a clarified liquid, depleted of approximately 90 percent of its organic matter. However, the clarified liquid still contains an abundance of nitrates and phosphates, both known to enhance algal growth. While excessive algal growth is undesirable, secondary treatment of sewage in all municipalities within the United States would certainly be an important step toward improving the quality of the environment.

Tertiary treatment, of course, is ideal. It involves chemical treatment designed to remove inorganic minerals, especially phosphates

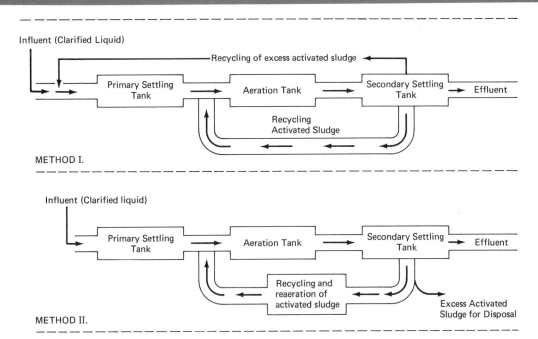

Figure 11.7. Two methods of operating an activated sludge digester.

and nitrates, from the effluent that emerges from the secondary process. After the minerals have been precipitated, the final clarified liquid is purged with chlorine gas and released to the receiving waterway—incapable of enhancing algal growth and of a quality that is microbiologically safe for human consumption. In some instances it is actually returned to the municipal drinking water supply.

Solid wastes—problems of disposal

It is difficult to define *solid waste*. A portion of it is trash and garbage, a portion is agricultural and industrial, and another portion is sludge, the undigested residue from sewage-treatment processes. Materials in each category continue to accumulate in the ecosystem, while methods for disposal are inadequate to deal with them. Thus, solid wastes can become a serious health hazard if not properly disposed of.

In most municipalities, trash and garbage are collected from homes and various other institutions in an orderly manner and disposed of by one or a combination of incineration, landfill, and recycling. The aims of solid-waste-disposal methods are to return the materials to the ecosystem in a manner that will not be hazardous to human health. Microorganisms have the ability to degrade the garbage and many other organics by methods previously described, but a major problem is to handle the bulk and separate the heterogeneous components in a manner appropriate for each kind of constituent. If not handled in a sanitary manner, solid waste can become a reservoir for numerous kinds of pathogens and a breeding ground for rodents.

The greatest immediate problem that surrounds the disposal of solid waste is the increasing volume of materials that must be handled. The ecological and health hazards usually reflect themselves as indirect effects. For example, emissions from improperly operated incinerators are major contributors to air pollution. In many cities incinerators have been forced to stop operating for failure to comply with air pollution standards. In such instances the alternative method has been to use landfills. This process consists of spreading the waste in thin layers in a previously opened ditch (approximately 6 feet deep), after which it is compacted and covered with about 2 to 3 feet of soil, which is the final seal. When properly constructed the waste is decomposed anaerobically without odor and without causing health problems. Land that has been used for a properly constructed landfill is not suited for the construction of buildings but is highly suitable for use as parks and playgrounds. Improperly constructed landfills may become serious hazards. Methane gas from anaerobic fermentation processes may escape and cause fires, drainage through such waste may cause underground water pollution, and the waste itself may become a breeding ground for rodents and other vectors.

Many kinds of solid-waste materials can be disposed of by recycling them through special facilities designed for that purpose. But because of heterogeniety of solid-waste materials, sorting and recovery operations have proved to be extremely costly. Consequently, the potential effectiveness of recycling processes has not been fully explored.

It is important to note that the efficiency of solid-waste disposal in a community can often be enhanced by handling the various kinds of waste in a selective manner. For example, sludge is a solid waste that can be disposed of in several ways. Most commonly, this infectious material is disposed of by incineration or by burial in landfills. However, some communities have experienced a remarkable degree of success at producing marketable fertilizers and soil conditioners from sludge. Income from the sale of such products helps to reduce the overall cost of waste disposal. Other communities have experimented with burning sludge, combined with papermill residues, in furnaces designed to generate power for the sewage-treatment facility.

Some types of agricultural and industrial solid waste also lend themselves to alternative methods of disposal, such as composting. Large-scale composting is not widely practiced in the United States, but some European countries have used it with a remarkable degree of success. Basically, composting consists of storing compostable materials (e.g., paper products, sawdust, grass, weeds), combined with slaughter-house waste, in large piles on concrete platforms or in shallow ditches, in order for the materials to undergo natural microbial decomposi-tions, after which the residue is used as a soil conditioner. If compost piles are left undisturbed, microbial processes will be predominantly anaerobic. However, aerobic decomposition can be enhanced by fre-quent turning or mixing of the materials. Usually composting does not produce serious health hazards, because heat generated in the piles is adequate to kill most pathogens. However, thermophiles, such as *As-pergillus fumigatus,* can become aerosolized and produce infections. For many years official agencies charged with the responsibility of maintaining the quality of the environment showed less concern for the development of methods for solid-waste disposal than for the de-velopment of programs for the control of air and water pollution. Con-sequently, open dumps were widespread, incinerators were inade-quate, and sanitary landfills were poorly constructed. However, since the inception of the Environmental Protection Agency in 1970, we have seen a great deal of progress in the field of solid-waste manage-ment. Currently, many new incineration processes are under develop-ment, and a few innovative recycling programs are being federally funded as demonstration projects. At last there seems to be an aware-ness of the fact that one form of pollution in the environment cannot be controlled independently of another. Excessive solid-waste pollution will eventually lead to air and water pollution.

The environment in which we live is not impervious to continued pollution with excessive amounts of toxic and nonbiodegradable mate-rial. Our resources are finite, and they are contained within the land, water, and air that we pollute. Furthermore, all ecosystems within the biosphere are interconnected, and all living things are interdependent. Human organisms are not in control of nature, we are a part of it. Therefore, if the world population continues to grow at the present rate, if technology continues to expand and pollute at the present rate, catastrophic environmental effects are inevitable.

Microorganisms as tools for detecting specific pollutants

For any test system to be useful for the purpose of detecting an environmental contaminant, it must be sensitive, reproducible, and preferably easy to perform. In regard to these factors, microorganisms possess many qualities that can be exploited by the prudent investigator. These include: rapid growth, the ability to be cultivated in compact units (small vessels), continuous growth, and the ability to utilize a wide variety of substrates (especially those containing carbon). In addition, their genetic complement can be modified and controlled by the investigator.

Bacterial and protozoan assays

Microorganisms require the same essentials for growth (carbon, nitrogen, energy, and micronutrients) as do all other forms of life. Therefore, organisms deficient in a specific enzymatic step or pathway that is needed for the biosynthesis of a particular growth factor will grow only when that factor is present in its environment (assuming that all other condition are optimal). Thus, a medium can be prepared to include all essential requirements except the growth-limiting factor. Then, if the growth-limiting factor is added in increasing concentrations to a series of tubes containing the medium, the test organism will grow in proportion to the concentration of the growth-limiting factor. Tubes with the least amount of limiting factor will have less growth than tubes with higher concentrations. This is the principle of the *microbiological assay*. When the same organism is inoculated into the medium without growth-limiting factor but with an impure substance suspected of containing the growth-limiting factor, its growth will occur only if that factor is present. Thus, the concentration of that factor in the unknown can be determined by comparing the growth response of the test organism in that substance with the growth curve prepared with known concentrations. Microbiological assays are inexpensive, highly sensitive, and more specific than most chemical and physical methods. Microbiological assay techniques have wide application in the quantitative determination of antibiotics, vitamins, amino acids, and a wide variety of other chemicals, based on the sensitivity of the test organism to these agents.

In a recently described bioassay, luminescent bacteria were used to measure air pollution. Cells of bioluminescent bacteria, *Photobacterium phosphoreum,* were treated with pollutants generated by irradiation of a gas mixture of *cis*-2-butane and nitrogen oxide. Then at 10-minute intervals, the gas mixture was replaced with nitrogen. The resulting anaerobic conditions reduced the bacterial luciferin and the enzyme reduction produced a flash of light, the intensity of which was measured with great sensitivity. As the available oxygen was quickly removed, the light intensity decayed logarithmically. When visible air pollution was evident, portions of the polluted air were passed over the cells and similar results were obtained, but when clean ambient air was passed over the cells, no significant changes in luminescence of the cells were detected.

In another study, a protozoan, *Paramecium aurelia,* was used to measure toxicity in the gas phase of cigarette smoke. The protozoan was exposed intermittently to cigarette smoke (simulating human smoking) and the survival time of the organism was noted. The microorganism survived longer in the smoke from cigarettes that contained an activated charcoal filter than in the smoke from the nonfiltered cigarettes. This system proved to be a convenient model for testing materials designed to reduce the toxicity of cigarette smoke. Cigarette smoke is known to be a damaging source of air pollution. It contains many substances: benz (a) pyrene, carbon monoxide, nitrogen dioxide, hydrogen cyanide, acrolein, aldehydes, and phenols. These compounds are present in very high concentrations, but the smoker survives because he puffs intermittently and most of the time breathes air that is not heavily contaminated. However, many smokers in a poorly ventilated room may produce pollutants in concentrations that are damaging to health. Cigarettes that contain sodium nitrate produce smoke that is less carcinogenic to mice, and contains less benz (a) pyrene, nicotine, and phenols than smoke from standard cigarettes. These effects are attributed partly to the thermal decomposition of the nitrate into oxygen and nitrogen oxides, the former enhancing combustion of tobacco and the latter inhibiting free-radical reactions which lead to the formation of benzo (a) pyrene. Cigarettes containing sodium nitrate burned at 778°C, while normal cigarettes burned at 850°C. Thermal decomposition of the additive may increase concentrations of nitrogen oxides in smoke and lead to the formation of other compounds that may be toxic.

A great deal of progress has been made in the development of bioassay methods for the assessment of specific air pollutants. The photodynamic effect of the carcinogen, 3:4-benzpyrene was tested on *Paramecium caudatum*. This photodynamic bioassay reflects the ability of various compounds to sensitize cells of the protozoan to the otherwise nontoxic effects of a specific wavelength of ultraviolet light. Since the results are available in a few hours, this method shows promise of yielding useful presumptive evidence of the carcinogenic potential of trace amounts of materials extracted from the air.

Tissue culture assays

Just as the bacterium *Escherichia coli* has been exploited by biochemists for the elucidation of many cellular processes, tissue culture methods have many applications in air pollution research. Their use is very appropriate to studying the effects of air pollutants at the cellular level, because all cells, both microbial and macrobial, have many features that are similar. Any data obtained from controlled experimentation with microorganisms and tissue culture systems could be used to supplement data obtained from animal exposures and may aid significantly in understanding the relationship between air pollution and human health.

Techniques of tissue culture are now well established, and media and equipment are so readily available at commercial supply houses that many biochemists, microbiologists, pharmacologists, physiologists, and other biological scientists use the mammalian cell culture as a biochemical tool. Most pharmaceutical companies use tissue culture systems in their research department. Since viruses multiply only in living cells, a tissue culture assay is the method of choice for the initial screening of antiviral drugs.

It is believed that many of the difficulties encountered in eliciting measurable responses of whole animals to low concentrations of specific air pollutants might be eliminated through the use of tissue culture assays. At least three animal cell lines (strain L mouse fibroblasts, mouse liver cells, and HeLa cells) have been tested for their sensitivity to sulfur dioxide (SO_2) and nitrogen dioxide (NO_2).

All test systems consist of growing the various cell lines in an appropriate medium and then exposing them to a continuous flow of gases at different concentrations. All cells in each line died when exposed to sulfur dioxide at a concentration of 25 ppm, but when the concentration was reduced to 10 ppm and 5 ppm, the cells survived for several days.

Similar studies were conducted with nitrogen dioxide. Mouse liver cells and HeLa cells are grown in an appropriate medium and exposed to various concentration of the gas. All cells in both lines died rapidly when exposed to 100 ppm of nitrogen dioxide, but when the concentration was reduced to 10 ppm and 5 ppm, cells in both lines survived for several days.

With the increasing concern for new complex chemical substances that are being detected as pollutants and assumed to be harmful, the need for assays to demonstrate the specific effects that low concentrations have on living cells has become crucial. Microorganisms, which have the ability to reflect many changes within a short period of time, might prove to be useful tools for monitoring the environment for pollutants that will produce a specific effect. In this regard a number of tests that employ microorganisms to detect mutational changes produced by chemical mutagenic substances are now being developed under the auspices of the Environmental Mutagen Society.

Key Words

aerosol   A gaseous medium that contains suspended particles.

ambient   Pertains to the natural atmosphere, especially the outdoor environment.

bioluminescent   The production of light in an ecosystem by an organism.

carcinogenic   The ability to produce or cause cancer.

dimorphic fungus   A fungus that has the ability to exhibit alternative growth forms: to grow in the yeast phase under one set of environmental conditions and to grow in the mycelial phase under a different set of environmental conditions.

emphysema   A respiratory disease in human beings characterized by a loss of elasticity in the alveoli or lung sacs.

septicemia   Pertains to an infection (replication) or microorganisms in the blood of human beings or other animals.

sludge   The residual mass of solids that remains in each sedimentation tank after the liquid portion of sewage has passed through a municipal sewage-treatment facility.

sludge digestion   The biological degradation of the solid residue from sewage in a municipal sewage-treatment facility; usually takes place in tanks designed specifically for this purpose.

trickling filter   The apparatus in a municipal sewage-treatment facility that contains a bed of rocks upon which sewage is sprayed and subsequently degraded by aerobic organisms.

Selected Readings

1. Biersteker, K. 1969. Air pollution and smoking as a cause of bronchitis. *Arch. Environ. Health* 18:513–535.

2. Clarke, N. A., and S. L. Chang 1975. Removal of enteroviruses from sewage by bench-scale rotary-tube trickling filters. *Appl. Microbiol.* 30:223–228.

3. Council of the Environmental Mutagen Society. 1975. *Environmental Mutagenic Hazards.* Science 187:503–514.

4. Ehrlich, P. R., A. H. Ehrlich, and J. P. Holdren. 1973. *Human Ecology,* San Francisco: W. H. Freeman and Company.

5. Gilbert, R. G., R. C. Rice, H. Bouwer, C. P. Gerba, C. Wallis, and J. L. Melnick, 1976. Wastewater renovation and reuse: virus removal by soil filtration. *Science* 192:1004–1005.

6. Hughes, J. M., et al. 1975. Outbreaks of waterborne diseases in the United States. *J. Inf. Dis.* 132:336–338.

7. King, D. E. 1970. The role of carbon in eutrophication. *J. Water Pollution Control Fed.* 42:2035–2051.

8. Liken, G. E., and F. H. Borman, 1974. Linkages between terrestrial and aquatic ecosystems. *BioScience* 24:447–456.

9. Wade, N. 1973. Microbiology: hazardous profession faces new uncertainties. *Science* 182:566–567.

10. Wurster, C. F., 1969. DDT goes to trial in Madison. *BioScience* 19:809–813.

Comprehensive glossary

A

abiotic   Nonliving.

acid   Any substance that liberates hydrogen ions when placed in solution.

acidophilic   Growing optimally in acidic habitats.

aciduric   Having the ability to survive in acidic environments.

active dispersal   Movement of organisms and/or their reproductive structures from one habitat to another as a result of their own physiological processes.

active immunity   Immunity acquired by individuals as a result of their own production of antibodies in response to an antigenic stimulus.

active transport   Energy-requiring transport that causes ions and solutes to enter cells through the cell membrane from a lower to higher concentration gradient.

Adenosine triphosphate (ATP)   An important high-energy compound that consists of one molecule of adenine, one molecule of D-ribose, and three molecules of phosphoric acid.

aerobe   An organism that requires molecular oxygen for growth.

aerosol   A gaseous medium containing suspended particles.

aflatoxin   A special class of carcinogenic toxins produced by certain kinds of fungi, *Aspergillus flavus*.

akinete   A resting stage or spore in certain species of blue-green algae.

algal bloom   The accumulation of algae in bodies of water to the extent that the water appears to be colored.

allochthonous   Foreign or transient in a particular habitat or region in the biosphere.

allogenic succession   The sequential replacement of one population of organism by another as a result of abiotic changes in a given habitat.

ambient   Surrounding, as the atmosphere, especially the outdoor environment.

anabolic   Involving the synthesis of macromolecules and structures from simpler molecules.

anaerobe   An organism that does not require molecular oxygen for growth.

anion   A negatively charged atom.

antibiosis   An interaction in which one kind of microorganism excretes substances that retard the growth of another microorganism in a given habitat.

antibiotic   A chemical substance, excreted by microorganisms or synthetically produced, that has the capacity to inhibit or kill bacteria when applied in dilute solutions.

antibody   A soluble proteinaceous substance produced in the blood stream of animals in response to the introduction of an antigen; it will combine specifically with the antigen that stimulates its formation.

antigen   Any substance (usually foreign) that, when introduced into the body of an animal, will stimulate the formation of specific antibodies.

antimicrobial agent   Any chemical substance or physical agent that will inhibit the growth of or kill microorganisms.

aphotic zone   The region in natural bodies of water so deep that it receives no significant amounts of light.

apocrine gland   The type of sweat gland, located in specific areas of the body such as the arm pit, having excretory ducts that open into canals from which hair grows.

aqueous   Pertaining to water.

arbovirus   A virus that is transmitted from environmental reservoirs to susceptible host by arthropods.

arthropod   An invertebrate animal with horny segmented body and jointed limbs.

ascospore   A dormant stage, or sexual spore, of fungi belonging to the class Ascomycetes.

asexual reproduction   Reproduction without discrete sexual gametes (opposite mating types).

attenuate   To treat pathogens so as to render them less virulent, but without destroying their antigenicity.

autochthonous   Native or indigenous to a particular habitat or region in the biosphere.

autogenic succession   The sequential replacement of one population of organism by another as a result of organism-effected changes in a given habitat.

autotroph   An organism that obtains its energy from solar radiations or inorganic elements, its carbon from carbon dioxide.

B

bacillus   Any rod-shaped bacteria, as opposed to *Bacillus*, which refers to a specific bacterial genus.

bacteremia   A condition in which bacteria are present in the blood stream in detectable concentrations.

bacteriocidal   Able to kill bacteria.

bacteriocin   Antibiotic-like substance secreted by bacteria that will kill closely related bacteria, usually within the same species.

bacteriophage (phage)   A virus that infects bacteria.

bacteriostatic   Inhibiting the growth of bacteria, but not killing them.

barophile   An organism that grows optimally in regions of great hydrostatic pressure (i.e. in great depths of oceans).

base   Any substance that liberates hydroxyl (OH^-) ions when placed in solution.

base analog   Any of several structures that is chemically similar to one of the natural purines or pyrimidines found in nucleic acids and can be incorporated in genes in such a manner as to cause mutations.

bdellophage   The group of bacterial viruses that have a host-range specificity limited to bacteria of the genus *Bdellovibrio*.

benign   Self-limiting, mild, or nonlethal.

binary fission   A method of asexual reproduction in which a single parent cell splits or divides transversely into two new cells (daughters or progeny).

biochemical oxygen demand (BOD)   The demand for free oxygen in aquatic ecosystems by aerobic organisms.

biodegradable   Pertaining to substances that can be decomposed by physiologically active organisms or their metabolic products.

biofouling   Deterioration or pollution of a structure by living organisms and/or their metabolic products.

biogeochemical changes   Changes that involve biological, geological, and chemical reactions.

biological vector   A specific kind of invertebrate animal in which a particular kind of microorganism must undergo a part of its life cycle.

bioluminescent   Referring to the production of light by an organism.

biomass   The weight of living organisms in a designated habitat, region, or even a vessel.

biosphere   The life-supporting region of the universe.

biota   The sum total of living organisms within a region or habitat.

boil   An abscess that develops on the skin, usually as a result of staphylococcal infections.

bone marrow   The central tissue within bones that gives rise to blood cells.

broad spectrum antibiotic   Any chemotherapeutic antibiotic that will exhibit its antibacterial action on both gram-negative and gram-positive bacteria.

budding   A form of asexual reproduction in which a new cell is formed as a knoblike outgrowth on the parent cell.

buffer   Any substance that, in solution, will minimize changes in pH when acids or alkalis are added to the solution.

C

calorie   That quantity of heat required to raise the temperature of one gram of water one degree Celsius.

cancer   A general term used to characterize a clinical condition in which cells continue to grow in an uncontrolled manner.

capsid   The protein shell that surrounds the nucleic acid core of a virus.

capsomers   The protein subunits from which the capsid of viruses is constructed.

capsule   A gelatinous material that forms the outermost covering of certain kinds of microorganisms.

carbuncle   An abscess that develops in subcutaneous tissues, usually from the collective drainage of boils.

carcinogenic   Producing or causing cancer.

carcinoma   The generalized group of cancers that develop in epithelial cells.

carnivore   An organism that obtains the major portion of its nutrients from meat.

carrier   An individual harboring an infectious agent in its body, but exhibiting no apparent symptoms of the disease the agent may cause.

catabolic   Involving the degradation of macromolecules with the liberation of simpler molecules.

cation   A positively charged atom.

cell   The smallest functional unit in any organism.

cell membrane   A selective barrier that surrounds the cytoplasm and functions to regulate the entry of nutrients and the exit of cell products.

cellulose   A polysaccharide homopolymer in which all subunits are glucose molecules linearly connected; it is a major component of plant cell walls.

chemoautotroph   Any bacterium that obtains its energy from the oxidation of inorganic compounds, and obtains carbon from carbon dioxide.

chemosynthesis The process through which certain organisms obtain and utilize energy and carbon from inorganic substances.

chemotherapy Use of chemical substances to prevent the development of a disease.

chitin A polysaccharide found as a cell-wall constituent in fungi and in the outer coverings of many invertebrates.

chloroplasts The chlorophyll-containing organelles in photosynthetic eucaryotic cells.

chromosome The structure that contains the hereditary DNA in all cells.

ciliates Protozoans characterized by having hairlike (ciliary) appendages on their surface during a period of their life cycle.

coenzyme A compound that functions coordinately with those enzymes that catalytically removes an element or chemical group from a substrate; may be considered as a carrier molecule.

coliform The general term used to describe certain types of bacteria that live in the gastrointestinal tract of humans and other animals, for example, *Escherichia coli*).

colonization The ability of bacteria or other kind of microorganism to proliferate (or grow) in a given habitat.

colony A population of cells growing on the surface of a solid medium; can be seen with the naked eye.

commensalism The type of symbiotic relationship from which one partner in the association receives benefits, while the other partner is neither benefitted nor harmed.

communicable Pertaining to a disease, caused by pathogens or their toxic products, that can be transmitted from an environmental reservoir to a susceptible host.

congenital Acquired during development stages of the fetus, but not determined by heredity.

conjugation The transfer of genetic material through a mating process that requires cell-to-cell contact.

convalescent carrier An individual harboring a pathogenic agent without any apparent symptoms, subsequent to suffering from the disease caused by the agent.

cortex A structural component of bacterial endospores that lies beneath the spore coat.

covalent bond The force that holds molecules tightly together, resulting from the sharing of one or more electrons between two atoms.

cyanophages The algal viruses that have a host range specificity limited to the blue-green algae.

cyst A dormant or resting stage of certain kinds of organisms, often formed in response to unfavorable environmental conditions.

cytoplasm   The watery collection of substances in the cell's interior.

D

death phase   That period of the growth curve of unicellular organisms in which the organisms die at a logarithmic rate.

defaunation   The removal or clearing of animal life from a region or habitat, such as the removal of protozoan endosymbionts from the gut of termites.

denitrification   Process in which microorganisms degrade nitrogen-containing compounds with the liberation of molecular nitrogen.

deoxyribonucleic acid (DNA)   A heteropolymer composed of nucleotide subunits in which the sugar deoxyribose is a component; it is the carrier of the hereditary material (genetic code) in all cells.

dermis   The portion of the epidermal layer of the skin containing nerves, blood vessels, and hair follicles.

destructive parasite   Organism that lives symbiotically at the expense of its host and whose interaction is lethal for the host.

differential stain   A dye that stains certain kinds of cells or structures preferentially when applied in a specified manner.

dimorphic fungus   A fungus with the ability to exhibit alternate growth forms, exhibiting yeast-like cells at 37°C, and exhibiting filamentous (mold) growth at room temperature.

dinoflagellate   The algae belonging to the class *Dinophyceae* and having cells that are naked or covered with a layer of cellulose, sculptured in definite patterns (armored).

dipicolinic acid   A unique chemical substance that is synthesized by microbial cells during the formation of endospores, and is subsequently present within the endospore.

diploid   Pertaining to that stage in the reproductive life cycle of eucaryotic organisms in which the haploid number of chromosomes for a given organism is doubled.

dispersal   Spread or distribution of organisms in nature.

E

eccrine gland   A type of sweat gland, widely distributed over the body, having ducts that open on the outer surface of the epidermis.

ecosystem   The inclusive region or habitat within the biosphere where interactions between living organisms and nonliving components occur.

ectomycorrhizae The symbiotic association between certain kinds of fungi and the roots of higher plants in which the fungal partner grows on the outside surface of the plant root as a sheath.

electron The subatomic particle that surrounds the nucleus of each atom and has a negative charge of -1.

electron microscope A microscope that uses an electron beam instead of light and magnets instead of glass lenses; it is capable of great magnification.

emphysema A respiratory disease in human beings characterized by a loss of elasticity in the alveoli or lung sacs.

endemic Always present within a population of a given geographical area.

endergonic Referring to biochemical reactions that require inputs of energy to proceed.

endocarditis Inflammation of the mucous membranes of the heart.

endoenzyme An organic catalyst that is synthesized within cells, and participates in intracellular reactions.

endomycorrhizae The symbiotic association between certain kinds of fungi and the roots of higher plants in which the fungal partner lives entirely within tissues of plant roots.

endospore A specialized structure formed by certain kinds of bacteria; a dormant stage that enhances the organism's survival in adverse conditions.

endosymbiont A microbial partner that lives symbiotically within cells or tissues of its host.

endotoxins Harmful substances that are contained within the cells that produce them, or are integral constituents of cellular structure and are not released until the cell disintegrates.

enteric Pertaining to the intestinal tract; especially to microorganisms that live in that region of the body, such as *Salmonella* and *Shigella* species.

enucleated Having no discrete nucleus; the genetic material in enucleated cells is distributed throughout the cytoplasm in areas called nuclear regions.

enzootic Pertaining to a condition in which the agent that causes a particular zoonotic disease is always present in susceptible hosts in a given area.

enzyme An organic catalyst that participates in metabolic reactions as a mediator.

epidemic A sudden increase in frequency of a disease, above the normal expectancy, in a population of human beings.

epidemiology The field of science that concerns itself with the determinants of disease and with the factors that influence its distribution.

epidermis The outer layer of human skin, composed of a thin layer of epithelial cells.

episome An extrachromosomal genetic element (piece of DNA) that may exist autonomously in the cytoplasm of a bacterial cell or may become integrated into the chromosome.

epizootic A sudden increase in frequency of a disease, above the normal expectancy, in a population of lower animals.

essential nutrient A required substance that cannot be synthesized in the body and must be supplied in the diet.

established cell line The type of transformed cells from higher organisms that can be cultivated *in vitro* by serial passage for an indefinite period.

etiological Pertaining to the cause of a disease or abnormal condition.

eucaryon A cell that has a well-defined nucleus surrounded by a nuclear membrane, and contains other intracellular organelles.

eutrophication The nutrient enrichment of natural aquatic habitats caused directly or indirectly by human activities.

exergonic Referring to biochemical reactions that liberate energy.

exoenzyme (extracellular enzyme) An organic catalyst, synthesized within cells, but excreted to the cells' exterior where it participates in extracellular reactions.

exogenous Arising from the exterior surroundings of a given host or habitat.

exosporium The outermost structural component of bacterial endospores.

exotoxin Harmful substance produced within cells, but excreted from intact cells into the surrounding environment.

extramicrobial association A symbiotic interrelationship between a microorganism and a plant or animal.

F

facultative Having the ability to thrive in the presence or absence of a specific environmental factor.

fastidious Pertaining to the sensitivity of microorganisms, especially in respect to specific nutritional requirements.

F-factor The specific kind of transferable nonchromosomal genetic element that, when present in certain kinds of bacteria, confers upon them the property of maleness (F+).

fistula   Experimental, surgically constructed opening through the abdominal wall of ruminants from which rumen fluids can be withdrawn.

flagella   Appendages found on certain types of microorganisms that function as organs of locomotion.

fomite   An inanimate object (such as a pencil or towel) that, when contaminated, can transport microorganisms passively.

food chain   The various processes in ecosystems through which chemical elements are cycled, and through which energy flows unidirectionally.

free-living   Capable of self-perpetuation in nature without the aid of a host cell.

G

gamete   A mature germ cell with a haploid chromosome number (sperm or egg).

gamma globulin   The protein fraction of blood serum that contains antibodies.

gas vacuole   A hollow rigid structural component in certain kinds of microorganisms that functions to give buoyancy to the organism.

gene   A unit of genetic material that codes for a specific trait.

generation time   The amount of time required for a population of any kind of organisms to double in number.

genome   The complete set of hereditary genetic material present in an organism; one haploid set of chromosomes.

germ-free animal   Animal that is obtained by a caesarean operation in a sterile environment, and subsequently maintained in an isolator in which the internal environment is sterile. Such animals live in the complete absence of contact with microorganisms.

glycolysis   The group of anaerobic reactions that degrade glucose to pyruvic acid with the formation of lactic acid or alcohol and carbon dioxide.

gnotobiotic   Referring to the condition in which a known population of microorganisms is introduced into a germ-free environment.

gram-negative bacteria   Bacteria that do not retain the crystal violet stain in their cells, when stained by the procedure described by Christian Gram, and thus appear pink or red when viewed under the light microscope.

gram-positive bacteria   Bacteria that retain the crystal violet stain in their cells when stained by the procedure described by

Christian Gram, and thus appear blue or purple when viewed under the light microscope.

H

halophile   An organism that requires high concentrations of salt for optimal growth.

haploid   The number of unpaired chromosomes present in the nucleus of eucaryotic cells; the number is characteristic for a species.

heat-labile   Heat-sensitive; pertaining to chemical structures and/or properties that can be altered or destroyed by moderate heat.

heat-stable   Resistant to alteration or destruction by moderate heat.

hemolysin   Bacterial extracellular product that will lyse and/or destroy red blood cells.

herbivore   Animal that obtains the major portion of its nutrients from plants.

heterocyst   A specialized structure in certain species of blue-green algae.

heteropolymer   A macromolecule in which the subunits (monomers) are different.

heterothallic   Having the ability to produce hyphae that contain opposite sexual gametes.

heterotroph   An organism that requires organic materials for carbon and for energy.

holozoic   Pertaining to the process of ingesting nutrients; generally recognized as a distinct animal trait.

homopolymer   A macromolecule in which the subunits (monomers) are identical.

humus   Residual organic debris that results from the partial decomposition of organic matter.

hydrogen bond   A relatively weak bond between atoms in different molecules that always connects an atom of hydrogen to either an atom of oxygen or an atom of nitrogen.

hydrolytic reaction   The addition or insertion of a water molecule at the site where enzymes catalytically break chemical bonds.

hypersensitivity   An altered state or activity in an individual following contact with certain kinds of substances (inanimate or animate).

hypertonic solution   Solution in which the solute concentration of the extracellular environment is higher than the solute concentration of cell's cytoplasm.

hyphae   Filaments or threadlike structures that form the mycelium of a fungus.

hypotonic solution   Solution in which the solute concentration in the extracellular environment is lower than the solute concentration of cell's cytoplasm.

I

immunity   The state of being nonsusceptible to an infectious agent or to an antigenic stimulus.

immunoglobulin (Ig)   One of the specific class of proteins that contains antibodies.

immunosuppressant   A chemical compound or agent that retards the functioning of the immune system in animals.

indigenous   Originating from within or naturally inhabiting an area; not introduced.

infectious   Capable of causing a disease in a susceptible host.

insecticide   Any material or agent capable of killing insects.

interspecific competition   A condition in which different kinds of organisms or populations have a mutual requirement for a resource that is available in limited supply.

intraspecific competition   A condition in which organisms of the same kind and usually within the same species have a mutual requirement for a resource that is available in limited supply.

invasive ability   The ability of microorganisms to penetrate host's normal defense barriers (i.e. the skin).

in vitro   In vessels; apart from the living body.

ion   Any atom that has an electrical charge.

ionic (electrostatic) bond   The force that holds oppositely charged atoms together, and results from the complete transfer of an electron from one of the atoms to the other.

isotonic solution   Solution in which the solute concentration of the cell's interior is equal to the solute concentration of its exterior surroundings; a state of equilibrium between a cell and its aqueous exterior environment.

K

keratin   The major proteinaceous component of horny or outermost layer of the epidermis.

Krebs cycle   The collective series of aerobic enzymatic actions that degrade pyruvic acid to carbon dioxide and water with the liberation of energy.

l

lag phase   The first portion of the generalized growth curve for unicellular organisms in which there is no cell division, but cells are metabolically active.

LD$_{50}$ The concentration (lethal dose) of microorganism or other substances that is required to kill 50% of the test animals to which the substance or organism has been administered.

leukemia   An abnormal state in which white blood cells grow in an unregulated manner; cancer of white blood cells.

leukocyte   A white blood cell.

lichen   A distinct form that results from the mutualistic association of an algae and a fungus.

logarithmic phase   The portion of the generalized growth curve for unicellular organisms in which cells divide at their maximum rate by geometric progression.

lymph node   A lymph gland occurring in lymphoid tissue.

lysis   Process whereby cells are destroyed by dissolution.

lysogeny   A state in which certain bacterial cells harbor a piece of bacteriophage DNA integrated into the bacterial chromosome as a prophage.

lysozyme   A hydrolytic enzyme that catalyzes the degradation of certain components in bacterial cell walls.

M

macromolecule   A large structure that is composed of smaller subunits (monomers).

mechanical vector   A living organism, such as the common house fly, that transmits infectious agents passively on the surface of its body from one place to another.

medium   A nutrient-containing substance used for the growth and multiplication of microorganisms.

meiosis   A process that occurs at various points in the life cycle of eucaryotic organisms, reducing the chromosome number by half, from diploid to haploid.

mesophile   An organism that grows optimally within the temperature range of 25°C to 40°C.

mesosome   The membranous structure in the cytoplasm of bacteria associated with the formation of cross-walls during the process of cell division.

messenger RNA   The particular species of RNA that is complementary to a strand of DNA and functions to carry the genetic code from DNA to ribosomes, where it determines the sequence of amino acids that are assembled during protein synthesis.

metastasis The spread of a disease-producing agency within tissues, especially as occurs with malignant cancer cells.

microaerophile An organism that requires oxygen for growth, but at a tension lower than is present in the atmosphere.

microflora The sum total of microorganisms in a given habitat.

micrometer (μm) A unit of length that is 1/1000 of a millimeter.

mitochondria The organelle in eucaryotic organisms in which energy is produced for metabolic processes.

mitosis A form of nuclear division in eucaryotic organisms, characterized by complex chromosome movements and exact chromosome duplication.

molecule The smallest unit of any compound.

monomer The subunit molecule from which macromolecules are composed; they may be identical or different.

morphological Relating to the form, size, and shape of an organism.

mutagenic agent Any chemical substance or physical agent that is capable of enhancing the frequency of detectable mutants within a population of organisms.

mutation A sudden, usually rare, change in the genetic code of an organism and results in a change in function that is inheritable.

mutation frequency The speed or rate at which a specific kind of hereditary change (mutation) will appear in a given population of organisms.

mutualism The type of symbiotic relationship from which both partners in the association benefit.

mycelium Mass of threadlike filaments forming the vegetative structure of a fungus.

mycorrhiza The symbiotic association that exists between certain kinds of fungi and the roots of higher plants.

mycovirus Viruses with a host-range specificity limited to the fungi.

N

narrow spectrum antibiotic Antibiotic that acts preferentially on a specific group of microorganisms.

natural immunity The inherent state of nonsusceptibility to a disease or antigenic stimulus.

neutron The subatomic particle of any element; it has a zero electral charge.

niche   The functional role that a given organism or population plays in an ecosystem.

nonseptate   The type of fungal hyphae that have no cross walls or segmented divisions.

nonsymbiotic nitrogen-fixer   Free-living organisms that have the ability to incorporate molecular nitrogen into organic constituents for subsequent use in metabolic processes.

nosocomial infections   The type of infections that are acquired in hospitals.

nuclear region   An area within the cytoplasm of procaryotic cells where hereditable genetic material is concentrated, but not surrounded by a membrane.

nucleotide   Any of a class of compounds that are constituents of nucleic acids, and are composed of a pentose sugar, a purine or pyrimidine, and phosphoric acid.

nucleus   A membrane-enclosed structure that contains hereditary genetic material within the cytoplasm of eucaryotic cells.

O

obligate anaerobe   Any microorganism, having oxygen-sensitive enzymes, that can only proliferate in environments completely devoid of molecular oxygen.

omnivore   Animal that obtains its nutrients from both plants and animals.

oncogenic virus   A cancer-producing virus.

organelle   Any membrane-enclosed structure within eucaryotic cells, such as mitochondria and chloroplasts.

osmophiles   Organisms that grow optimally in habitats where osmotic pressure is relatively high.

osmosis   The diffusion of water across a semi-permeable membrane from a region of low solute concentration to a region of higher solute concentration.

osteomyelitis   An infectious disease involving the destruction of bone tissue.

oxidative phosphorylation   The process in which inorganic phosphorus is enzymatically coupled to a compound such as ADP to form ATP.

P

pandemic   Referring to a disease that occurs in world-wide proportions, not restricted by geographical barriers or other boundaries.

parasitism   The type of symbiotic association in which one partner (the parasite) receives benefits, and simultaneously inflicts harm on the other partner in the association (the host).

passive dispersal   The transport of organisms from one location to another by processes that are not organism-controlled such as by water and air.

passive immunity   Immunity caused by the natural or artificial transfer of preformed antibodies to a nonimmune individual.

pasteurization   The use of heat under controlled conditions to reduce the number of microorganisms that may be present in liquids or other food substances without changing the quality or texture of them; not a sterilization process.

pathogenic   Having the ability to produce or cause a disease.

pellicle   A slime layer or film on a liquid surface due to the growth of micoorganisms.

periplasmic space   The area between the cell wall and the plasma (unit) membrane of a bacterial cell.

peristalsis   The wave-like rhythmic muscular movement that forces foodstuff through the gastrointestinal tract.

peritoneal cavity   The cavity of the abdomen (in mammals) that lies between the viscera and the skin.

permease   Enzyme that functions to transport substances into cells across the unit (plasma) membrane.

phagocyte   A specific kind of white blood cell (a leukocyte that is capable of ingesting foreign materials).

photic zone   The region in natural bodies of water (lakes, oceans, etc.) that is penetrated by light rays.

photoautotroph   An organism that obtains its energy from solar radiation and its carbon from carbon dioxide.

photoorganotroph   Any bacterium that is a facultative heterotroph, capable of utilizing either radiant energy or energy from organic sources, but having an absolute requirement for organic carbon.

photosynthesis   Process in which certain kinds of organisms utilize solar energy and carbon dioxide to synthesize carbohydrates.

pili   A specific kind of appendage present on certain types of bacteria; they are not organs of locomotion.

pioneer organism   The first to arrive in a virgin habitat, or the first species of a kind to become colonized in a region that contains other kinds of living cells.

plasmid   An extrachromosomal piece of DNA that may be present in certain kinds of bacteria, but it is not essential to growth of the host bacterial cell.

poly-β-hydroxic butyric acid (PHB)   A homopolymer composed of β-hydroxybutyric acid subunits: a major intracellular nutritional reserve (storage product) in bacterial cells.

polymer   A macromolecule composed of distinct small molecules as subunits.

polypeptide   A series of amino acids linked together by peptide bonds to form a chain.

potable water   Water that is safe for human consumption.

primary consumers   Heterotrophic organisms in the first stage of food chains that obtain their nutrients preformed from autotrophs.

primary immune response   The initial detectable increase in antibodies in the blood following an antigenic stimulus.

primary producers   Autotrophic organisms that utilize energy from solar radiation or inorganic chemicals, and carbon from carbon dioxide to synthesize organic compounds.

procaryon   An undifferentiated cell that has its genetic material concentrated in nuclear regions of the cytoplasm, not surrounded by a membrane, and reproduces itself asexually.

prophage   A piece of DNA from a bacteriophage (phage) which is integrated into the bacterial chromosome.

proteolytic enzyme   An enzyme that is capable of catalyzing reactions that degrade proteins.

proton   The subatomic particle of any element that has a positive charge of +1.

prudent parasite   A parasite that maintains a symbiotic relationship with its host in such a manner that only minimal damage is inflicted on the host; as opposed to a *destructive parasite,* which kills the host quickly.

pseudomycelia   False mycelia.

psychrophile   An organism that grows optimally within the temperature range of 0°C to 20°C.

pure culture   A culture of microorganisms in which all cells are of a single type.

pus   Exudates formed during an infection, consisting of bacteria, white blood cells, dead tissue cells, and other foreign materials from the blood stream.

R

resolving power   The ability of light microscopes to project clear and distinct images, to discriminate two separate points in close proximity.

R-factor   A class of transferable plasmids that carry genes for resistance to antimicrobial substances; found in many kinds of bacteria.

ribonucleic acid (RNA)   A heteropolymer composed of nucleotide subunits in which the sugar ribose is a component. It is the carrier of the hereditary material (genetic code), and plays a major role in protein synthesis.

ribosomes   Intracellular granules or particles that are composed of proteins and RNA, and function in protein synthesis.

S

saccharophile   An organism that requires sugar in high concentrations for optimal growth.

sanitary sewer   The collective group of pipes in a sewage system that transports human fecal waste to the treatment facility.

saprophyte   Any organism that requires and utilizes preformed nutrients from dead or decaying organisms.

sarcoma   Cancer that affects soft tissues and bones.

secondary immune response   The sudden increase in antibody production in an animal in response to an antigenic stimulus from a substance to which the animal has been previously stimulated.

septate   The type of fungal hyphae that have subdivisions (cross walls).

septicemia   The invasion and multiplication of microorganisms in the blood stream.

serial passage   The sequential transfer of living organisms or agents from one culture media to another at various intervals, such as the propagation of animal viruses in tissue culture systems.

sex pilus   A specific kind of appendage present on (F+) cells; the structure that forms the conjugation tube through which genetic material passes during conjugation.

sexual reproduction   A method of reproduction that requires the interaction and fusion of opposite sexual gametes.

sheath   An external structure, outside of cell wall, of some microorganisms.

silent mutation   A genetic change in one or two nucleotides of the triplet code that does not express itself as a functional change.

simple stain   A single dye that can be used to increase the contrast in a specimen by making the cells darker than the surrounding medium. Simple stains react equally with all kinds of bacteria.

slime layer   The gelatinous outermost covering of certain bacteria; unlike capsules, which are tightly bound to cell walls.

sludge   The residual mass of solids which remains in sewage sedimentation tank after the liquid portion of sewage has passed through a municipal sewage treatment facility.

sludge digestion   The biological degradation of the solid residue from sewage in a municipal sewage treatment facility; it usually takes place in tanks designed specifically for this purpose.

spore coat   The structural component of bacterial endospores that lie between the exosporium and the cortex.

symbiosis   The living together of two or more dissimilar organisms with some degree of constancy.

symbiotic nitrogen-fixer   The incorporator of molecular nitrogen into specific cells that result from the symbiotic association of two organisms (i.e. *Rizobium* and a legume).

stationary phase   That portion of a unicellular life cycle (growth curve) in which cells are dividing and dying at an equal rate.

sterilize   To free from all forms of life.

storm sewer   The collection of pipes in a sewer system that transports all water, except that portion that contains human waste, to the sewage treatment facility.

submicroscopic   Too small to be seen with a light microscope; some submicroscopic structures are visible with an electron microscope.

superinfection   An infection that develops subsequent to an already existing infection; most often involving strains that are resistant to antimicrobial agents.

SV-40   The designation for a Simian virus strain originally isolated as a contaminant in a tissue culture system of monkey kidney cells used for the propagation of polioviruses. It is an *in vitro* tumorigenic virus strain.

T

taxis   The movement of an organism toward or away from a source of stimulation.

temperate phage   The type of bacterial virus that, after penetration of host cells, exists as a prophage integrated into the bacterial chromosome.

thermophile   An organism that grows optimally within the temperature range of 45°C to 60°C.

toxemia   Condition caused by a toxin in the blood.

toxigenic   Producing a toxin.

toxin   A substance produced by a living organism that injures tissues or alters the functions of another organism.

toxoid   A toxin modified (weakened) but still retaining the ability to function as an antigen.

transcription   The process during protein synthesis in which the genetic code in DNA is transcribed to a complementary strand of messenger RNA.

transduction   The transfer of a gene or of several genes from a donor bacterial cell to a recipient bacterial cell by an intermediate agent called a bacteriophage or bacterial virus.

transfer RNA   The species of ribonucleoprotein that functions to carry amino acids to the ribosome during protein synthesis; each transports a specific amino acid.

transformation   The process through which a portion of bacterial chromosome (DNA) is incorporated into a recipient bacterial cell in the absence of cell-to-cell contact, and without the aid of a bacterial virus.

translation   The process during protein synthesis in which the genetic code in messenger RNA directs the assembly of amino acids into a polypeptide chain.

trickling filter   The apparatus in a municipal sewage treatment facility consisting of a bed of rocks upon which sewage is sprayed and subsequently degraded by aerobic organisms.

trypsin   A specific kind of proteolytic enzyme.

U

ultramicrotome   An instrument used to cut cells and other materials into very thin slices for observation under the transmission electron microscope.

V

vaccine   Any suspension of organisms and/or their products that is used for the purpose of inducing immunity to an infectious agent or disease.

valence   The combining capacity or electron affinity of any atom.

vector   A living organism that has the ability to transmit infectious organisms from one locale to another.

vegetative cell   A cell that is actively growing; in bacteria, replicating, as opposed to surviving as an endospore.

vehicle   Any contaminated inanimate material, such as water or food, through which microorganisms can be transported or dispersed to a susceptible host.

villi Hair-like projections present on mucous membranes such as those in the interior of the intestinal tract.

virion A complete extracellular virus particle.

virulence The degree of pathogenicity or disease-producing capacity of an organism.

virulent phage The kind of bacterial virus that, after penetration of a host cell, will undergo replication and subsequently cause lysis of the host cell with the release of complete viruses.

virus An infectious agent that contains either RNA or DNA in its core surrounded by a protein shell, is able to alternate between intracellular and extracellular states, and replicates only when present in living cells.

W

weathering The disintegration of substances such as rocks by physical, chemical, and biological processes.

X

xerophile An organism with enzyme systems that enable it to grow optimally in dry places (deserts).

Z

zoonoses (zoonotic diseases) The group of diseases primarily confined to lower animals (wild and/or domestic) but which are transmissible to human beings directly or indirectly.

zooplankton The group of microscopic animals found free floating in a body of water.

Index

Explanation of symbols: f = figure; t = table.

A

Acetobacter, 142
Acholeplasma, 26
Achromobacter, 184f
Acid-fast stain, 15, 239
Acidity, 59f, 104
Acids, definition of, 58
 organic 63t
Adenosine triphosphate, 81–82, 81f
Aedes aegypti, 280t, 282
Aerobic respiration, 85–88
Aerosol, definition of, 317
Aflatoxins, 301–302
Agaricus bisporus, 163
Air indoor, 107–108
 microflora of, 107, 108t
 outdoor, 107
 microflora of, 107
Algae, blue-green, 25–26, 27f
 eucaryotic, 35–36, 37f
Algal blooms, 300
Allochthonous, 106
Amebiasis, 260–261
Amino acids, 63t, 64, 66t, 66f
Anabaena, 27f, 196, 297, 298, 298t
Anaerobic respiration, 88
Animal starch, 64. *See also* Glycogen
Anopheles, 230, 284t, 285–286
Anthrax, 272–274, 295t
 animal outbreaks of, 272
 clinical features of, 273–274
 prevention of, 273–274
 transmission of, 272–273, 272t
Antibiosis, definition of, 133
Antibiotics, 133–139
 definition of, 135
 nonmedical use of, 139t
 sensitivity testing of, 138f

Antibodies, 217–221. *See also* Immunoglobulins
 classes of, 218t
 immune response curve, 219f
Antigens, 218–219
Aphanizomenon flos-aquae, 297, 298t, 299, 306
Arboviruses, group A, 280t, 280–281
 group B, 280t, 281–283
Aspergillosis, 321
Aspergillus flavus, 301
Aspergillus fumigatus, 258t, 321, 322f
Assays, microbiological, 334–336
Athlete's foot, 207. *See also* Dermatomycoses
Atoms, 54–56
Autochthonous, 106
Autotrophs, 74
Azotobacter, 7, 196, 197t

B

Bacillus, 19, 197t
 anthracis, 5, 12t, 133, 272, 272t, 273, 295t, 323
 Calmette-Guerin (BCG) vaccine, 241
 cereus, 141t
 lentimorbus, 311–312
 megaterium, 140, 141t
 popilliae, 311–312
 subtilis, 20, 21, 22f, 23f
 stearothermophilus, 100
 thuringiensis, 296–297, 310–311
Bacteria, 9–25
 capsules of, 20–21

Bacteria *(continued)*
  cell dimensions of, 11t
  cell structures of, 18–19, 19f
  endospores of, 21–23, 22f, 23f
  flagella of, 18f, 19
  Gram reaction of, 14–15, 15t
  intracellular structures of, 18f,
    19f
  morphological types, 9f, 10f, 11f,
    20f
  pili of, 20
  smear preparation of, 12, 14f
Bacteriocins, 140, 141f
Bacteriophages. *See* Viruses, bac-
    teriophages
Bacteroides, 210
Baltimore, D., 48
Bases, 57–60
Bdellophage, 161, 163f
Bdellovibrio, 160–163, 162f
  bacteriovorus, 161, 161f
Beauveria bassiana, 312
Beggiatoa, 200
Beijerinck, M., 6
Biochemical oxygen demand
    (BOD), 131
Biological environments, 116–117
Biosphere, 98, 181
Blastomyces dermatitis, 258, 258t,
    259f
Blastomycosis, 258–259
Boils, 234
Bonds, chemical, 56–57
Botulism, 295t. *See also*
    Clostridum botulinum
Brucella, 272t, 323
  abortus, 274, 288
  melitensis, 274
  suis, 274
Brucellosis, 272t, 274–276
  clinical features of, 276
  human cases of, 274, 275f
  prevention of, 276, 288
  reservoirs of, 274t
  transmission of, 274
Bubonic plague, 272t, 276–277
  clinical features of, 276–277
  human cases of, 277

prevention of, 277, 289
transmission of, 276

C

Caedobacter, 160t
Calorie, 81
Cancer, definition of, 46
Candida, 211
  albicans, 30f, 31f, 32, 138–139,
    212, 258, 258t, 268
Candidiasis, 258
Carbon cycle, 193–195, 194f
Carbon sources, 72
Carbuncles, 234–235
Cellobiose, 64, 65f
Cells, classification of, 8f
  eucaryotic, 28–40
  procaryotic, 7–28
Cellulose, 64, 65f
Chain, E. B., 135
Chemoautotrophs, 75
Chemoorganotrophs, 75
Chemotherapy, 266–268
Chlorella, 35
Chlorobiaceae, 200
Chloroplasts, 29, 37f
Cholera, 244–246, 295t
  cases of, 246
  clinical features of, 245
  pandemics of, 245, 325
  prevention of, 246
Chromatiaceae, 200
Claviceps purpurea, 307–308, 307f
Clostridium, 196, 197t, 210
  botulinum, 23, 115, 234t, 294,
    295t
    dimensions of, 12t
    food poisoning from, 116t, 304,
      305f
    toxins, types of, 304t
  perfringens, 10f, 233t, 304, 305f
  tetani, 115, 232, 294, 295t
Coagulase, 233t, 234
Collagenase, 233t

Composting. *See* Waste disposal, solid
Conjugation, 24
Corynebacterium acnes, 117, 207
Corynebacterium diphtheriae, 43, 156, 210, 232, 234t, 294, 295t
Coxiella burnetii, 11
Cryptococcosis, 320–321
Cryptococcus neoformans, 258t, 320–321, 320f, 321f
Culex, 280–282, 280t
Cyanobacteria. *See* Algae, blue-green
Cyanophages, 164
Cytophaga, 158, 160t

D

Dental caries, 213
Deoxyribonucleic acid (DNA), 69, 71f, 77t, 90–93
  components of, 70f
  helical structure of, 71f, 91f
Dermatomycoses, 256–257
Desulfovibrio, 88, 199f, 200
Diphtheria. *See Corynebacterium diphtheriae*
Dobereiner, J., 196
Dulbecco, R., 48
Dysentery, bacillary, 295t. *See also* Amebiasis

E

Ecosystems, 98, 98f
Electromagnetic spectrum, 82–83. *See also* Radiation, solar
Electron transport system, 86, 87f
Elements, chemical, 54, 62f, 73t
Encephalitides, 280t, 280–282, 282f, 289
  Eastern equine encephalitis (EEE), 280–282, 280t

St. Louis encephalitis (SLE), 280–282, 280t
  Venezuelan equine encephalitis (VEE), 280–282, 280t
  Western equine encephalitis (WEE), 280–282, 280t
Energy, kinds of, 72, 83f
Entamoeba histolytica, 260
Enterobacter cloacae, 141t
Environmental Mutagen Society, 336
Environmental Protection Agency (EPA), 324
Enzymes, 76–80
  classes of, 77, 77t
  cofactors of, 79
  model reactions of, 78–80, 78f, 79f,
  properties of, 77t
Epidemiology, 226–231
  definition of, 226
  investigations, kinds of, 230–231
  triangle of causation, 226f
Epidermophyton, 256. *See also* Dermatomycoses
Ergotism, 307–308
Escherichia coli, 100, 101f, 104, 140, 141t, 163f, 185, 210, 335
Eucaryotic organisms, 28–39
Euglena gracilis, 37f

F

Fermentation, 83–85, 86t
Fleming, A., 133–135
Florey, H., 135
Food chains, 190–192, 191f
Food poisoning. *See* Botulism, *Clostridium botulinum, Clostridium perfringens,* Salmonellosis, *and Staphylococcus aureus*
Fracastoro, G., 4
Fungi, 30–35, 34f, 35f. *See also* Dermatomycoses

G

Gastrointestinal tract, 208–210, 208f, 213f
Genetic code, 90–93
German measles. *See* Measles, rubella
Germ-free animals, 221–223, 222f
Giardia lamblia, 38f, 261–262
Giardiasis, 261–262, 327t
Glycogen, 64, 65f
Glycolysis, 84–85, 85f
Glycoside bond, 64, 65f
Gnotobiotic, 222
Gonorrhea, 262–263
  cases of, 265f
  clinical features of, 262
  prevention of, 146, 263
Gonyaulax catenella, 298t, 299–300, 306
Gonyaulax monilata, 298t, 300
Gram, C., 14. *See also* Stains, Gram's procedure for
Great Salt Lake, 182
Growth, 70–74
  essential trace elements, 73, 73t
  micronutrients for, 73–74
  unicellular organisms, 183–189
    binary fission, 19f, 184
    generation time for, 23, 185
    measurements in nature, 187–189
    phases in liquid media, 185–186, 185f
    on solid media, 186, 187f
Gymnodinium breve, 298t, 300

H

Habitats, 106–117
  interactions in, 98f
  kinds of, 108–117
  linkages of, 120f

microbial dispersal in, 117–120, 118t
Haemophilus influenzae, 12t
Hemolysins, 233t, 235, 236f
Hepatitis, 251–253
Herpetosiphon geysericola, 11
Heterotrophs, 75–76
Histoplasma capsulatum, 258t, 318f, 318, 320
  vectors for, 319f
Histoplasmosis, 318–320
Hooke, R., 4

I

Immunity, 220–221
Immunizations, 269
Immunoglobulins, 218, 218t. *See also* Antibodies
Incineration. *See* Waste disposal, solid
Influenza, 253–256
  causative agents, 253
  clinical features, 253–254
  susceptibility to, 254
  pandemics, 253
  prevention, 254
Insecticides, microbial, 308–312
  bacterial cells, 311–312
  bacterial toxins, 310–311
  desirable attributes of, 310t
  fungal cells, 312
Interference, microbial, 123
Intestinal microflora. *See* Gastrointestinal tract

K

Kircher, A., 4
Koch, R., 4, 5, 12, 239–240, 245, 272

Koch's postulates, 5, 183, 239
Kreb's cycle, 86, 87f

L

Lake Erie, 109
Landfill. *See* Waste disposal, solid
Leeuwenhoek, A. van, 4, 11
Legume, 174–177, 175f, 176f, 177f
Leucidins, 233t
Lichens, 126, 154–155, 155f
Limulus assay, 187–188
Limulus polyphemus, 188
Lipids, 66–68
    beta-hydroxybutyric acid, 68
    complex, 67
    phospholipids, 67, 69f
    simple, 67
    triglyceride, 68f
Lipopolysaccharides, 68, 187–188
Lipoproteins, 68
Lyngbya, 165, 186
    majuscula, 298t, 299
Lysergic acid, 308
Lysogeny, 42, 43f, 155–156
Lyticum, 160t

M

Macrocystis, 35
Macromolecules, 64–69, 62f
Malaria, 284t, 285–287
    cases, civilian, 285f
        military, 285f
    causative agents, 284t, 286
    clinical features of, 286
    endemic areas, 285
    pathogens, life cycle of, 286
    prevention/control, 286–287
Malpighi, M., 4
Maltose, 64, 65f
Martian soil, 181

Measles, 249–251
    rubella, 251
    rubeola, 249–250
Meresmopedia, 27f
Mesophiles, 100, 101f
Metabolism, 76–92
    aerobic respiration, 85–88, 87f
    anaerobic respiration, 88
    glycolysis, 84, 85f
    fermentation, 83–85
    interactions, combined, 87f
    lipid synthesis, 89–90
    polysaccharide synthesis, 89
    protein synthesis, 90–92, 91f
Metarrhizium anisopliae, 312
Metcalf, E. S., 100
Methanobacterium, 88
Micrococcus roseus, 12t
Microcystis aeruginosa, 297, 298,
        298t, 306
Microscopes, electron, 18–20, 20f
        scanning, 20, 20f
        transmission, 16–18, 16f
    light, 11–16, 13f
        compound, 11–16
        simple, 4, 11
Microsporium canis, 256, 257f. *See
        also* Dermatomycoses
Mitochondria, 28, 31f
Molds. *See* Fungi
Molecules, 55
Morita, R. Y., 104
Mosquito-borne disease. *See* En-
        cephalitides, Malaria, *and*
        Yellow fever
Mosquitoes, 280t, 280–282
Mutagenic agents, 93
Mutations, 92–93
Mycobacterium, 15
    avium, 239
    bovis, 239
    leprae, 5
    tuberculosis, 185, 217, 239
Mycoplasma, 16, 26–28, 210, 301
    pneumoniae, 28, 29f
Mycorrhizae, 172–174
Myxobacter, 144, 145f

N

Neisseria gonorrhoeae, 5, 12t, 234t, 263, 263f
  penicillin resistance, 146, 263
Nitrobacter, 197, 197t
Nitrogen, kinds of, 72
Nitrogen cycle, 195–197, 196f
Nitrogen-fixation, 174–177
Nitrosomonas, 197, 197t
Nobel prizes, 48, 135
Nostoc, 27f, 145, 145f, 196
Nucleic acids, 68–69, 71f. *See also* Deoxyribonucleic acid *and* Ribonucleic acid
Nucleotide, 68

O

Olsen, R. H., 100
Oral infections. *See* Candidiasis *and* Streptococcus pyogenes
Organelles. *See* Eucaryotic organisms
Osmosis, 102–103

P

Para-aminosalicylic acid (PAS), 241. *See also* Tuberculosis
Paramecium, 38, 38f
  aurelia, 157–160, 159f, 160t
  bursaria, 157–158
  caudatum, 335
Parasites, definition of, 178
Pasteur, L., 4, 5, 133, 272, 283
Pathogenicity, 232–234, 233t, 234t
Penicillium, 34f, 35f, 187f, 301
  crysogenus, 164
  notatum, 135
Peridinium polonicum, 298t

Phagocyte, 216, 217f
Phagocytosis, 216–217
Phormidium, 165, 186
Phosphorus cycle, 198, 198f
Photoautotrophs, 74
Photobacterium phosphoreum, 334
Photoorganotrophs, 75
Photosynthesis, 82–84
Plague, 295t. *See also* Bubonic plague
Plasmids, 24–25, 267–268
  c-factor, 140
  F-factor, 24
  R-factor, 24, 267–268
Plasmodium falciparum, 286
  malariae, 286
  ovale, 286
  vivax, 286
Plectonema, 165
Pneumonia, 237–239, 238f
Polio, 247–249
  causative agents, 247
  clinical features, 248
  epidemic, definition for, 249
  prevention, 249
  vaccines, 248–249
Pollution, definition of, 315
Pressure, hydrostatic, 103–104
  osmotic, 102–103, 103f
Procaryotic organisms, 7–28
Proteins, 64–66, 66t
Protozoa, 36–38
Prynesuim parvum, 300
Pseudomonas, 197t
  aeruginosa, 138, 138f, 141t, 185, 212, 268
  fluorescens, 141t
Psychrophiles, 100, 101f

R

Rabies, 280t, 283–284
  clinical features of, 283
  prevention, 283–284

recent human cases of, 284
vaccines for, 4, 283
Radiations, solar, 83, 83f, 99, 101
Recycling. *See* Waste disposal, solid
Reed, W., 282–283. *See also* Yellow
fever
Rhizobium, 6, 151, 174, 175f, 176f,
177f, 182, 196, 197t
Rhizopus, 34f, 35f. *See also* Fungi
Rhodotorula infirmoninata, 101f
Ribonucleic acid (RNA), 69, 77t,
90–93
components of, 70t
kinds of, 90–91, 91f
Ribosomes, 91, 91f
Ringworms, 256
Rous, P., 47
Ruminants, 165–167, 166f, 167f

S

Sabin, A., 248
Saccharomyces cerevisiae, 31–32,
84, 142
life cycle of, 33f
St. Louis encephalitis. *See* Encephalitides
Salk, J., 248
Salmonella, 210, 272t, 277, 305f
paratyphi, 244
schottmulleri, 244
typhi, 234t, 242t, 277
Salmonellosis, 272t, 277–279, 327
clinical features of, 277–278
prevention, 278–279
transmission of, 272t, 278, 279f
Sclerotia, 307, 307f
Serratia marcescens, 141t
Sewage, 327
treatment, 328–331, 329f
Shigella, 210
dysenteriae, 234t, 260, 294, 295t
exotoxin of, 295t
food poisoning from, 305f
flexneri, 143–144

Shigellosis, 325, 327t
Skin, 204–208
appendages, 207
glands, 206
microflora, 207–208
structure, 205–208, 205f
Sludge, 328
digestion of, 329–331, 331f
Smallpox, 246–247
causative agents, 247
clinical features, 247
endemic areas, 246
prevention, 247
Smith, H. S., 308
Snow, J., 244–245
Soils, 112–116. *See also* Terrestrial
environments
formation, 112–114, 113f, 114f
microflora of, 115–116
Sphaerotilus, 186
natans, 161, 162f
Spirillum lipoferum, 176–177, 196
Spirillum volutans, 11f, 12t
Sporotrichosis, 259–260
Sporotrichum schenkii, 258t, 259–
260, 260f
Stains, differential, 13
acid-fast, 15, 239
Gram's procedure for, 14–16,
15t
simple, 13
Staphylococcus aureus, 12t, 134,
138f, 187f, 207, 233t, 234,
267, 296
exotoxin of, 295t
food poisoning from, 304, 305f
Starch, 64, 65f
Stigeoclonium, 186
Streptococci, oral, 212–215
Streptococcus lactis, 11f
Streptococcus miteor, 213, 214f
Streptococcus mutans, 212
Steptococcus pneumoniae, 21, 210,
217, 233t, 237, 238f
Streptococcus pyogenes, 12t, 233t,
235–37, 236f
Streptococcus salivarius, 212

Streptomyces, 144
Succession, ecological, 125–129, 129f
  pioneers in, 125
  types of, 126–127
Sulfur cycle, 199–200, 199f
Swine influenza. *See* Influenza
Symbiosis, 150, 151–153, 165–170, 174
Syphilis, 263–265
  cases of, 265f
  clinical features, 264
  prevention of, 265

T

Tectobacter, 160t
Temin, H. M., 48
Terrestrial environments, 112–116
Tetanus, 295t. *See also Clostridium tetani*
Thermophiles, 100, 101f
Thermoplasma acidophilum, 26
Thermus aquaticus, 100, 101f
Thiobacillus, 88
  ferrooxidans, 200
  thiooxidans, 104, 200
Thrust, oral, 258
Tissue culture, 39–40
  assays, 335–336
Tolypothrix, 186
Toxins, algal, 297–301
    blue-green toxin producers, 297–299, 298t
    eucaryotic toxin producers, 298t, 299–301
  bacterial, 294–297
    endotoxins, characteristics of, 296–297
    exotoxins, characteristics of, 295–296, 295t
  fungal, 301, 307–308, 312
Toxoplasma gondii, 284t, 287, 287f
Toxoplasmosis, 284t, 287–288
Transduction, 24, 42

Transformation, 24, 39
Treponema pallidum, 12t, 264f
Trichomonas vaginalis, 211
Trichophyton, 256. *See also* Dermatomycoses
Tuberculosis, 239–242
  causative agents, 239
  clinical types, 240
  skin test for, 240–241
  treatment of, 241
  vaccine for, 241
Typhoid fever, 242–244, 327t
  carriers, 243
  causative agent, 242
  clinical features, 242–243

U

Ulothrix, 36f, 186
Urea, 61

V

Vectors, 230
Vehicles, 229
Venereal diseases (VD), 262–265
Vibrio cholerae, 12t, 234, 294, 295t
Viking I, 181
Virulence, 232
Viruses, 40–48
  animal, 44–48
    cancer hypothesis, 46–48
    classification of, 45t
    enveloped, 44–45, 45f, 45t
    naked, 44–45, 45f, 45t
  bacteriophages, 42–44, 43f
  nucleic acids of, 42, 44–45, 45t
  structural components, 40
Volcanic rocks. *See* Soils
Volcanoes, 112–114, 112f, 113f, 114f
Volvox, 36f

W

Waksman, S. A., 135
Waste disposal, liquid. *See* Sewage,
    treatment
  solid, 331–333
Water, drinking, 324–327, 327t
    chlorination of, 326
    coliform test, 326
    filtration process, 326
  salt, 110–112
    ions in, 111t
Waterborne diseases. *See* Amebi-
    asis, Cholera, Giardiasis, Sal-
    monellosis, Shigellosis, *and*
    Typhoid fever
Winogradsky, S., 6
Wohler, F., 61
Woods Hole Harbor, 188
  detection of microbes in, 189f
Wool-sorter's disease. *See* Anthrax

Y

Yeasts, 30–33, 84, 85f, 142. *See also*
    Dermatomycoses
Yellow fever, 280t, 282–283
    clinical features of, 282–283
    prevention, 282
    vector for, 280t, 282
Yersinia pestis, 12t, 141t, 272t, 276,
    295t

Z

Zobell, C. E., 104
Zoogloea, 186
Zoonoses, 271